Concepts and Recent Advances in Generalized Information Measures and Statistics

Andres M. Kowalski, Raúl D. Rossignoli
Departamento de Física–IFLP, Universidad Nacional de La Plata
and Comisión de Investigaciones Científicas
La Plata
Argentina

Evaldo M. F. Curado
Centro Brasileiro de Pesquisas Físicas
and National Institute of Science and Technology for Complex Systems
Rio de Janeiro
Brazil

CONTENTS

Part II: Applications

Foreword by Fernando Dantas Nobre

The concept of entropy appears in many areas of knowledge, like thermodynamics, statistical mechanics, information theory, biology, economy, and human sciences. From the historical point of view, it was introduced in 1865 by Rudolf Clausius through an elegant formulation of the second law of thermodynamics. A nice historical review of how the concept of entropy emerged in physics is presented in Chapter 1, written by the editors of this book. According to the second law, the total entropy of an isolated system can never decrease in thermodynamical transformations, and particularly in the case of irreversible transformations, it always increases. Since most natural processes are irreversible, entropy has been associated with the "arrow of time". Some years later (1872), Ludwig Boltzmann wrote down an equation (known as Boltzmann equation) to describe the evolution of the single-particle distribution of a rarefied gas. Considering this distribution, he defined a quantity H and proved the famous H-theorem, by showing that H always decreases in time. Boltzmann realized that for a perfect gas in equilibrium, the quantity H was related to Clausius' entropy S (apart from a minus sign and some multiplicative constants). This identification led to the definition of statistical entropy, *i.e.,* the entropy defined in terms of a probability distribution, $P(\vec{x}, t)$, associated with the occurrence of a given physical quantity $\vec{x}$ (in the case of gases $\vec{x}$ may represent the position, the velocity, or both position and velocity, of a molecule) at a time t. Moreover, the H-theorem yielded a microscopic interpretation of the second law of thermodynamics.

The dynamical approach of Boltzmann, together with the elegant theory of statistical ensembles at equilibrium proposed by Josiah Willard Gibbs, led to the Boltzmann-Gibbs theory of statistical mechanics, which represents one of the most successful theoretical frameworks of physics. The knowledge of the equilibrium distribution associated with a given statistical ensemble allows one to calculate average values to be related with thermodynamic quantities. This theory is based on a fundamental assumption, namely, the ergodic hypothesis, which requires that the system will pass through all its microstates after a sufficiently long time. Only if ergodicity holds is that one can replace a given time average (defined within the Boltzmann framework) by the corresponding average over a statistical ensemble (defined at equilibrium).

In Chapter 2 (also written by the editors of this book) the essentials of information theory are introduced and discussed. This theory was created in 1948 by Claude Shannon through the definition of a measure of information following a form similar to the quantity H introduced by Boltzmann. Consequently, from this measure of information one constructs a statistical entropy, sometimes called Boltzmann-Gibbs-Shannon entropy. In 1957, E. T. Jaynes introduced the Maximum Entropy Principle, which allows one to obtain equilibrium distributions by maximizing the statistical entropy under given constraints. In this way, one can derive the equilibrium distributions of statistical mechanics from the entropy maximization procedure, by considering the Boltzmann-Gibbs-Shannon entropy under the constraints suitable for each statistical ensemble.

The interest in a wide variety of complex systems, which are usually characterized by a large number of constituents that interact through long-range forces, and may present peculiar features, like strong correlations, long-time memory, and breakdown of ergodicity, has increased in the latest years. Therefore, due to these characteristics, the complex systems are expected to exhibit collective behaviors very different from those of the rarefied gas considered by Boltzmann in 1872. Many experiments, as well as computational studies in complex systems, have shown properties in disagreement with the predictions of Boltzmann-Gibbs statistical mechanics, suggesting the need of a more general theory for their description. A breakthrough occurred in 1988 with the introduction of a generalized entropy S_q by Constantino Tsallis; this proposal is discussed in detail in Chapter 3. It is characterized by a real index q, such that in the limit $q \to 1$ one recovers the Boltzmann-Gibbs-Shannon entropy. The entropy S_q, and more particularly the distribution that comes from its maximization, has been very successful in describing many situations where Boltzmann-Gibbs statistical mechanics fails.

Although the entropy S_q has been the most successful for describing complex systems so far, other generalized entropic forms have also been proposed in the literature, as discussed in Chapter 4 by S. Thurner and R. Hanel. This is what the present book is about, containing interesting chapters in history, theory, computational methods, experimental verifications, and applications. The book is addressed not only to physicists, but to researchers in a large diversity of fields, like biology, medicine, economics, human sciences, and to all those interested in understanding the mysteries within the realm of complex systems.

Fernando Dantas Nobre

Centro Brasileiro de Pesquisas Físicas

Rio de Janeiro - RJ

Brazil

Foreword by Héctor Vucetich

In the fourth century BC, Aristotle stated that he disposed of an infallible method to find the truth, namely, the inductive one. And in the XVII century, Leibniz proposed the construction of a "Calculus of Ideas" that would end vain debates. Neither project succeeded, but each of them started several projects that enriched science: logic and statistics.

In the XIX century a new, powerful idea was developed, namely, that of entropy: an ever-growing physical magnitude that measured the degree of decay of order in a physical system. This powerful idea not only explained quantitatively the behavior of gases, dilute solutions and chemical reactions but also explained the philosophical problem of decay that Aristotle attributed to earthly matter. With the introduction of entropy, thermodynamics became a model of theoretical science.

In 1948 Shannon developed a "Statistical theory of communication" taking ideas from both fields that in turn opened new paths for research. The powerful notion of *information entropy* played a major part in the development of new statistical techniques, overhauled the Bayesian approach to probability and statistics and provided powerful new techniques and approaches on several fields of science.

Later on, several generalizations of the concept of information entropy were introduced, that extended and shed new light on the field. These generalizations are already applied in statistical problems and may have interesting applications in fields of science such as critical behavior or neuroscience.

These and related topics are treated in this book that is a review of an old subject from a young point of view. Starting from its historical roots, the book proceeds with the mathematical foundations, generalizations, properties and applications to different branches of mathematics and natural science of the powerful notion of *information entropy.* And as such, it gives a state-of-art perspective of the subject in the second decade of the XXI century.

Reader: enjoy!

Héctor Vucetich
Observatorio Astronómico
Universidad Nacional de La Plata
La Plata
Argentina

Preface

Since the introduction of the concept of information entropy by Claude Shannon in his famous 1948 article [1], quantifiers based on information theory have played an increasingly fundamental role in several fields. Different generalizations of the Shannon entropy have been developed, among them the Rényi and Tsallis entropies, which have found important applications not only in physics but also in quite distinct areas such as biology, economy, cognitive sciences, *etc.* In addition, other information measures such as the Fisher information, which predates the Shannon entropy, and the more recent statistical complexities, have also proved to be useful and powerful tools in different scenarios, allowing in particular to analyze time series and data series independently of their sources.

It is our goal to expose in this book, in a broadly accessible level, the basic concepts and some of the latest developments in the field of generalized information measures, understanding as such all those quantities which allow to obtain and quantify information from a probability distribution. Addressed not only to physicists, but also to researchers in other fields like biology, medicine, economics, *etc.*, it offers through its chapters an overview of the main measures and techniques, together with some recent relevant applications which illustrate their potential. Its scope ranges from generalized entropies and the majorization based concept of disorder to complexity measures and metrics in probability space. It includes methods for extracting probability distributions from general data series and applications ranging from quantum entanglement to biology and brain modeling. A comprehensive list of references is also contained.

Reference

[1] Claude E. Shannon, "A Mathematical Theory of Communication", Bell System Technical Journal 27, 379–423 and 623–656 (1948).

Andres M. Kowalski & Raúl D. Rossignoli
Departamento de Física–IFLP
Universidad Nacional de La Plata
and Comisión de Investigaciones Científicas
La Plata
Argentina

Evaldo M. F. Curado
Centro Brasileiro de Pesquisas Físicas and National
Institute of Science and Technology for Complex Systems
Rio de Janeiro
Brazil

List of Contributors

Josep Batle
Universitat de les Illes Balears
Mallorca, Spain

Xavier Calbet
EUMETSAT
Eumetsat Allee 1, D-64295 Darmstadt, Germany

Norma Canosa
Departamento de Física-IFLP, CONICET,
Universidad Nacional de La Plata,
C.C. 67, 1900 La Plata, Argentina

Laura C. Carpi
Civil, Surveying and Environmental Engineering.
The University of Newcastle.
University Drive. Callaghan NSW 2308, Australia
and
Departamento de Física, Instituto de Ciências Exatas.
Universidade Federal de Minas Gerais,
Av. Antônio Carlos, 6627 - Campus Pampulha.
31270-901 Belo Horizonte - MG, Brazil

Montserrat Casas
Late Professor of Universitat de les Illes Balears
Mallorca, Spain
and
IFISC-CSIC of Spain

Evaldo M. F. Curado
Centro Brasileiro de Pesquisas Físicas
and
National Institute of Science and Technology for Complex Systems,
Rua Xavier Sigaud 150
22290–180 Rio de Janeiro, - RJ, Brazil

Mabel D'Arrigo
Facultad de Ciencias Bioquímicas y Farmacéuticas,
Universidad Nacional de Rosario.
Suipacha 531, 2000 Rosario, Argentina

Jesus S. Dehesa
Institute "Carlos 1" for Computational and Theoretical Physics,
University of Granada, Granada, Spain and
Department of Atomic, Molecular and Nuclear Physics,
University of Granada, Granada, Spain

Luciana De Micco
Facultad de Ingeniería,
Universidad Nacional de Mar del Plata.
Juan B. Justo 4302. 7600 Mar del Plata, Argentina
and
CONICET

Rodolfo O. Esquivel
Institute "Carlos 1" for Computational and Theoretical Physics,
University of Granada, Granada, Spain
and
Department of Chemistry,
Metropolitan Autonomous University, Mexico D.F., Mexico

Gustavo Luis Ferri
Facultad de Ciencias Exactas y Naturales, Universidad Nacional de La Pampa
Uruguay 151, Santa Rosa, La Pampa, Argentina

Alejandra Figliola
Instituto del Desarrollo Humano.
Universidad Nacional de General Sarmiento.
Juan María Gutiérrez 1150, Los Polvorines, Buenos Aires, Argentina

Claudio M. González
Facultad de Ingeniería,
Universidad Nacional de Mar del Plata.
Juan B. Justo 4302. 7600 Mar del Plata, Argentina

Rudolf Hanel
Section for Science of Complex Systems, Medical University of Vienna.
Spitalgasse 23, A-1090, Austria

Ana María Korol
Facultad de Ciencias Bioquímicas y Farmacéuticas,
Universidad Nacional de Rosario.
Suipacha 531, 2000 Rosario, Argentina
and
Applied Optical Group, Instituto de Física Rosario, CERIDER-CONICET.
2000 Rosario, Argentina

Andres M. Kowalski
Departamento de Física-IFLP,
Universidad Nacional de La Plata,
C.C. 67, 1900 La Plata, Argentina
and
Comisión de Investigaciones Científicas (CIC), Argentina

Pedro W. Lamberti
Facultad de Matemática, Astronomía y Física,
Universidad Nacional de Córdoba and CONICET,
Ciudad Universitaria, 5000 Córdoba, Argentina

Hilda A. Larrondo
Facultad de Ingeniería,
Universidad Nacional de Mar del Plata.
Juan B. Justo 4302. 7600 Mar del Plata, Argentina

Ricardo López-Ruiz
DIIS and BIFI, Facultad de Ciencias, Universidad de Zaragoza,
E-50009 Zaragoza, Spain

Ana P. Majtey
Instituto Carlos I de Física Teórica y Computacional and
Departamento de Física Atómica Molecular y Nuclear,
Universidad de Granada, Granada, E-18071, Spain

Héctor Mancini
Department of Physics and Applied Mathematics, Institute of Physics,
Universidad de Navarra, 31080 Pamplona, Spain

María Teresa Martín
Departamento de Física-IFLP, Facultad de Ciencias Exactas,
Universidad Nacional de La Plata.
C.C. 727, 1900 La Plata, Argentina

Fernando Montani
Instituto de Física La Plata, CONICET
and Universidad Nacional de La Plata.
115 y 49 - C.C: 67 - 1900 La Plata, Argentina

Flavia Pennini
Departamento de Física, Universidad Católica del Norte.
Av. Angamos 0610, Antofagasta, Chile

Angelo Plastino
Instituto de Física, Facultad de Ciencias Exactas,
Universidad Nacional de La Plata (UNLP).
C.C. 727, 1900 La Plata, Argentina

Angel Ricardo Plastino
Instituto Carlos I de Fisica Teorica y Computacional
Universidad de Granada, Spain
and
CeBio y Secretaría de Investigación,
Universidad Nacional Buenos Aires-Noroeste UNNOBA,
Junin, Argentina

Bibiana Riquelme
Facultad de Ciencias Bioquímicas y Farmacéuticas,
Universidad Nacional de Rosario.
Suipacha 531, 2000 Rosario, Argentina
and
Applied Optical Group, Instituto de Física Rosario, CERIDER-CONICET.
2000 Rosario, Argentina

Raúl D. Rossignoli
Departamento de Física-IFLP and Facultad de Ingeniería,
Universidad Nacional de La Plata,
C.C. 67, 1900 La Plata, Argentina
and
Comisión de Investigaciones Científicas (CIC), Argentina

Osvaldo A. Rosso
Chaos & Biology Group, Instituto de Cálculo,
Facultad de Ciencias Exactas y Naturales, Universidad de Buenos Aires.
Pabellón II, Ciudad Universitaria. 1428 Ciudad de Buenos Aires, Argentina
and
Departamento de Física, Instituto de Ciências Exatas.
Universidade Federal de Minas Gerais,
Av. Antônio Carlos, 6627 - Campus Pampulha.
31270-901 Belo Horizonte - MG, Brazil

Patricia M. Saco
Civil, Surveying and Environmental Engineering.
The University of Newcastle.
University Drive. Callaghan NSW 2308, Australia
and
Departamento de Hidráulica,

Facultad de Ciencias Exactas, Ingeniería y Agrimensura.
Universidad Nacional de Rosario.
Avenida Pellegrini 250, 2000 Rosario, Argentina

Pablo Sánchez-Moreno
Institute "Carlos 1" for Computational and Theoretical Physics,
University of Granada, Granada, Spain
and
Department of Applied Mathematics,
University of Granada, Granada, Spain

Simon R. Schultz
Department of Bioengineering, Imperial College London.
South Kensington, London SW7 2AZ.

Eduardo Serrano
Centro de Matemática Aplicada, ECyT.
Universidad Nacional de San Martín.
Irigoyen 3100, San Martín, Buenos Aires, Argentina

Stefan Thurner
Section for Science of Complex Systems, Medical University of Vienna.
Spitalgasse 23, A-1090, Austria
and
Santa Fe Institute
1399 Hyde Park Road, Santa Fe, NM 87501, USA

Constantino Tsallis
Centro Brasileiro de Pesquisas Físicas
and National Institute of Science and Technology for Complex Systems,
Rua Xavier Sigaud 150
22290–180 Rio de Janeiro,–RJ, Brazil
and
Santa Fe Institute
1399 Hyde Park Road, Santa Fe, NM 87501, USA

Introduction - Summary of Contents

The goal of this book is to offer an updated overview on generalized information measures and statistics, including the basic concepts as well as some recent relevant applications.

The book begins with an historical introduction describing the fascinating development of the concepts of heat and entropy. Starting from the ideas of the ancient Greece, an account of the main historical breakthroughs is provided, which allows to appreciate the fundamental contributions of Nicolas Sadi Carnot, Rudolf Clausius, Ludwig Boltzmann, Josiah Willard Gibbs and others. It ends with the seminal works of Claude Shannon, which led to the foundation of Information Theory, and Edwin Jaynes, which provided the connection of the latter with Statistical Mechanics.

The second chapter is a basic tutorial on the essentials of information entropy, describing in an accessible level the concepts and quantities used in the rest of this book. The Shannon entropy and its associated measures such as the conditional entropy, the mutual information (a measure of correlations) and the relative entropy (also known as Kullback-Leibler divergence, a measure of the discrepancy between two probability distributions) are all presented, together with their main properties and the most important proofs. We also provide the main features of the Fisher Information, which can be presented in terms of the relative entropy between two slightly displaced distributions, and the associated Cramer-Rao bound. Other topics include the definition of entropy in quantum systems, the fundamental property of concavity and a brief introduction to the maximum entropy approach and its connection with statistical mechanics. It contains finally the Shannon-Khinchin axioms leading to the uniqueness theorem for the Shannon entropy together with an introduction to the concept of generalized entropies.

Chapters 3, 4 and 5, are devoted precisely to the generalized entropy concept. Chapter 3 presents a review by Constantino Tsallis of the famous non-additive entropy S_q which is known by his name, together with the associated generalized statistical mechanics and q-distributions. As described there, such generalized framework allows for the possibility of an extensive thermodynamic entropy in strongly correlated systems where the standard additive Boltzmann-Gibbs entropy is non-extensive. The chapter includes a comprehensive list of relevant recent applications of the formalism in the most diverse fields, together with the concomitant references. It also comments on recent relevant results related with the connection of S_q with the theory of numbers through the Riemann *zeta* function.

Chapter 4 presents an axiomatic approach for deriving the form of a generalized entropy. After describing in full detail the four Shannon-Khinchin axioms, it considers the situation where just the first three are conserved, together with the requirement of two newly discovered scaling laws which the generalized entropy should fulfill. It is shown that this leads to a general form of entropy depending essentially on two parameters, which define entropic equivalence classes. The connection with the Shannon, Tsallis, Rényi and other entropies is described in detail, together with the associated distribution functions and some related aspects. It includes an appendix containing the technical details and the demonstration of four associated theorems.

Chapter 5 discusses the relation between generalized entropies and the concept of majorization. The latter is a powerful and elegant mathematical theory for comparing probability distributions, which leads to a rigorous concept of mixedness and disorder. This chapter describes first the concept of majorization in an accessible level. It then considers its connection with entropy, and shows that by means of generalized entropies it is possible to express the majorization relation in terms of entropic inequalities. It also describes the majorization properties of the probability distributions determined by the maximization of general entropic forms, and the concept of mixing parameters, *i.e.,* parameters whose increase ensure majorization. Finally, the concept of majorization in the quantum case, *i.e.,* for density operators, is also examined. As application, the problem of quantum entanglement detection is considered, where it is shown that majorization leads to a generalized entropic criterion for separability, which is much stronger than the standard entropic criterion.

In chapter 6, the notion of distance measures for probability distributions is reviewed. An overview of the most frequently used metrics and distances like Euclidean metrics, Wootters's distance, Fisher metric and Kullback-Leibler divergence is made, centering the analysis in the distance known as the Jensen-Shannon divergence both in their classical and quantum versions. Application of the latter as a measure of quantum entanglement is also discussed. This chapter is related to the next two chapters, which are devoted to Statistical Measures of Complexity, because of the dependence of these measures with distances in probability space.

There is no universally accepted definition of complexity, nor of quantifiers of complexity. An extensive list of relevant contributions can be found in the introduction of the chapter 8, as well as in the references of chapter 7. A comparative classification of various complexity measures, by Wackerbauer, Witt, Atmanspacher, Kurths and Scheingraber, can be found in the reference [10] of chapter 8. We will here consider just a particular class of complexity measures based on information theory, which are essentially a combination of an entropy with a distance measure in probability space. Such measures vanish when the probability distribution implies either full certainty or complete uncertainty, being maximum at some "intermediate" distribution. This approach is precisely adopted in chapter 7, where Ricardo López-Ruiz, Hector Mancini and Xabier Calbet introduce the well-known measure of complexity known by their surnames (LMC Statistical Measure of Complexity). Its properties are discussed in full detail and some interesting applications (gaussian and exponential distributions, and complexity in a two-level laser model) are also provided.

In chapter 8 the properties of a Generalized Statistical Complexity Measure are discussed. The authors adopt the functional product form of the LMC Statistical Measure of Complexity, but consider different entropic forms and different definitions of distance between distributions of probability, beyond the Shannon Entropy and Euclidean distance used in chapter 7. In particular, the use of the Jensen divergence introduced en chapter 6, together with the Shannon Entropy (Shannon Jensen Statistical Complexity) is analyzed in depth. Another important aspect considered in this chapter is the methodology for the proper determination of the underlying probability distribution function (PDF), associated with a given dynamical system or time series. We should also mention here the Statistical Complexity of Shiner, Davison and Landsberg (ref. [13] of chapter 8).

In chapters 9-15, different applications of generalized information measures are considered. Chapter 9 deals with the Fisher Information, whose basic properties were introduced in Chapter 2. The chapter describes its use in radial probability distributions associated with quantum states, determining the related Cramer-Rao inequalities and the explicit expressions of the Fisher information for both ground and excited states of D-dimensional hydrogenic systems. It then considers its application to some physico-chemical processes, showing that the Fisher information can be a valuable tool for detecting the transition rate and the stationary points of a chemical reaction.

Chapters 10 and 11 deal with problems in physics while chapters 12-15 with applications in others fields. In particular, chapters 14 and 15 are devoted to biological applications.

In chapter 10, the links between the entanglement concept (see also chapter 5) and the information entropy are analyzed, as represented by different measures like the Shannon, Renyi (see chapter 2 and 4) and Tsallis (see chapter 3) ones. In chapter 11, the authors review the difference between quantum statistical treatments and semiclassical ones, using a semiclassical Fisher Information measure built up with Husimi distributions.

Chapter 12 deals with the use of Information theory tools for characterizing pseudo random number generators obtained from chaotic dynamical systems. The authors make use of the conjunction between Entropy and the Shannon Jensen Statistical Complexity introduced in chapter 8, to evaluate the quality of pseudo random number generator. It is done by quantifying the equiprobability of all its values and statistical independence between consecutive outputs by means the comparison of a Shannon Entropy calculated with a Histogram PDF and a Shannon Jensen Statistical Complexity calculated with a Symbolic Bandt–Pompe PDF (see chapter 8) in an Entropy–Statistical Complexity plane.

In chapter 13, the authors employ different information measures such as Shannon Entropy, Fisher Information measure (see chapter 2) and the Shannon Jensen Statistical Complexity (introduced in chapter 8), to analyze sedimentary data corresponding to the Holocene and so characterize changes in the dynamical behavior of ENSO (El Niño/Southern Oscillation) during this period.

In chapter 14, the authors present an application of wavelet-based information measures to characterize red blood cells membrane viscoelasticity. Relative Energy, Shannon Entropy, Shannon Jensen Statistical Complexity calculated with a Wavelet PDF, technic introduced in chapter 8, together with an Entropy–Complexity plane, are used to analyzing a human haematological disease.

Finally, in chapter 15 the authors apply an information theoretic approach to analyze the role of spike correlations in the neuronal code. By considering certain brain structures as communication channels, application of Information Theory becomes feasible, allowing in particular to investigate correlations through the pertinent mutual information. It is a nice example of the important role played by Information theoretical methods in current problems of Theoretical Neuroscience. The chapter also includes a comprehensive list of references on the subject.

Concepts and Recent Advances in Generalized Information Measures and Statistics, 2013, 3-29

CHAPTER 1

Heat and Entropy: A Brief History

Evaldo M. F. Curado [a,*], **Andres M. Kowalski** [b,c] **and Raúl D. Rossignoli** [b,c]

[a] Centro Brasileiro de Pesquisas Físicas and
National Institute of Science and Technology for Complex Systems
Rio de Janeiro, Brasil

[b] Departamento de Física-IFLP, Facultad de Ciencias Exactas,
Universidad Nacional de La Plata,
C.C. 727, 1900 La Plata, Argentina

[c] Comisión de Investigaciones Científicas (CIC), Argentina

Abstract

We provide a brief description of the historical development of the concept of Entropy, from its origins related to heat to its modern interpretation as a special measure of information.

Keywords: Heat; Entropy; Information Theory; Maximum Entropy Principle.

1 A brief history of Heat

Entropy is one of the most fundamental and controversial concepts in theoretical physics, historically related with heat. We will therefore begin this short history with the development of the concept of heat, which receives a clear definition only in the middle of the XIXth century. We will see the context in which the concept of entropy arises, some of the discussions it provokes, the traps it is involved with - mainly its relation with heat and with information theory - and our current understanding about it. It is impossible to cover a long history like that in a few pages without committing injustices with some thinkers and ideas. We are aware we may have omitted some of them and inaccurately described others. Nevertheless, we hope to provide a historical view to the reader with some of the most important actors and with some special points of view, of one of the most important

* Corresponding author. E-mail: evaldo@cbpf.br

Andres M. Kowalski, Raul D. Rossignoli and Evaldo M. F. Curado (Eds)

concepts of modern theoretical physics that is also applied in many different areas of science such as chemistry, biology, and, using related notions, in economy, social sciences and human sciences, showing the broad interest the concept of entropy provokes.

A long time ago. Historically, the entropy concept could be considered as a son of the notions, developed by many civilizations, of fire and heat. In the ancient Greece, it is known that Heraclitus of Ephesus (540-480 BC) defended the idea that the universe is in equilibrium between two opposites, in a perpetual state of tension, in an unstable state. In his universe, fire, as an agent of change, has the primordial place. However fire was not considered by him as an element of nature, it was rather the transforming process itself, symbolizing the continuous change. Empedocles (born in 1490, in the greek city of Acragas, today Agrigentum, in Sicily and died in 1430 BC) conceived the world as composed by four "roots", earth, air, water and fire, and two forces, equivalent to attraction and repulsion forces. Leucippus and Democritus, Vth century BC, the greek atomists, proposed a materialist doctrine where the world is composed of atoms and empty space. The atoms are solid matter and are so small to be visualized. The substances are different due to different atoms they are composed or due the manner they are combined. The fire, for the atomists, was also composed by spherics and very fast atoms. Note that for Empedocles, fire was one of the basic elements and for the atomists it was composed by special atoms, both views indicating the notion of materiality of heat, contrarily to Heraclitus. We could see Heraclitus as an ancestor of the modern concept of heat, as for him heat was related to a process of change, and Empedocles, supported by Plato and Aristotle (with an adapted theory of the elements[1]), and the atomists, as ancestors of the phlogiston and caloric theories, due the materiality of heat they consider [1, 2, 3]. Therefore, since then and for two thousand years, fire was considered as one of the basic elements of nature. The alchemists followed these ideas and based their "experiments" on this classification until very recently, in centuries XVI and XVII AC, with some additional properties. In fact, from their experiences in working with metals, the alchemists had introduced, during the centuries, three more properties, the "sulfur", the "mercury" and "salt". The property "sulfur" was related to the combustibility of the materials, besides their colors, affinity *etc.* The metals could be transmuted one into another by varying these properties [3]. After

[1]Aristotle introduced four properties besides the four elements of Empedocles: hot, cold, dry and wet. The basic elements could change one into another if their properties are modified.

some time, the sulfur property becomes accepted as having a substantial nature, related to fire.

Renaissance. The reintroduction of the classical texts in Europe had an important impulse with the muslim scholar Ibn Rushd, most known as Averroes in the European literature, a master of Aristotelian philosophy, born in Cordoba, Spain, in the XIth century AC. Later, the conquest of Constantinople by the turks, in 1453, led many scholars from Constantinople to Europe, also reintroducing many texts from classical greek culture into the occident. The spread of these classical greek manuscripts have been largely improved by the printing revolution, introduced in Europe by Johannes Gutenberg, a german publisher, goldsmith and printer, in the XVth century AC (in fact, it seems that the invention of printing was happened in China: the first known movable type system was invented by the Chinese in the XIth century and in Korea, with metal movable type, in the XIIIth century), and has enormously contributed to the renaissance cultural movement. In science, this period consisted, first, in rediscovering old scientific texts, mainly from the classical greek period, and later, to overpass them, building the basis of the new scientific methods, starting the scientific revolution. Works by Nicolaus Copernicus (*De Revolutionibus Orbium Coelestium*, XVIth century), proposing a new cosmology, with the sun in the center of the universe, contradicting Claudius Ptolemy; the anatomist Andreas Vesalius (XVIth century) and his *De Humani Corporis Fabrica*, going beyond Galen; Tycho Brahe (XVIth century), collecting a huge amount of data on the position of stars and planets, data that were used by Joahnnes Kepler (XVIth and XVIIth centuries) to establish, his three laws about the movement of planets, the two first laws in the book *Astronomia Nova* and finally the three laws in the work *Epitome of Copernican Astronomy*, laws showing that the best fitting trajectories for the planets were given by ellipses, having the sun at one of the focus of the ellipse, and not by circles (which were the geometrical figures used in all ancient texts), are good examples of the new winds blowing in Europe. The knowledge recovered and revered from the ancient texts was becoming surpassed.

Dawn of science. Galileo Galilei, born in Pisa, at that time belonging to the Duchy of Florence, in 1564, and died in Arcetry, Tuscany, in 1642, laid the solid foundations of the new science, as we know it today, in proposing that the experiments to be performed should be expressed by means of mathematical relationships among the physical variables.

This recipe, leading to the use of quantitative methods in science, is in the origin of the unstoppable development of science until our days. Galileo wrote in his book *Il Saggiatore*, published in 1623 by the Academia dei Lincei: *"La filosofia scritta in questo grandissimo libro che continuamente ci sta aperto innanzi a gli occhi (io dico l'universo), ma non si pu intendere se prima non s'impara a intender la lingua, e conoscer i caratteri, ne' quali scritto. Egli scritto in lingua matematica, e i caratteri son triangoli, cerchi, ed altre figure geometriche, senza i quali mezzi impossibile a intenderne umanamente parola; senza questi un aggirarsi vanamente per un oscuro laberinto"* [2]. This mathematical approach affected many areas of science, in particular in physics and chemistry.

Many new fundamental instruments were invented in the XVIth and XVIIth centuries that have largely contributed to the realization of new quantitative experiments and to the establishment of new quantitative relationships - or laws - among the physical variables. The invention of the thermometer (Galileo invented one of the first thermometers, in 1593), was an innovation of major importance. The invention of the barometer, by Evangelisto Torricelli, in 1643, and the air pump, by Otto von Guericke, in 1650, allowed Robert Boyle to discover in 1662 that, for sufficiently diluted gases, the pressure is inversely proportional to volume (at constant temperature), an important quantitative relation, today called Boyle's law, according with Galileo's prescription.

Terra Pinguis and Phlogiston. The study of heat has then experienced its first scientific approaches, starting by the "mise au point" of several instruments capable to measure the temperature, giving place to the proposition, in the XVIIIth century, of several temperature scales such as Römer, Farenheit, Celsius, Réaumur[3]. It was then possible to measure a quantity related to heat, but what about its ultimate nature? In the first half of the XVIIIth century, two main approaches existed in order to explain heat. The first one, put forward firstly by Johann J. Becher, in 1667, in which he proposed a different classification for the basic elements other than the classic four earth, air, fire and water, eliminating air

[2]Philosophy is written in this grand book which stands continually open to our gaze (I mean the universe), but no one can understand it unless one first learns to comprehend the language and to know the characters in which it is written. It is written in the language of mathematics, and its characters are triangles, circles, and other geometrical figures, without which it is impossible to understand a single word of it; without these, one is wandering around in a dark labyrinth.

[3]Isaac Newton invented one of the first temperature scales and is considered by many authors as having inspired Celsius to create his temperature scale.

and fire and expanding earth in three elements, *terra lapidea, terra fluida* and *terra pinguis*. Observing that even sulfur burns, Becher proposed the existence of a combustible principle associated with combustion, called terra pinguis, that was involved in any process with fire. Later, in 1703, Georg E. Stahl used the concept of terra pinguis of Becher to propose a theory where he postulated the existence of a substance, called by him *phlogiston*, coming from the greek, meaning burning up, present in all flammable materials, that was without color, taste, odor or mass. In burnings, this substance was liberated in the air and the fire was over when there was no anymore phlogiston in the material [3, 4]. Thus, like the propositions of Empedocles, Democritus and Becher (*terra pinguis*), in the theory of Stahl heat was associated with an existing element of nature. The other proposition was made by Daniel Bernoulli, in his book published in 1738, *Hydrodynamique*, were he considered heat as a consequence of the mechanical movement of the particles of air, and the pressure as caused by the impact of these particles on a surface, a proper kinetic theory of gases, more than one century before it was accepted. The Bernoulli theory was practically ignored, and the phlogiston theory was mostly accepted during large part of the XVIIIth century. One of the reasons why the Bernoulli kinetic theory was not accepted, besides the fact that the accepted theory of phlogiston explained heat in a different way, as a material substance, was that the idea of air at that time was dominated by a hypothesis formulated by Isaac Newton. In Newton's model, air could be constituted by very small particles having repulsive forces acting among them, a force proportional to the inverse of the distance between two particles. In this view, the particles of air had a fixed position in space, an equilibrium position taking into account the repulsive forces, only oscillating around these positions. With this picture, Newton was able to derive the Boyle's law. The authority of Newton and the fact that he was able to explain the Boyle's law led this model of air to be accepted by most scientists of the XVIIIth century. The Bernoulli model, with particles moving fast in empty space, without forces acting on them except during the impact with others particles, was unlike the Newton's view and only will be accepted in the middle of the XIXth century.

Joseph Black. An important work during this period, that largely contributed to the clarification of the concept of heat, was made by Joseph Black, in the second half of the XVIIIth century, a Scottish chemist and physicist, who noted that an understanding of heat demanded two different and independent concepts, that of intensity of heat, the temperature, and that of quantity of heat. Black also introduced other important concepts. One of them is the concept of *latent heat*, coming from the observation that when heat is added to ice at its melting point the temperature of the mixture ice/water does not increase but rather it increases the relative amount of water. The same happens at the boiling point of water, where heat only changes the amount of water but not raise its temperature. The term "latent heat" means that the heat added to the mixture was in a latent state. Other form of heat was the "sensible heat", the amount of heat used to increase the temperature of the body. Other important concept also introduced by Black is the notion of *heat capacity* [5, 6], showing that each material needs a different amount of heat to increase in one degree its temperature. The works of Black[4] had a broad and fast diffusion in the scientific community[5]

Caloric. The phlogiston theory remained dominant until the work of Antoine Lavoisier, in 1783, *Réflexions sur le phlogistique*, were he shown that the combustion requires a gas that has weight, the oxygen, and that there were no flammable material mixed within the bodies. The reason why the phlogiston theory was so long time accepted is perhaps because the phlogiston is a kind of "anti-oxygen"; where there are a large amount of phlogiston there are few amount of oxygen and vice-versa, and therefore many experiments could be explained fairly well by the phlogiston theory. With the work of Lavoisier, there was no more space to phlogiston.

Lavoisier then proposed a new theory of heat, supposing that heat was a self-repelling fluid, called caloric[6], that flows from hotter to colder bodies. Each body has some amount of caloric and the quantity of this substance is constant through universe, only being exchanged among the bodies. The elements of any material are supposed to be not in contact one with another, the space between them being filled with the caloric fluid. There was an attracting force between the elements of the material and a repulsive force due to the self-repelling caloric fluid. With the relative strength between these two attracting and repelling forces, Lavoisier was able to explain many phenomena related to heat. In particular the three states of the matter, solid, liquid and gas, were associated with a smaller or larger separation between the elements of the material, caused by a small or large amount of caloric fluid.

[4]The carbon dioxyd was also discovered by him.

[5]We note the the word heat had not a precise meaning among the scientists until the middle of the XIXth century, when the modern concept was adopted.

[6]The term "caloric" existed already in the scientific literature at that time. In 1785, for example, it was used by Louis-Bernard Guyton de Morveau (1737-1816) to denote heat or fire.

The three states of matter were rather, in the caloric theory, a physical phenomena than a chemical one. Heat was then a material substance, which was conserved, could not being destroyed nor created. A hot material contains more caloric than a cold material and the caloric fluid goes from the hotter to colder. Also, the expansion of air could be explained by the self-repelling property of the caloric fluid: more caloric implying expansion. The quantity of caloric was a state variable, like in Black. There was a heat function - state variable - associated with each material, depending on the parameters of the system, such as volume, temperature, *etc.*, giving the amount of heat (caloric) contained in that material[7] [5, 7, 8].

The caloric theory was largely accepted by the scientific community and among the scientists that have used it we can mention, besides Lavoisier, Pierre-Simon Laplace, Siméon Denis Poisson, Joseph Fourier[8], Benoit Paul-Émile Clapeyron *etc.*

Soon, problems started to appear for the theory. In 1789, Benjamin Thompson, Count Rumford, was minister of war and police in the German state of Bavaria. He noted that heat could be generated apparently without limit during the work of boring cannons of the army. According to the theory of caloric, heat should flow into the cannon from its surrounding, which should become colder. It was observed that the surroundings of the cannon become even hotter, instead of colder. Caloric, according with Lavoisier's theory, could not be created nor destroyed. Thompson concluded that some mechanical work done on the cannon was being converted into heat. This was not a definitive point against caloric theory as it could be explained, for example, as the transformation of some latent heat, stored in the cannons, in sensible heat. But the issue of conversion of work into heat was then raised [5, 7, 9].

Sadi Carnot. Nicolas Léonard Sadi Carnot, using the concept of caloric as a fluid going from a hotter system to a colder one, published in 1824 the book entitled *Réflexions sur la puissance motrice du feu sur les machines propres à développer cette puissance* [10], a major work in the history of thermodynamics, studying a generalized theory of heat engines. This

[7]We note that at that time there were in fact many different interpretations of the caloric among the scientific community, as also happened with the phlogiston theory. We mention here only the basic aspects of these theories.

[8]Although the work of Fourier on heat diffusion can be formulated independently of the caloric theory, and he never said explicitly that he adopted this theory, many of his ideas are compatible with the caloric theory [7].

book is considered as one of the cornerstone works of thermodynamics. In this book, written without equations, it was stated that the power of the engines operating between a hotter and a colder reservoir comes from the flow of caloric from the hotter to the colder. In modern language, he studied the efficiency of a reversible heat engine, showing that the efficiency of this idealized machine depends only on the two temperatures of the reservoirs[9]. One of his great contributions was to propose an idealized model of a heat engine, representing the steam engines existing at that time, allowing exact calculations and getting general results, applied to all machines. All the concepts introduced in his book were explained without equations, in a literary explanation. Concepts well known nowadays, such as Carnot cycle, Carnot efficiency, Carnot heat engine, reversible process and others, came from this small book. Also, the text contains a preliminary sketch of the second law of thermodynamics [7]. Perhaps because there were many new ideas and no equations, his book remained forgotten by almost ten years.

Equivalence work-heat. In 1834, the French mining engineer Benoit Paul-Émile Clapeyron examines the Carnot's book in his *Mémoire sur la puissance motrice du la chaleur* [11] where he, for the first time, placed in quantitative terms what is now known as Carnot cycle. The work of Carnot was also recognized by William Thomson, later Lord Kelvin, in his famous work (1848) [12] about the absolute temperature scale, where he used the Carnot's theory. Few years before, and inspired by the work of Count Rumford, Julius Robert von Mayer [13], James Prescott Joule [14] and Ludwig August Colding [15], independently, carried out a series of experiments intending to show that heat and work were equivalent. Several numerical values for the mechanical equivalent of heat were published in these years by Mayer, Joule and Colding, and, due to experimental difficulties, these values were not exactly the same but they were consistently close. The word thermodynamics appears, for the first time, in the Thomson's work *Account of Carnot's theory*, published in 1849: *"A perfect thermo-dynamic engine of any kind, is a machine by means of which the greatest possible amount of mechanical effect can be obtained from a given thermal agency"* [16].

After the effort of many scientists, and due mainly to the recent experiments that had shown the equivalence heat - work, the first explicit assertion of the first law of thermody-

[9] The maximum efficiency is equal to $(T_h - T_c)/T_h$, where T_h is the temperature of the hotter reservoir and T_c is the temperature of the colder one, an expression that Carnot did not write. This mathematical expression was due, very likely, to Lord Kelvin.

namics was apparently made by Rudolf Clausius, in the famous paper *Über die bewegende Kraft der Wärme und die Gesetze, welche sich daraus für die Wärmelehre selbst ableiten lassen*, 1850. Clausius, considering an infinitesimal Carnot cycle, calculated the work produced and the *"amount of heat which must be imported to change the gas by a definite process from any given state to another"* [17], called by him Q. He shown, first, that Q cannot be a function of the volume and the temperature and that *"The equation can be brought under the form of a 'complete differential' thus:*

$$dQ = dU + AR\frac{a+t}{v}dv,$$

where U denotes an arbitrary funcion of v and t" [17]. In this equation dU is the transfer of energy during the process, $a+t$ is the Kelvin temperature, v is the volume and A expresses the equivalent of heat for the unit of work. It can be rewritten as $dQ = dU - dW$, where $dW = -pdv$. In the same paper Clausius put forward the argument that in any irreversible process a small amount of heat is dissipated. At the same time he showed that there exist no heat function and set down on solid basis the first and second laws of thermodynamics.

The equivalence between work and heat signaled the end of the caloric theory, the long-living hypothesis of the materiality of heat and the idea that heat was a state variable.

In a series of papers published between 1851 and 1854, contained in the article *On the dynamical theory of heat* in his *Mathematical and Physical Papers, Volume I, Art. XLVII*, Thomson delineated the new theory, based on the experiments of equivalence of work and heat, and remarked that *"The recent discoveries made by Mayer and Joule, of the generation of heat through the friction of fluids in motion, and by the magneto-electric excitation of galvanic currents, would either of them be sufficient to demonstrate the immateriality of heat"*, and, *"heat is not a substance, but a dynamical form of mechanical effect, we perceive that there must be an equivalence between mechanical work and heat, as between cause and effect"* [18, 19].

Revival of kinetic theory. The alternative hypothesis of the caloric theory, the kinetic theory, proposed on scientific grounds since Daniel Bernoulli, in 1738, and supported by few scientists since then, as Mikhail Vasilyevich Lomonosov[10], Georges-Louis Le Sage[11], John

[10]His two main manuscripts about heat are *Meditations on the Cause of Heat and Cold*, published in 1750 and *Meditations on the solidity and liquidity of bodies*, published in Russian as a separate publication, in August 1760 [20].

[11]He proposed a kinetic theory of gravity in the manuscripts *Essai sur l'origine des forces mortes* (1748),

Herapath[12], John James Waterston[13], among others, and whose works remained practically ignored by the scientific community, saw, then, a revival. This theory was based on the hypothesis that all things are made by atoms and molecules, that are very small objects that we can not see, even with the help of microscopes. Gases are made of atoms and molecules traveling fast among empty space. This hypothesis had many opponents in the scientific community and the acceptation of any theory based on this hypothesis was not easy as well.

The scientific program was to explain the conservation of energy, including heat, as stated by Clausius and Kelvin, by means of the motion of atoms. In 1857 August Krönig, who apparently knew the work of Waterston, published a very simple model of kinetic theory of gases (*Grundzüge einer Theorie der Gase*) where only the translational motion of the particles were considered [23]. Clausius, in 1857, apparently not aware of the work of Krönig, proposed a more complex kinetic theory, taking into account, besides the translational motion, the rotational degrees of freedom. He could then estimate, using heat data, the fraction of total energy that could be associated with the translational motion. He also introduced the concept of mean free path of a particle of gas. This work inspired James Clerk Maxwell to conceive the idea that the particles of a gas have not the same velocity[14], obeying in fact a distribution of velocities, that he was able to calculate and is nowadays known as Maxwell-Boltzmann's distribution. It was not anymore necessary to calculate the motion of each individual particle using Newton's law (by the way a hopeless task, given the extremely large number of molecules involved), it was sufficient to know the density of particles having a particular speed. All the particle motions were represented in a single equation, obeying a statistical law. He communicated these results in a meeting in 1859 and published the paper, *Illustrations of the Dynamical Theory of Gases*, one year later [24]. This was the first time probability enters in a physical theory. The time of determinism

Essai de Chymie Méchanique (1758), and *Lucrèce Newtonien* (1784).

[12]He considered the momentum of a particle as proportional to the temperature and obtained that the product of the pressure by the volume was proportional to the square of the temperature. In 1821 he presented his ideas in the paper *On the Causes, Laws and Phenomena of Heat, Gases, Gravitation* [21].

[13]He was able to derive the relation $PV/T =$ constant based only on the kinetic hypothesis. The Waterston's work was submitted to the Royal Society in 1845 but it was rejected; after his death his paper was finally published in the Philosophical Transactions of the Royal Society in 1892 [22].

[14]Clausius assumed that, at some temperature, all molecules, of any type, have the mean velocity; he did this not because he believed in that but because he did not know how to do otherwise.

and certainties in physical theories was over.

2 Entropy

Clausius. Clausius faced the challenge of putting in quantitative terms this new quantity, that was related to heat and that was referred in the literature only in qualitative terms. In a series of papers, starting in 1854[15], followed by papers in 1856, 1862[16], and finally in 1865, with the paper: *Ueber verschiedene für die Anwendung bequeme Formen der Hauptgleichungen der mechanischen Wärmetheorie* [25], he achieved his objective. In this last paper, also, the word entropy, substituting the word previously used by him, "equivalence-value", was invented. In his words: *"I prefer going to the ancient languages for the names of important scientific quantities, so that they mean the same thing in all living tongues. I propose, accordingly, to call S the entropy of a body, after the Greek word [τροπη], 'transformation'. I have designedly coined the word entropy to be similar to energy, for these two quantities are so analogous in their physical significance, that an analogy of denominations seems to me helpful".* This quantity, for any reversible process, was shown in his Eqs. (59-60) (in Clausius notation):

$$dS = \frac{dQ}{T} \quad \text{and} \quad S = S_0 + \int \frac{dQ}{T} \,,$$

where dQ is the amount of heat in an infinitesimal process, T is the absolute temperature proposed by Kelvin and S_0 is a constant[17]. He ended this seminal paper with the famous sentences:

1) Die Energie der Welt ist constant.

2) Die Entropie der Welt strebt einem Maximum zu.[18].

This new quantity, S, is a state variable, depending on the physical parameters of the system, and independent of the physical process realized. It is now called the thermodynamic entropy. It is not a conserved quantity, like energy is. In an isolated system this quantity or remains constant or it increases, it never decreases. It is also strongly related to the heat

---

[15]In this paper (*Ann. de Pogg. XCIII (1854) 481*) the first mathematical formulation of entropy, called by him "equivalence-value" and until then just explained by words, was presented.

[16]Where he states quantitatively the second law of thermodynamics: "The algebraic sum of all the transformations occurring in a cyclical process can only be positive, or, as an extreme case, equal to nothing."

[17]For an irreversible process the change of entropy is always higher than the change of heat divided by T, *i.e.*, $dS > dQ/T$.

[18]1) The energy of the world is constant. 2) The entropy of the world tends to a maximum.

exchanged in a physical process, where the absolute temperature T works, mathematically, as an integrating factor. Heat, in modern terms is a form of energy, existing only during a process, it is not a state variable, as it was supposed before. This is one of the reasons of the difficulties the nonexpert and even scientists have to understand modern thermodynamics. The technical scientific meaning of heat differs in many aspects from the intuitive notion of heat we have. By other way round, it is astonishing as entropy has many properties in common with the popular and not precise meaning of heat. Heat, for a nonexpert, is a state variable (each body has a quantity of heat associated with), it flows from a hotter body to a colder one and is an increasing function of temperature, meaning that higher temperatures mean hotter bodies. All these properties are valid for the thermodynamic entropy. We can speak about an entropy flow, going from a system with higher temperature to a system with lower temperature; it increases with the temperature and it is a state variable, we can associate an amount of entropy for a body, even when it is at rest. Besides that, it is a quantity that can be created but never destroyed. Clearly, entropy is not heat, which is a form of energy existing during a process, but the nonexpert can think in the common sense of heat he has, if he wants to understand, in a hand-waving way, the behavior of entropy in a physical system. This new quantity, entropy, with a new name, not known before had, since the beginning, an element of mystery. A new name doesn't bring any previous concept or intuition behind, the new intuition has to be built.

Clausius proposition of the new quantity entropy was based on macroscopic considerations, but it remained to understand this new quantity microscopically, in the framework of the kinetic theory. This step was worked mainly by Ludwig Eduard Boltzmann, who tried to find a mechanical interpretation for the entropy and to explain why it always increases. Soon, influenced by Maxwell's work, he recognized that it was necessary to introduce statistics to treat the large number of molecules of a gas.

Boltzmann. Boltzmann proposed in 1872, in a paper with 95 pages, *Weitere Studien über das Wärmegleichgewicht unter Gasmolekülen* [26], a partial differential equation to describe the dynamics of an ideal gas, or more precisely, the time evolution of the distribution function (a density function), $f(x, t)$, which determines the number of molecules of a gas having energy (*"vis viva"*) x at time t[19]. He could then show that if the distribution function

[19] This equation is defined in an one-particle 6-dimensional phase space (μ-space), and it is known in textbooks as Boltzmann's equation.

is given by the Maxwell's formula the time derivative of $f(x,t)$ will vanish, showing that: *"once this velocity distribution has been reached, it will not be disturbed by collisions"*. This only shows that the Maxwell distribution is a stationary distribution for the time evolution equation he derived (Boltzmann equation). To show that this distribution is also a stable distribution, Boltzmann proposed a new quantity, E[20], and shows that $dE/dt \leq 0$, being zero only for the Maxwell distribution[21]. In his words, chosen among ten pages of the demonstration of the theorem: *"However, we shall now consider the problem more generally. We shall assume that the distribution of kinetic energy is initially a completely arbitrary one, and we shall ask ourselves how it changes in the course of time. Its variation is determined by the partial differential equation (16) [Boltzmann's equation]... We shall first give the proof of a theorem which forms the basis of our present investigation: the theorem that the quantity*

$$E = \int_0^\infty f(x,t)\left\{\log\left[\frac{f(x,t)}{\sqrt{x}}\right] - 1\right\} dx$$

can never increase, when the function $f(x,t)$ that occurs in the definite integral satisfies the differential equation (16) ... E does not depend on x, but only on t. Since t does not appear in the limits of the integral, we can obtain the time derivative dE/dt... Hence, the quantity under the integral sign, and therefore the entire integral, is necessarily negative. Therefore E must necessarily decrease. $\cdots$ Since, as we shall see later on, E cannot go to negative infinity, it must approach a minimum with increasing time, and at the minimum we have $dE/dt = 0$, hence equation $\cdots$ holds. This equation implies ...,

$$\frac{f(x,t)f(x',t)}{\sqrt{x}\sqrt{x\prime}} = \frac{f(\xi,t)f(x+x'-\xi,t)}{\sqrt{\xi}\sqrt{x+x'-\xi}}$$

In order that this equation may be satisfied for all values of the variables x, x', and ξ, it is necessary, as can easily be shown, that

$$f(x,t) = C\sqrt{x}e^{-hx}$$

It has thus been rigorously proved that, whatever may be the initial distribution of kinetic energy, in the course of a very long time it must always necessarily approach the one found by Maxwell" [26, 27]. In order to make clear the above statement, he used a "trick" that

[20]Later, this quantity was called H by Samuel Hawksley Burbury, in 1890 [28]. Both notations are not happy because E remembers energy and H remembers Hamiltonian, in modern notation.

[21]This is known nowadays as Boltzmann's H-theorem.

becomes clear only many decades later, with the advent of quantum mechanics and its associated energy discretization. This trick was necessary because the phase space in classical mechanics is a continuum space, where each point represents a microscopic state of the system. Therefore, any small volume of the phase space contains an infinite number of states, and the combinatorial reasoning used by Boltzmann could not be applied. This coarse-grained phase space adopted by Boltzmann (then with 28 years-old) reveals an extremely happy intuition, corroborated only almost thirty years later, with the work of Max Planck in 1900. As written in the summary of Boltzmann's 1872-paper [27], *"To clarify the reasoning involved in these proofs, we replace the continuous energy variable x by a discrete variable which can take only the values ϵ, 2ϵ, 3ϵ, ...; we then show that the same result can be derived by taking the limit $\epsilon \to 0$"*. Although Boltzmann had used this discretization as a mathematical tool, it was the first time energy was considered as assuming only discrete values. He thought then that, based only on dynamics, he had shown the approach to equilibrium and the second law.

The Boltzmann's H-theorem, stating that the new quantity, $-\int f(x,t)\log f(x,t)dx$, subject to constraints, can never decrease in function of time and corresponds to the second law, was commented by Johann Josef Loschmidt, in 1876. He noted that the microscopic evolution laws are reversible and it was, then, impossible to obtain a macroscopic irreversibility based only on microscopic reversible laws. As a counter-example, he mentioned that if, in an instant of time during a process where the new Boltzmann's quantity (entropy) was increasing, we reverse the velocity of all particles, the system will follow exactly the inverse trajectory in phase space, implying thus the new quantity introduced by Boltzmann should decrease [29], breaking in this way the Boltzmann's statement. This is known as the reversibility paradox ("umkehreinwand") and represented a serious argument against the purely mechanical interpretation of the second law of thermodynamics. This important remark forced Boltzmann to review his statements and to remark, for the first time, the connection between entropy and probability. The Boltzmann's answer appeared one year later, in 1877: *"Since the entropy would decrease as the system goes through this sequence in reverse, we see that the fact that entropy actually increases in all physical processes in our own world cannot be deduced solely from the nature of the forces acting between the particles, but must be a consequence of the initial conditions. Nevertheless, we do not have to assume a special type of initial condition in order to give a mechanical proof of the second law, if we are willing to accept a statistical viewpoint. While any individual non-uniform state (corresponding to low entropy) has the same probability as any individual uniform state (corresponding to high entropy), there are many more uniform states than non-uniform states. Consequently, if the initial state is chosen at random, the system is almost certain to evolve into a uniform state, and entropy is almost certain to increase"* [30]. The important sentence *"...the system is almost certain to evolve into a uniform state"* recognizes that the second law is a statistical law, not a deterministic rule. In order to clarify this argument, he published in the same year the paper, *Über die Beziehung zwischen den zweiten Hauptsatze der mechanischen Wärmetheorie und der Wahrscheinlichkeitsrechnung respektive den Sätzen über das Wärmegleichgewicht,* [31] where he discussed an arbitrary and generalized model. His arguments can be summarized in the following way. Coming back to his mathematical model, he assumed that each molecule could present only p discrete values of the kinetic energy ϵ, 2ϵ, $3\epsilon, \cdots, p\epsilon$. In a collision, two molecules could exchange their kinetic energies, however, both before and after the interaction, the two molecules could have only energies belonging to these p values of energy. The total energy of the system is $\lambda\epsilon$, where λ is an integer (very large). He looked the way to distribute N molecules among the p discrete energy levels. Suppose the N molecules are distributed amongst the p-levels, with ω_1 molecules with energy ϵ, ω_2 with energy 2ϵ, ω_3 with energy 3ϵ and so on, until the level p. To this specific microscopic configuration, called by him "complexion", we can associate a state (also called macrostate) consisting of the set $\{\omega_i\}$ of occupation numbers of the energy levels ($\sum_{i=1}^{p} \omega_i = N$ and $\sum_{i=1}^{p} \omega_i \, i\epsilon = \lambda\epsilon$), and a corresponding number P, called by Boltzmann *"Permutabilitätsmass"*, permutability,

$$P = \frac{N!}{\omega_1! \, \omega_2! \, \cdots \, \omega_p!} \, ,$$

which is the number of possible permutations of the N particles, discarding the permutations of particles within each energy level. This number is proportional to the probability W of the macrostate associate with the set $\{\omega_i\}$ of molecules among the levels,

$$W = \frac{P}{\sum_{states} P} \, ,$$

where the sum is over all possible macrostates, *i.e.*, all different sets of occupation numbers $\{\omega_i\}$ satisfying the fixed values of energy and number of particles. As the logarithm is a monotonically increasing function, the maximum of P among the possible states $\{\omega_i\}$ can be reached maximizing $\log P$, taking the constraints into account. Using the Stirling approximation and taking into account the constraints of number of molecules and energy, Boltzmann found that the probability for a molecule to have kinetic energy equal to $j\epsilon$ is given by the expression $\omega_j \propto \exp(-j\epsilon/\bar{\epsilon})$, where $\bar{\epsilon} = \lambda\epsilon/N$ is the average kinetic energy per molecule, which is a kind of Maxwell distribution for discrete values of energy. After considering the continuum limit for the energy he found that maximizing $\log P$ leads to the Maxwell distribution. He also related $\log P$ with the entropy he proposed in the 1872's paper and with the thermodynamical entropy [31, 32, 33]. No dynamical aspects were considered by Boltzmann in this new approach but, nevertheless it led to a relation between the thermodynamic entropy and the probability W_{eq} of the equilibrium state in a system whose energy is constant. Thus, identifying the probability of a configuration (macrostate) with the number of likelihood microscopic configurations associated with this macrostate, divided by the total number of microscopic configurations, it is possible to see that the macrostate associated with the thermal equilibrium is the most-probable macrostate. When the number of particles is huge, as it is in physical systems of macroscopic size, the number of microscopic states associated with the macrostate of thermal equilibrium is much higher than the number of microscopic states associated with macrostates of non-equilibrium such that, in practice, the system stays in the equilibrium state forever (*law of large numbers*).

Boltzmann started his research trying to demonstrate the second law with only mechanical considerations but ended with a relation between the second law and probability only, without needing to consider any mechanical considerations for the equilibrium state. The meaning of the second law started to become clear: the system evolves in direction of the macrostates with larger probability; once reached these states the system remains there, since the probability to go to macrostates with lower probability is extremely small if the number of particles is very large. In the words of Maxwell: *"the second law is drawn from our experience of bodies consisting of an immense number of molecules. ... it is continually being violated, ..., in any sufficiently small group of molecules ... As the number ... is increased ...the probability of a measurable variation...may be regarded as practically an impossibility"*, or Josiah Willard Gibbs: *"In other words, the impossibility of an uncompensated decrease of entropy seems to be reduced to an improbability"* [34, 35, 36, 37].

Therefore, the second law could be regarded as an improbability, not as an impossibility, and the thermodynamic entropy as a quantity related to the number of microscopic states (or to the volume in phase space) associated with a macroscopic equilibrium configuration of the system, *i.e.*, with the physical parameters on which the experiment is realized.

Heat-entropy. In spite of many similarities between entropy and the common sense of heat, as remarked before, now we can understand better these similarities and differences. When there is a heat flux into a system, the system is receiving energy (heat is a form of energy), increasing its internal energy. Higher energy implies a higher number of microscopic accessible states (or, equivalently, a larger volume in phase space), implying an increase of entropy. Also, when heat flows from one system to another, one system looses internal energy, lowering its entropy, and the other increases its internal energy, increasing its entropy. We can then speak, in a metaphoric sense, in flow of entropy, following the common sense of flow of heat - considered a long time as a real material fluid. Looking another similarity, when the temperature of the system increases, it means that the system has a higher internal energy than before, implying that its entropy has also increased (there is a larger number of accessible microscopic states than before, *i.e.*, a larger volume in the phase space). Finally, when the heat flow ceases, the system reaches a new thermodynamical equilibrium, with a new energy value, and with a new number of microscopic accessible states, *i.e.*, with a new value of the entropy, which is a state variable. There is always an amount of entropy associated with each body - related to the number of accessible states, or volume occupied in the phase space -, like the common sense of heat says that each body has an amount of heat. The old intuition of heat has a new actor as protagonist, the entropy; the old actor, heat, is now, technically, completely understood and has a smaller rôle. As always happens in history of science, and in physics in particular, an old concept gives place to a new one, presenting some similarities with the old one but being more complex and much more abstract.

We know that in a reversible process, the change in entropy is related to the change in heat during the process, divided by the temperature. But, in an irreversible process this connection is not straightforward and we, using the interpretation brought to us by Boltzmann, can see a clear difference between the entropy and heat concepts. The discovery of Boltzmann, associating entropy with the number of possible microscopic states - or volume in phase space - compatible with the external constraints, helps us to better understand the adiabatic expansion of a gas for example, where there is no heat involved but the entropy increases. It increases because the phase space has been enlarged when the volume suddenly increases, and this larger phase space is immediately occupied, implying a larger number of possible configurations, leading to a higher value for the entropy. No heat is involved but the entropy is changed. The concept of entropy is becoming clearer. We could, at this point, have realized that a measure of the quantity of microscopic states associated with a particular configuration (information) should have a strong relation with entropy, but it was too early for that. In fact this connection was recognized, but only more than seventy years later.

Boltzmann and Gibbs entropies. Boltzmann also proposed another very fruitful idea [32], later extensively developed by Gibbs [38], that statistical mechanics could not only describe a single object, but an ensemble of similar objects. The collection of the states of all these objects gives a density in the $6N$-dimensional phase space, also called Γ space. Gibbs discussed several ensembles, the microcanonical, canonical and grand-canonical ones. In particular, for a microcanonical ensemble he postulated that all the microscopic states having fixed values for the energy (with a small fluctuation around) and all other external variables (volume, number of particles, *etc.*), are equally likely to happen, having the same probability [38]. Note that with the concept of ensemble and this hypothesis of equal probability for states satisfying the same external constraints, it is not necessary to know anything more about the particular dynamics of the system, the macroscopic equilibrium properties of the system could be deduced based only on the probabilities of the microscopic states. It is then possible to avoid many problems related to the dynamics, but only the equilibrium states can be studied. The entropy, that depends only on the probability of the states (or complexions), could be deduced without reference to the dynamics. However, Boltzmann has also used another derivation in order to find the probability of a volume Ω in the phase space of a physical system based on the dynamics, instead of a statistical approach. He looked the trajectory one system describes in a phase space as time goes on. The probability to find the system in the volume Ω, W_Ω, is proportional to the amount of time the system spends in this volume, t_Ω, divided by the total evolution time, T, when the evolution time goes to infinity, $W_\Omega = \lim_{T\to\infty} t_\Omega/T$ [39, 40]. In a ergodic system this probability W_Ω should be the same as if ensemble methods are used. In the beginning of the XXth century Einstein made important remarks about this point.

Entropy is then rather a statistical concept related to the system. Later, Gibbs introduced an analogous definition of entropy, $S_G = -\int f_G(\Gamma)\log f_G(\Gamma)\,d\Gamma$, similar to the Boltzmann one, but defined on the phase space of N-particles, *i.e.*, the phase space of

$6N$-dimensions, usually called Γ-space. $f_G(\Gamma)$ is the probability distribution in the Γ-space. The distribution of particles, $f(x,t)$, in an one-particle space, a 6-dimensional space, used by Boltzmann in his entropy definition, can be obtained from the general distribution function f_G in the Γ-space, by means of the integration over the other $N-1$ particles. Clearly, when we integrate over the other $N-1$ particles, we loose the information contained in the correlations among them, that there are present in f_G, and then the following inequality $S_{Boltzmann} \geq S_{Gibbs}$ should be satisfied, the equality happening only for independent particles [41].

The now famous expression $S = k_B \log W$, where W is the number of allowed states, which is written in the Boltzmann's tombstone, in Vienna, was written for the first time by Max Planck (with $\Re_0$ in the place of W) [42], in 1906, where also, for the first time, the Boltzmann constant k_B was presented.

These great contributions of Boltzmann, leading the fundamentals of statistical mechanics, were not shared by the whole scientific community at that time and actually a large number of dispute with other scientists occurred, involving many technical aspects of his achievements. Besides the reversibility paradox, already discussed, there were also the questions of ergodicity, the discussion on initial conditions, the Maxwell's demon, the Poincaré recurrence theorem (wiederkehreinwand), the Gibbs paradox, coarse-grained and fine-grained phase space, the hypothesis of molecular chaos (Stosszahlansatz), the existence of atoms and molecules *etc.* [5, 7, 27, 32, 39, 43]. Nowadays, the ideas and interpretations introduced by Boltzmann are solidly stated. They are part of modern science and constitutes the base of modern statistical mechanics.

Einstein remark on dynamics. Einstein, in the beginning of the XXth century, remarked that the statistical properties of a physical system have to be derived from the dynamics of the system, being at variance with the statistical methods, without dynamical considerations, proposed by Boltzmann and Gibbs [44, 45]. The point raised by Einstein is that the probability of a system to be in a specific volume in phase space comes from the dynamics of the system, and only after the dynamics fill the phase space in such a way the density at each region of the phase space no longer changes, it is possible to attribute a probability to a specific volume in the phase space. This probability cannot be given a priori, it comes as a consequence of the equations of motion of the system. In the Einstein's words: *"Usually W equals the number of complexions. In order to compute W one needs*

a complete (molecular-mechanical) theory of the system. Therefore it is dubious whether the Boltzmann principle has any meaning without a complete molecular-mechanical theory or some other theory which describes the elementary processes. In equilibrium, the expression $S = k \ln W$ *+ constant, seems without content from a phenomenological point of view, without giving in addition such an elementary theory"* [45, 46]
Certainly, in Hamiltonian systems presenting chaotic behavior, the phase space can be filled completely and the statistical approach works well, as it is well-known. The question if it works also for all complex dynamics is a current dispute, with no clear answer until now.

Quantum entropy. With the advent of quantum mechanics, in the twenties of the XXth century, the mathematical space where the physical systems are represented is not anymore the classical phase space of positions and momenta but the Hilbert space. The discretization of energy, used by Boltzmann as a mathematical trick, is now accepted as a fundamental property of nature. It was then necessary to adapt the entropy formula proposed by Boltzmann (that worked in a classical phase space) to the new mathematical Hilbert space. This was done by John von Neumann in 1927, where, first, he defined the density operator $\hat{\rho}$, acting on the Hilbert space and representing the macrostate of a physical system (it is the analogous, in quantum mechanics, of the probability distribution on the phase space in classical mechanics). It is important to note that this density matrix has also been proposed, independently, by Lev Landau, who was trying to overpass the difficulty to describe a subsystem of a composite quantum system by the state vector formalism [47]. John von Neumann then defined the quantum entropy as

$$S = -k_B Tr \hat{\rho} \log \hat{\rho} \,,$$

where Tr is the trace on the Hilbert space of the system [48]. The quantum entropy $S(\hat{\rho})$ is additive and presents many properties in common with the classical entropy, not all of them however, due the intrinsic properties of the quantum formalism (the Shannon additivity is not a valid property for the quantum entropy). He showed that at equilibrium the quantum entropy is equal to the thermodynamical entropy. In quantum statistics the particles are indiscernible, the permutation of two particles does not lead to a new state, unlike classical statistics. This particular property of quantum particles solves the Gibbs paradox; it was necessary to have a new microscopic theory to understand it. The quantum entropy is also known in the literature as the von Neumann entropy.

Information Theory. At this time, the relations between the entropy concept and a measure of amount of accessible states were not strange. In the forties, in large part due the second war, the communication science was being rapidly developing. Claude Elwood Shannon, in 1948, created the information theory in his seminal paper, *A Mathematical Theory of Communication* [49], where he studied the best way to encode an information to be transmitted. He was faced to quantify the uncertainty in a message. In this paper, section 6, entitled "Choice, Uncertainty and Entropy", he wrote: "*... Can we define a quantity which will measure, in some sense, how much information is produced by such a process, or better, at what rate information is produced?*

Suppose we have a set of possible events whose probabilities of occurrence are p_1, p_2, $\cdots, p_n$. *These probabilities are known but that is all we know concerning which event will occur. Can we find a measure of how much choice is involved in the selection of the event or of how uncertain we are of the outcome?*. He then required three reasonable properties such a measure $H(p_1, p_2, \cdots, p_n)$ should obey: i) "*H should be continuous in the p_i*"; ii) "*If all the p_i are equal, $p_i = 1/n$, then H should be a monotonic increasing function of n*; iii) he required what it is now called in the literature as the Shannon additivity. With these three assumptions he proved that the only measure satisfying his three requirements is:

$$H = -K \sum_i p_i \log p_i \,,$$

where K is a positive constant. Still Shannon: "*Quantities of the form $H = -\sum p_i \log p_i$ (the constant K merely amounts to a choice of a unit of measure) play a central role in information theory as measures of information, choice and uncertainty. The form of H will be recognized as that of entropy as defined in certain formulations of statistical mechanics ... H is then, for example, the H in Boltzmann's famous H theorem*". Finally he gave the name of his measure of information, choice and uncertainty: "*We shall call $H = -\sum p_i \log p_i$ the entropy of the set of probabilities $p_1, \cdots, p_n$*"[22]. The letter H was chosen, without doubt, due the H-theorem by Boltzmann. Note that in this work he was building a theory of information, there was no physical substance there, no heat as well. Besides of that, based only on the mathematical similarities between the measure of information/uncertainty[23] he

[22]The entropic form $S = -\sum p_i \log p_i$ is nowadays known as Boltzmann, Boltzmann-Gibbs or Boltzmann-Gibbs-Shannon entropy in the scientific literature.

[23]Information is a measure of the decrease of uncertainty at a receiver. They are concepts dialectically related.

proposed the same expression used in statistical mechanics, he gave the same name given by Clausius to a state variable in thermodynamics, a state variable related to a physical substance. It is unnecessary to say that this choice of name for his measure of uncertainty in a set of probabilities $\{p_i\}$'s, which is a very beautiful and abstract concept, is the origin of some confusion even today. There are many measures of information/uncertainty [50, 51], each one with its own properties, but there is only one entropy, a physical quantity associated with a physical system and strongly related with temperature (they are thermodynamically conjugate variables). The physical entropy has to be extensive and related to the exchange of heat in reversible transformations. The Shannon entropy, used in information theory, has none of these properties. The Clausius entropy can be seen as a measure of information but a measure of information, in general, is not a physical entropy. According to some authors, the name entropy was suggested by von Neumann to Shannon, who had said once [52]: *"My greatest concern was what to call it. I thought of calling it information, but the word was overly used, so I decided to call it uncertainty. When I discussed it with John von Neumann, he had a better idea. Von Neumann told me, 'You should call it entropy, for two reasons. In the first place your uncertainty function has been used in statistical mechanics under that name. In the second place, and more important, no one knows what entropy really is, so in a debate you will always have the advantage' "*.

Statistical mechanics and information theory. One key Shannon's remark quoted above is when he noted that the important and sufficient point to use his measure is to know the probabilities p_i's, it doesn't matter what they are. Like, in a physical system, we can associate a probability distribution with a classical phase space or a density operator with a quantum system, the connection between information theory and thermodynamics was in view. This connection came in 1957 with Edwin Thompson Jaynes, when he published two seminal papers entitled *Information theory and Statistical Mechanics*, [53]. Jaynes: *"Previously, one constructed a theory based on the equations of motion, supplemented by additional hypotheses of ergodicity, metric transitivity, or equal a 'priori' probabilities, and the identification of entropy was made only at the end, by comparison of the resulting equations with the laws of phenomenological thermodynamics. Now, however, we can take entropy as our starting concept, and the fact that a probability distribution maximizes the entropy subject to certain constraints becomes the essential fact which justifies use of that distribution for inference... In freeing the theory from its apparent dependence on physical hypotheses of the above type, we make it possible to see statistical mechanics in a much more general light. Its principles and mathematical methods become available for treatment of many new physical problems."*. In fact, the approach used by Jaynes is actually used in practically all statistical mechanics courses, lectures, researches *etc.*, allowing physicists to treat a broad number of problems, including interdisciplinary ones, otherwise unimaginable. It enormously simplifies the task, but by means of an abstract formalism, an inference approach. Entropy is in its more pure state, depending only on the probabilities of configurations.

In the last years, some questions were raised concerning the extensivity of the Boltzmann-Gibbs entropy for physical systems presenting long range interactions. Some mathematical examples exhibiting long range correlations have been shown, where the entropic measures which are extensive are not the Boltzmann-Gibbs one but other entropic forms, as for example the Tsallis entropy, $S_q = k(1 - \sum_i p_i^q)/(q - 1)$, which, depending of the type of correlations present in some simple models, is extensive for some specifics values of q [51]. According to some authors, if this happens for a physical system, the information measure that we could call entropy, in the Clausius sense, should be the one which is extensive . This question is related to the Einstein's remark about the dynamical content of the probabilities in phase space and has to be answered by looking the dynamics of strongly correlated systems, which are very hard to study. The final word will come, as always in physics, from the experiments, from nature.

Conflict of interest: None.

Acknowledgments

E.M.F.C. thanks J. P. Gazeau for a critical reading of the manuscript and acknowledges the financial support from the brazilian scientific agencies CNPq, FAPERJ and CAPES. A.M.K. and R.D.R. acknowledge support from CIC of Argentina.

References

[1] R. Taton, *Histoire Générale des Sciences*, Presses universitaires de France, 1957.

[2] C. Ronan, *Histoire mondiale des sciences*, Éditions du Seuil, 1998.

[3] J. R. Partington and D. McKie; *Historical studies on the phlogiston theory*, Annals of Science 2 (1937) 361-404; 3 (1938) 1-58 and 337-71; 5 (1939) 113-49.

[4] J. Priestley, *Considerations on the doctrine of phlogiston and the decomposition of water*, Philadelphia, Thomas Dobson, 1796 (http://web.lemoyne.edu/~giunta/phlogiston.html);
J. H. White, *The History of the Phlogiston Theory* (Arnold, London, 1932); http://en.wikipedia.org/wiki/Phlogiston_theory.

[5] S. G. Brush, *The Kind of Motion we call Heat*, Amsterdam, North Holland, 1976.

[6] M. K. Barnett, The Scientific Monthly, 62 (1946) 165-172, 247-257.

[7] C. Truesdell, *The tragicomical history of thermodynamics 1822-1854*. Springer, New York, 1980.

[8] R. J. Morris, http://faculty.poly.edu/~jbain/heat/readings/72Morris.pdf

[9] M. Fowler, *Early attempts to understand the Nature of Heat*, http://galileo.phys.virginia.edu/classes/152.mf1i.spring02/What%20is%20Heat.htm

[10] S. Carnot, *Réflexions sur la puissance motrice du feu et sur les machines propres à déveloper cetter puissance*, Bachelier, Paris,1824.

http://books.google.com/books?id=YcY9AAAAMAAJ&printsec=frontcover&hl=pt-BR#v=onepage&q&f=false
http://www.history.rochester.edu/steam/carnot/1943/Section2.htm
http://www.history.rochester.edu/steam/carnot/1943/Section3.htm

[11] E. Clapeyron, Journal de l'Ecole Polytechnique, XXIIIéme Cahier, Tome XIV, 1834. Reprint: *Mémoire sur la puissance motrice de la chaleur*, Ed. Jacques Gabay, Paris, 2006.

[12] W. Thomson - Proceedings of the Cambridge Philosophical Society 1 (1848) 66-71.

[13] J. R. Mayer, Annalen der Chemie und Pharmacie 42 (1842) 233-240; Heilbronn Dechsler, (1845); both papers are in *Die Mechanik der Wärme*, Stuttgart (1867), by J. R. Mayer (http://www.philosophiebuch.de/waerme.htm).

[14] J. P. Joule, Philosophical Magazine 26 (1845) 369-383; 27 (1845) 205-207; Manchester Courier, May 5 and 12 (1845); J. P. Joule and W. Thomson, Philosophical Magazine 4 (1852) 481-492. Also: *Joule's Scientific Papers*, Vol. 1, 1887.

[15] L. A. Colding, *Nogle saetninger om kraefterne* ("Theses concerning moving forces"), read at the Royal Danish Academy of Sciences and Letters, 1843, published in L. A. Colding, *Scientific reflections on the relationship between intellectual life's activity and the general forces of nature*, 1856. Also: P. F. Dahl, (1963), *Ludvig A. Colding and the Conservation of Energy.* Centaurus, 8 (1963) 174-188.

[16] W. Thomson, Transactions of the Royal Society of Edinburgh 16 (1849) 541-574.

[17] R.Clausius, Annalen der Physik und Chemie 79 (1850) 368-397; R. Clausius, Philosophical Magazine II (1951) 1-21.

[18] W. Thomson, Transactions of the Royal Society of Edinburgh 20 (1851) 261-268, 289-298.

[19] W. Thomson, *Mathematical and Physical Papers*, Vol. I, Cambridge University Press, 1882. Also in:
http://gallica.bnf.fr/ark:/12148/bpt6k95118c/f5.image.r=william+thomson.langPT.

[20] M. V. Lomonosov, Novi Commentarii Academiae scientiarum imperialis Petropolitanae 1 (1750) 206-229. Also in: M. A. Lomonosov, Collected Works III, 1952.

[21] J. Herapath, Annals of Philosophy 9 (1821) 273-293.

[22] J.J. Waterston, Philosophical Transactions of the Royal Society of London A 183 (1892) 179.

[23] A. Krönig. Annalen der Physik, 99 (1857) 315-322.

[24] J. C. Maxwell, Philosophical Magazine XIX (1860) 19-32 and XX (1860) 21-37.

[25] R. Clausius, Annalen der Physik und Chimie 7 (1865) 23.

[26] L. Boltzmann, Wiener Berichte 66 (1872) 275-370.

[27] S. G. Brush, *Kinetic theory, Vol. II*, Pergamon Press, Oxford, 1966, p.88.

[28] S. Burbury, Philosophical Magazine 30 (1890) 298-317.

[29] J. Loschmidt, Wiener Berichte 73 (1876) 139.

[30] L. Boltzmann, Wiener Berichte 75 (1877) 67.

[31] L. Boltzmann, Wiener Berichte 76 (1877) 373-435.

[32] S. Dahmen, Revista Brasileira de Ensino de Física 28 (2006) 281-295.

[33] E. G. D. Cohen, *Entropy, probability and dynamics*, in the book *Boltzmann's Legacy*, Eds. G. Gallavotti, W. L. Reiter and J. Yngvason. European Mathematical Society, 2008, Freiburg, Germany.

[34] J. C. Maxwell, *Theory of Heat*, p. 308: Taits Thermodynamics, Nature 17, 257 (1878). Quoted in M. Klein, ref. 4d.

[35] J. W. Gibbs, Connecticut Academy Transactions 3 (1875) 229, reprinted in The Scientific Papers, 1, 167 Dover, New York 1961).

[36] L. Lebowitz, arXiv:math-ph/0010018v1.

[37] J. Uffink, *Rereading Ludwig Boltzmann.* http://www.phys.uu.nl/igg/jos/publications/oviedoboltzmann.pdf.

[38] J. W. Gibbs, *Elementary principles in statistical mechanics* (Yale University Press, 1902); also (Dover Publications, NY, 1960).

[39] P. Ehrenfest and T. Ehrenfest-Afanassjewa, *The conceptual Foundations of the Statistical Approach in Mechanics*, New York, Cornell University Press, 1959 (original edition 1912).

[40] E. G. D. Cohen, Physica A 305 (2002) 19-26.

[41] E. T. Jaynes, American Journal of Physics 33 (1965) 391-398.

[42] M. Planck, *The theory of heat radiation*, Dover, 1951, p. 119.

[43] S. T. R. Pinho and R. F. S. Andrade, *Evolução das Ideias da Termodinâmica e da Mecânica Estatística*, In J. F. Rocha (Org.) *Origens e Evolução das Ideias da Física* (Chapter II), EDUFBA, Salvador, 2002.

[44] A. Einstein, Ann. der Physik 17 (1905) 132-148.

[45] A. Einstein, Annalen der Physik 33 (1910) 1275-1298.

[46] Translation of: A. Pais, *Subtle Is The Lord*, Oxford University Press, New York, 1982 (Chapter II).

[47] http://en.wikipedia.org/wiki/Von_Neumann_entropy.

[48] J. von Neumann, *Mathematical Foundations of Quantum Mechanics*, first published in 1932. Princeton Univ. Press. 1996 edition.

[49] C. E. Shannon, *A mathematical theory of communication*, Bell System Technical Journal 27 (1948) 379-423 and 623-656.

[50] A. Wehrl, Review of Modern Physics 50 (1978) 221-260.

[51] C. Tsallis, *Introduction to Nonextensive Statistical Mechanics*, Springer, 2009.

[52] M. Tribus and E. C. McIrvine, Scientific American 224 (1971) 178-184.

[53] E. T. Jaynes, Physical Review 106 (1957) 620-630; 108 (1957) 171-190.

Send Orders for Reprints to reprints@benthamscience.net

CHAPTER 2

Essentials of Information Entropy and Related Measures

Raúl D. Rossignoli [a,b,*], Andres M. Kowalski [a,b] and Evaldo M. F. Curado [c]

[a] Departamento de Física-IFLP, Facultad de Ciencias Exactas,
Universidad Nacional de La Plata,
C.C. 67, 1900 La Plata, Argentina

[b] Comisión de Investigaciones Científicas (CIC), Argentina

[c] CBPF and
National Institute of Science and Technology for Complex Systems,
Rio de Janeiro, Brasil

Abstract

This introductory chapter provides a basic review of the Shannon entropy and of some important related quantities like the joint entropy, the conditional entropy, the mutual information and the relative entropy. We also discuss the Fisher information, the fundamental property of concavity, the basic elements of the maximum entropy approach and the definition of entropy in the Quantum case. We close this chapter with the axioms which determine the Shannon entropy and a brief description of other information measures.

Keywords: Shannon Entropy; Mutual Information; Relative Entropy; Fisher Information; von Neumann Entropy.

1 Shannon Entropy

Let us first consider a discrete probability distribution

$$\boldsymbol{p} = (p_1, \ldots, p_n), \;\; p_i \geq 0\,, \;\; \sum_{i=1}^{n} p_i = 1\,, \tag{1}$$

where p_i indicates the probability of a certain event i in a random experiment. The Shannon entropy [1] is a measure of the lack of information associated with the probability

* Corresponding author. E-mail: rossigno@fisica.unlp.edu.ar

distribution and is defined as

$$S(\boldsymbol{p}) = -\sum_{i=1}^{n} p_i \log p_i\,, \tag{2}$$

where the base in the logarithm can in principle be any number $a > 1$ ($\log_a x = \ln x/\ln a$, where $\ln x = \log_e x$ is the natural logarithm). The choice of a determines the unit of information. The most common unit is the bit, in which case we should set $a = 2$ such that $S(\boldsymbol{p}) = 1$ for $\boldsymbol{p} = (1/2, 1/2)$, *i.e.*, for an experiment with just two possible and equally likely outcomes (a random bit). Other units include the nat ($a = e$), equivalent to $\log_2 e \approx 1.44$ bits, and the ban ($a = 10$), also denoted as hartley or dit, equivalent to $\log_2 10 \approx 3.32$ bits. In what follows we will use the notation $\log x$ in all expressions valid for any $a > 1$. Let us also notice that if $p_i = 0$ for some i, the corresponding term in (2) should be understood as 0, in agreement with the limit $\lim_{p\to 0^+} p \log p = 0$ (Fig. 1).

As stated above, Eq. (2) is a measure of the lack information associated with the discrete probability distribution $\boldsymbol{p}$, quantifying the uncertainty about the possible outcome of the random experiment. It can be also considered as the average information gained once the outcome is known, as well as a measure of the disorder associated with $\boldsymbol{p}$. It satisfies

$$S(\boldsymbol{p}) \geq 0, \tag{3}$$

with $S(\boldsymbol{p}) = 0$ *if and only if* there is no uncertainty, *i.e.*, if there is just one event (say j) which occurs with probability 1, the others having then probability 0:

$$S(\boldsymbol{p}) = 0 \text{ if and only if } p_i = \delta_{ij}, \tag{4}$$

for some $j \in \{1, \ldots, n\}$, where δ_{ij} is the Kronecker symbol ($\delta_{ij} = 1$ for $i = j$ and 0 otherwise). Eqs. (3)–(4) are apparent as $-p\log p$ is strictly positive for $p \in (0, 1)$ and vanishes at $p = 0$ and $p = 1$ (Fig. 1).

On the other hand, $S(\boldsymbol{p})$ is maximum when there is maximum uncertainty, *i.e.*, when all n events are equally likely and hence $p_i = 1/n\ \forall\ n$:

$$S(p_1, \ldots, p_n) \leq \log n\,, \tag{5}$$

with

$$S(p_1, \ldots, p_n) = \log n \text{ if and only if } p_i = 1/n,\quad i = 1, \ldots, n. \tag{6}$$

This result is also apparent, as maximization of Eq. (2) with respect to the $p_i's$, with the normalization constraint $\sum_{i=1}^{n} p_i = 1$, directly leads to the condition $\log p_i = C$, with C

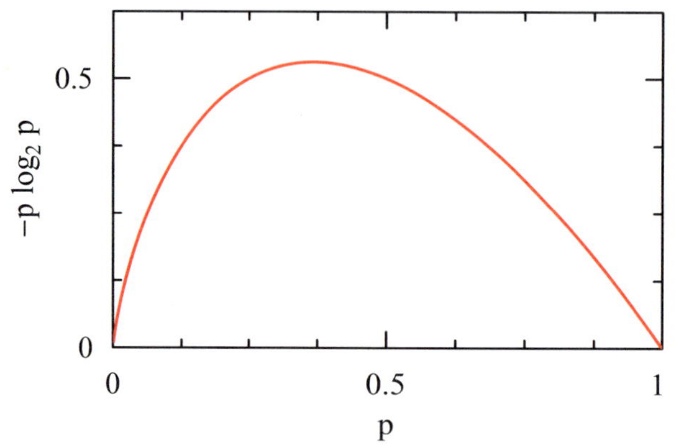

Figure 1: The function $-p\log_2 p$ for $p \in [0,1]$, which is maximum at $p = e^{-1}$.

constant $\forall\ i$, whence $p_i = 1/n$. This important result can actually be derived from the strict concavity of the function $-p\log p$ (Fig. 1) and the ensuing concavity of the entropy (2), which ensures a single maximum for any linear constraint (see sections 9–10).

Notice that the maximum value $\log n$ is, accordingly, an increasing function of n, such that a constant distribution for a subset of $m < n$ events has a smaller entropy $\log m < \ln n$. The maximum entropy distribution is fully symmetric under permutations (*i.e.*, it is the only completely symmetric distribution), in agreement with the symmetry against permutations of Eq. (2) and the uniqueness of its maximum.

It is frequently convenient to associate with $\boldsymbol{p}$ a discrete random variable X, such that $p_i = P(X = x_i)$ is the probability of X taking the value x_i ($x_i \neq x_j$ for $i \neq j$). We may then define the entropy of the discrete random variable X as

$$S(X) = -\sum_i p_i \log p_i \,. \tag{7}$$

Hence, $S(X) = 0$ if and only if X takes a single value with probability 1, while $S(X)$ is maximum when all values x_i are equally likely. The entropy of a binary random variable, *i.e.*, one taking just two possible values x_1 and x_2 with probabilities $p_1 = p$ and $p_2 = 1-p$, is plotted in Fig. 2. The case of continuous random variables is considered in section 6.

The entropy (2) or (7) can also be considered as the *average surprisal* [2]. The surprisal (also denoted as self-information) associated with an event i is just $-\log p_i$. The less probable the event, the larger the surprisal, with no surprisal at all if the event always occurs ($p_i = 1$), and a large surprisal if a very unlikely event ($p_i \approx 0$) occurs. We now recall that the average of the random variable X (also denoted as expectation value $E(X)$

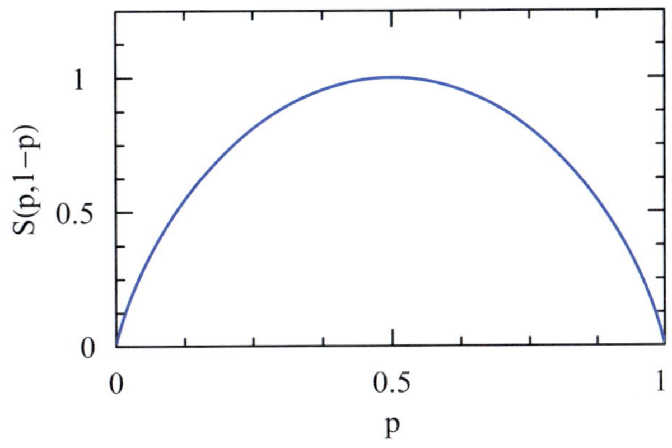

Figure 2: The entropy of a binary random variable. It is maximum at $p = 1/2$, where $S(p, 1-p) = 1$ (for $\log x = \log_2 x$).

or mean value $\overline{X}$) is just

$$\langle X \rangle = \sum_i p_i x_i, \tag{8}$$

where $p_i = p(X = x_i)$, and that the average of the random variable $G(X)$ (taking the value $G(x_i)$ when $X = x_i$) is just $\langle G(X) \rangle = \sum_i p_i G(x_i)$. Hence, by considering the surprisal as a random variable which takes the value $-\log p_i$ if the event i occurs, we may view the entropy (2) or (7) as its expectation value: $S(X) = \langle -\log p \rangle = -\sum_i p_i \log p_i$.

If the surprisal is assumed to be a decreasing continuous function $s(p)$ of the event probability p that satisfies additivity, *i.e.*, $s(pq) = s(p) + s(q)\ \forall\ q, p \in [0, 1]$ (such that the surprisal associated with the simultaneous occurrence of two independent events is the sum of surprisals) then $s(p) = -k \log p$, with $k > 0$. This provides a basic justification of the choice (2). A more rigorous justification based on the Khinchin axioms [3] is provided in section 11.

2 Joint Entropy and Additivity

Given two discrete random variables X and Y characterized by a *joint* probability distribution

$$p_{ij} = p(X = x_i, Y = y_j),\ i = 1, \ldots, n,\ j = 1, \ldots, m, \tag{9}$$

where $(X = x_i, Y = y_j) \equiv (X = x_i) \bigcap (Y = y_j)$ is the event where X and Y take simultaneously the values x_i *and* y_j respectively, the joint entropy is the entropy associated the joint distribution:

$$S(X,Y) = -\sum_{i,j} p_{ij} \log p_{ij} \,. \tag{10}$$

It is obviously symmetric: $S(X,Y) = S(Y,X)$.

The corresponding marginal distributions are

$$p_i^X = \sum_j p_{ij}, \;\; p_j^Y = \sum_i p_{ij} \,, \tag{11}$$

such that $p_i^X = p(X = x_i)$ is the probability of X taking the value x_i irrespective of the value taken by Y and $p_j^Y = p(Y = y_j)$ that of Y taking the value y_j irrespective of the value taken by X. The corresponding marginal entropies are then

$$S(X) = -\sum_i p_i^X \log p_i^X, \quad S(Y) = -\sum_j p_j^Y \log p_j^Y \,. \tag{12}$$

Two random variables are said to be *independent* if and only if $p_{ij} = p_i^X p_j^Y \; \forall i,j$. In such a case, Eq. (10) becomes

$$S(X,Y) = S(X) + S(Y) \quad (p_{ij} = p_i^X p_j^Y \;\forall\, i,j). \tag{13}$$

This is the fundamental *additivity property* fulfilled by the Shannon entropy, valid for independent random variables. It is, of course, a very natural and expected property of a measure of lack of information: The lack of information about two independent random variables or observables is just the sum of the lack of information about each random variable.

In the general case, the joint entropy is *smaller* than the sum of marginal entropies, such that for two arbitrary random variables X, Y we have

$$S(X,Y) \leq S(X) + S(Y) \,, \tag{14}$$

with $S(X,Y) = S(X) + S(Y)$ only if X and Y are independent. This inequality is demonstrated in section 4 and is usually denoted as the *subadditivity* property of the Shannon entropy.

3 Conditional Entropy

For the general case where X and Y are not necessarily independent, we may define the *conditional entropy* $S(X|Y)$ [4, 5] as the average entropy associated with the conditional distribution

$$p_{j/i}^{Y} = p(Y = y_j / X = x_i) = p_{ij} / p_i^X \,, \tag{15}$$

which, we recall, satisfies $p_{j/i}^{Y} \geq 0$ with $\sum_j p_{j/i}^{Y} = 1$, and represents the probability of outcome y_j given that the event $X = x_i$ has occurred. It is obviously defined for $p_i^X > 0$. The conditional entropy is then defined as

$$S(Y|X) = \sum_i p_i^X S(Y/X = x_i), \tag{16}$$

where

$$S(Y/X = x_i) = -\sum_j p_{j/i}^{Y} \log p_{j/i}^{Y} \,,$$

is the entropy associated with the conditional distribution (15). $S(Y|X)$ is therefore the average lack of information about Y given the knowledge of X. By replacing Eq. (15) in (16), we can also express the conditional entropy in terms of the joint entropy and the marginal entropy of X:

$$\begin{aligned} S(Y|X) &= -\sum_{i,j} p_{ij} \log(p_{ij}/p_i^X) \\ &= S(X,Y) - S(X). \end{aligned} \tag{17}$$

Eq. (17) turns out to be a *fundamental* property of the Shannon entropy, being essential for providing a uniqueness theorem (see section 11). Notice that $S(Y|X)$ differs in general from $S(X|Y) = S(X,Y) - S(Y)$, with $S(Y|X) - S(X|Y) = S(Y) - S(X)$.

If X and Y are independent then $p_{j/i}^{Y} = p_j^Y \ \forall i$ and

$$S(Y|X) = S(Y) \quad (X,\ Y \text{ independent})\,, \tag{18}$$

in agreement with Eqs. (13) and (17). In the general case, Eqs. (14) and (17) lead to

$$S(Y|X) \leq S(Y)\,, \tag{19}$$

i.e., the conditional entropy cannot be larger than the marginal entropy, being equal only if Y and X are independent. In the opposite case where Y is a function of X, such that for all x_i with $p_i^X > 0$ there is a *single* possible value y_i of Y such that $p_{i/i}^{Y} = 1$, $S(Y/X = x_i) = 0$ $\forall\ i$ and

$$S(Y|X) = 0 \quad (Y = F(X)). \tag{20}$$

In any other case, $S(Y|X) > 0$ (for discrete random variables).

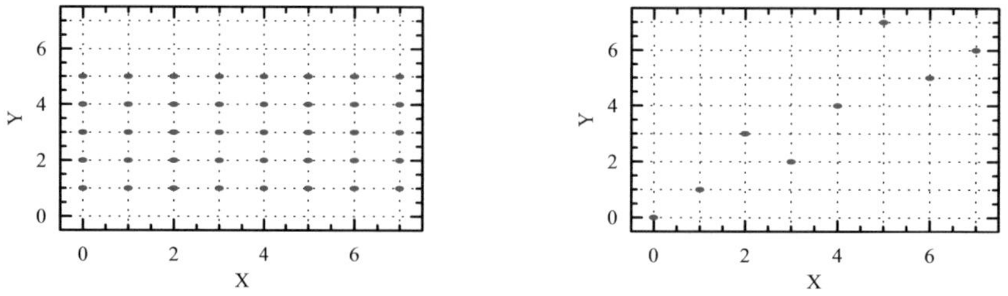

Figure 3: Independent (left) and fully correlated (right) random variables. On the left, $p_{ij} = p_i^X p_j^Y$, implying $S(X,Y) = S(X) + S(Y)$, $S(X|Y) = S(X)$, $S(Y|X) = S(Y)$ and $I(X;Y) = 0$. On the right, $S(X,Y) = S(X) = S(Y) = I(X;Y)$, with $S(X|Y) = S(Y|X) = 0$.

Therefore, in the discrete case $S(X,Y) = S(X)$ if and only if $S(Y|X) = 0$, *i.e.,* if and only if $Y = F(X)$ for the given joint distribution. Similarly, $S(X,Y) = S(Y)$ if and only if $X = G(Y)$. If $S(X,Y) = S(X) = S(Y)$ then $Y = F(X)$ and $X = G(Y)$, with $G = F^{-1}$. In this case knowledge of X provides full knowledge of Y and viceversa (fully correlated case).

As stated above, in the present discrete case we always have $S(Y|X) \geq 0$, $S(X|Y) \geq 0$. Eq. (17) implies, therefore, a fundamental property of the joint entropy:

$$S(X,Y) \geq S(X)\,, \quad S(X,Y) \geq S(Y)\,. \tag{21}$$

In other words, the lack of information associated with the joint distribution cannot be less than that associated with any of its marginals. We should remark that when applied to composite physical systems (such that X refers to a system A and Y to a system B), this inequality is characteristic of *classical* systems, *i.e.,* those that can be described by probability distributions. In contrast, *quantum* systems are described by *density matrices* rather than by probability distributions (see section 8), and the quantum version of Eq. (21) does not necessarily apply, as discussed in detail in chapter 5, being in fact broken in all *entangled* pure states.

4 Mutual Information

The mutual information [4, 5] can be defined as

$$I(X;Y) = S(X) + S(Y) - S(X,Y), \tag{22}$$

and is a measure of the correlation between X and Y. It measures how much information is shared by X and Y. If X and Y are independent then $S(X,Y) = S(X) + S(Y)$ and hence

$I(X;Y) = 0$. Moreover, we have in general

$$I(X;Y) \geq 0, \tag{23}$$

with

$$I(X;Y) = 0 \text{ if and only if } X \text{ and } Y \text{ are } \textit{independent}. \tag{24}$$

The demonstration of this fundamental property (which leads to Eq. (14)) is given shortly below, but let us remark that it is conceptually apparent: $S(X) + S(Y)$ represents the lack of information about the system when just the information about the marginal distributions is available, so that it cannot be smaller than that associated with the full distribution p_{ij} (which determines the marginals but also all correlations between X and Y). Moreover, since by Eq. (17) $S(X,Y) = S(X|Y) + S(Y)$, we have

$$\begin{aligned} I(X;Y) &= S(Y) - S(Y|X) \\ &= S(X) - S(X|Y). \end{aligned} \tag{25}$$

Hence, when Y is a function of X, $I(X;Y) = S(Y)$ while if X is a function of Y, $I(X;Y) = S(X)$. If X and Y are fully correlated $I(X;Y) = S(X) = S(Y)$ (Fig. 3). In all other cases $I(X;Y)$ is smaller than at least one of the marginal entropies. The mutual information is obviously symmetric: $I(X;Y) = I(Y;X)$.

Let us also notice that using the definitions (11)–(12) of the marginal distributions and entropies, Eq. (22) can be written explicitly as

$$I(X;Y) = -\sum_i p_i^X \log p_i^X - \sum_j p_j^Y \log p_j^Y + \sum_{i,j} p_{ij} \log p_{ij} \tag{26}$$

$$= -\sum_{i,j} p_{ij} \log \frac{p_i^X p_j^Y}{p_{ij}}. \tag{27}$$

Proof of Eqs. (23)–(24): From the strict concavity of the function $\ln x$, it follows that $\ln x \leq x - 1 \; \forall \; x > 0$ (*i.e.*, $\ln x$ is not greater than its tangent at $x = 1$, as illustrated in Fig. 4), with equality valid only if $x = 1$. Hence, using (27) we obtain (for $\log p = \ln p$, *i.e.*, $a = e$)

$$I(X;Y) = -\sum_{i,j} p_{ij} \ln \frac{p_i^X p_j^Y}{p_{ij}} \geq \sum_{i,j} p_{ij}\left(1 - \frac{p_i^X p_j^Y}{p_{ij}}\right) = 1 - 1 = 0. \tag{28}$$

The identity holds only if $p_i^X p_j^Y / p_{ij} = 1 \; \forall i, j$, in which case X and Y are then independent. The validity for general $a > 1$ follows immediately.

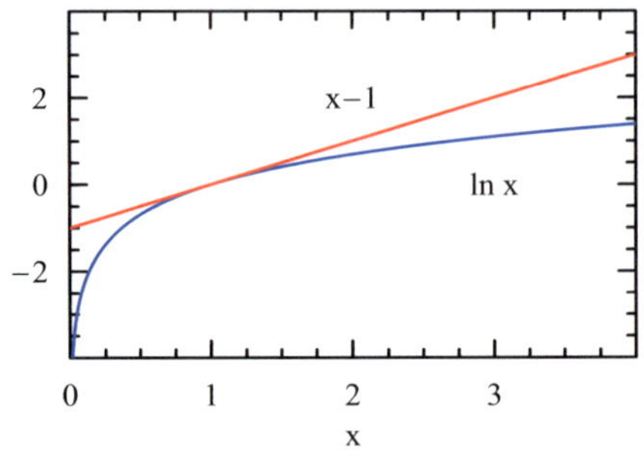

Figure 4: The function $\ln x$ and its upper bound $x - 1$.

5 Relative Entropy - Kullback-Leibler Divergence

Given now two probability distributions $\boldsymbol{p} = (p_1, \ldots, p_n)$, $\boldsymbol{q} = (q_1 \ldots, q_n)$ defined for the same set of events, the *relative entropy* [4, 6] is an entropic measure of the distance from $\boldsymbol{q}$ to $\boldsymbol{p}$ and is defined as

$$\begin{aligned} S(\boldsymbol{p}||\boldsymbol{q}) &= \langle -\log \boldsymbol{q}\rangle_{\boldsymbol{p}} - S(\boldsymbol{p}) \\ &= -\sum_{i=1}^{n} p_i \log \frac{q_i}{p_i}. \end{aligned} \tag{29}$$

It is also known as the *Kullback-Leibler divergence* [6]. Notice that $S(\boldsymbol{p}||\boldsymbol{q}) \neq S(\boldsymbol{q}||\boldsymbol{p})$. The *relative entropy* provides an entropic means for comparing the distribution $\boldsymbol{q}$ to $\boldsymbol{p}$, but is not a true distance since it is not symmetric. The first term $\langle -\log \boldsymbol{q}\rangle_{\boldsymbol{p}} = -\sum_i p_i \log q_i$ is also denoted as *cross entropy* ($S_c(\boldsymbol{p}, \boldsymbol{q}) = S(\boldsymbol{p}||\boldsymbol{q}) + S(\boldsymbol{p})$).

Eq. (29) satisfies the fundamental property

$$S(\boldsymbol{p}||\boldsymbol{q}) \geq 0, \tag{30}$$

with

$$S(\boldsymbol{p}||\boldsymbol{q}) = 0 \text{ if and only if } \boldsymbol{p} = \boldsymbol{q}. \tag{31}$$

Proof: Applying exactly the same upper bound employed in (28) (Fig. 4), we obtain, for $\log p = \ln p$,

$$S(\boldsymbol{p}||\boldsymbol{q}) = -\sum_i p_i \ln \frac{q_i}{p_i} \geq \sum_i p_i (1 - \frac{q_i}{p_i}) = 1 - 1 = 0\,, \tag{32}$$

with identity holding if and only if $q_i = p_i \ \forall\, i$.

In particular, the *mutual information* (22) can now be seen as *the relative entropy* between the product distribution $(\boldsymbol{p}^X \times \boldsymbol{p}^Y)_{ij} = p_i^X p_j^Y$ determined by the marginals and the actual joint distribution $\boldsymbol{p}^{XY}$ of elements p_{ij}, as is evident from Eq. (27):

$$I(X;Y) = S(\boldsymbol{p}^{XY}||\boldsymbol{p}^X \times \boldsymbol{p}^Y).$$

The mutual information (22) measures thus the distance from the "independent" to the actual joint distribution.

Similarly, denoting with $\boldsymbol{p}^X \times \frac{1}{n_Y}$ the special joint distribution of elements p_i^X/n_Y, where n_Y is the number of values taken by Y ($\sum_{i,j} p_i^X/n_Y = 1$), the conditional entropy (17) can be written in terms of the relative entropy as

$$S(Y|X) = S(X,Y) - S(X) = -S(\boldsymbol{p}^{XY}||\boldsymbol{p}^X \times \tfrac{1}{n_Y}) + \log n_Y, \tag{33}$$

an expression which we will use in section 9.

Notice that for two independent joint distributions $\boldsymbol{p} = \boldsymbol{p}^X \times \boldsymbol{p}^Y$ and $\boldsymbol{q} = \boldsymbol{q}^X \times \boldsymbol{q}^Y$, the relative entropy is *additive*, in the sense that

$$S(\boldsymbol{p}^X \times \boldsymbol{p}^Y||\boldsymbol{q}^X \times \boldsymbol{q}^Y) = S(\boldsymbol{p}^X||\boldsymbol{q}^X) + S(\boldsymbol{p}^Y||\boldsymbol{q}^Y)\,. \tag{34}$$

As mentioned before, the relative entropy is non-symmetric (Fig. 5). Nonetheless, for small *relative* deviations $|q_i - p_i|/p_i \ll 1\ \forall\ i$, we obtain, for $\log p = \ln p$ and recalling that $\ln(1+x) = x - \frac{x^2}{2} + O(x^3)$ for $x \to 0^+$,

$$S(\boldsymbol{p}||\boldsymbol{q}) \approx S(\boldsymbol{q}||\boldsymbol{p}) \approx \tfrac{1}{2}\sum_i \frac{(q_i - p_i)^2}{p_i}\,, \quad (|q_i - p_i|/p_i \ll 1) \tag{35}$$

if terms $O((q_i/p_i - 1)^3)$ are discarded.

On the other hand, a *symmetrized* version of (29), denoted as Jensen-Shannon divergence and defined as

$$D_{JS}(\boldsymbol{p},\boldsymbol{q}) = \tfrac{1}{2}[S(\boldsymbol{p}||\tfrac{1}{2}(\boldsymbol{p}+\boldsymbol{q})) + S(\boldsymbol{q}||\tfrac{1}{2}(\boldsymbol{p}+\boldsymbol{q}))]\,, \tag{36}$$

has also been introduced [7] (see chapter 6 for more details). This quantity obviously satisfies $D_{JS}(\boldsymbol{p},\boldsymbol{q}) = D_{JS}(\boldsymbol{q},\boldsymbol{p})$, and is non-negative (Eq. (31)), vanishing only if $\boldsymbol{p} = \boldsymbol{q}$. Its square root was shown to be a proper metric for probability distributions [8], satisfying the triangle inequality ($d(\boldsymbol{p},\boldsymbol{q}) \le d(\boldsymbol{p},\boldsymbol{r}) + d(\boldsymbol{r},\boldsymbol{q})$ for $d(\boldsymbol{p},\boldsymbol{q}) = \sqrt{D_{JS}(\boldsymbol{p},\boldsymbol{q})}$).

6 Entropy in the Continuous Case

Let us now discuss the extension of all previous quantities to the case of continuous random variables. A continuous random variable X is characterized by a probability density function

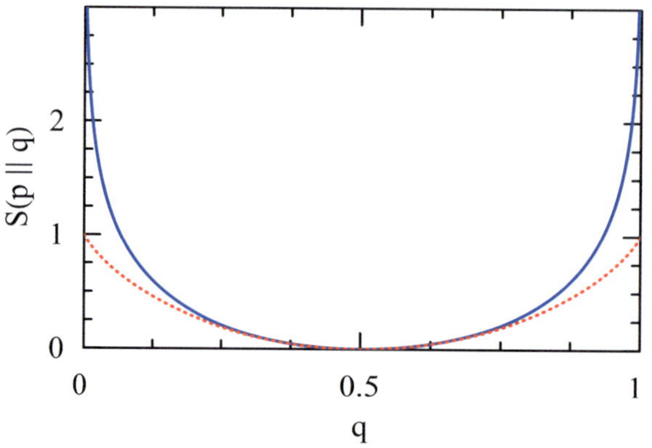

Figure 5: The relative entropy $S(\boldsymbol{p}||\boldsymbol{q})$ (solid line) between the binary distribution $\boldsymbol{q} = (q, 1-q)$ and the uniform binary distribution $\boldsymbol{p} = (\frac{1}{2}, \frac{1}{2})$, for $\log p = \log_2 p$. The dotted line depicts $S(\boldsymbol{q}||\boldsymbol{p})$. For $|q - \frac{1}{2}| \ll \frac{1}{2}$, $S(\boldsymbol{p}||\boldsymbol{q}) \approx S(\boldsymbol{q}||\boldsymbol{p}) \approx 2(q-\frac{1}{2})^2/\ln 2$ (Eq. 35).

(PDF) $p(x)$ (continuous or piecewise continuous), such that for $a < b$,

$$p(a \leq X \leq b) = \int_a^b p(x)dx\,. \tag{37}$$

The PDF should then satisfy

$$p(x) \geq 0, \quad \int_{-\infty}^{\infty} p(x)dx = 1\,. \tag{38}$$

In order to define the entropy, we may first consider the discrete probability distribution $p_i \equiv p(x_i \leq X \leq x_{i+1})$ associated with a discrete set of points x_i with spacing $\Delta x_i = x_{i+1} - x_i > 0$. If Δx_i is sufficiently small, we have

$$p_i = \int_{x_i}^{x_{i+1}} p(x)dx \approx p(x_i)\Delta x_i\,,$$

if $p(x)$ is continuous in $[x_i, x_{i+1})$. The associated entropy is

$$S(\boldsymbol{p}) \approx -\sum_i [p(x_i)\Delta x_i] \log[p(x_i)\Delta x_i] = -\sum_i p(x_i)\log[p(x_i)]\Delta x_i + C\,, \tag{39}$$

where $C = -\sum_i p(x_i)\Delta x_i \log \Delta x_i$ ($C \approx -\log\delta$ for a uniform spacing $\Delta x_i = \delta\ \forall\ i$). For $\Delta x_i \to 0^+\ \forall\ i$ we then obtain

$$\lim_{\Delta x_i \to 0^+} [S(\boldsymbol{p}) - C] = -\int_{-\infty}^{\infty} p(x)\log p(x)\,dx\,.$$

The *continuous entropy* (also denoted as *differential entropy*) of a continuous random variable X with probability density $p(x)$ is defined as this last limit:

$$S(p) = -\int_{-\infty}^{\infty} p(x)\log p(x)\,dx\,. \tag{40}$$

In contrast with the discrete case, Eq. (40) is no longer necessarily positive, since it omits the constant C in (39) (which approaches $+\infty$ for $\Delta x_i \to 0^+ \ \forall\ i$). Nonetheless, the difference between the entropies (39) associated with two continuous random variables with densities $p(x)$ and $q(x)$ approaches $S(p) - S(q)$ for $\Delta x_i \to 0^+$.

For instance, in the case of the *gaussian* distribution

$$p(x) = \frac{e^{-(x-\mu)^2/(2\sigma^2)}}{\sqrt{2\pi}\sigma}\,, \tag{41}$$

Eq. (40) becomes

$$S(p) = \log(\sqrt{2\pi e}) + \log(\sigma)\,, \tag{42}$$

where $\sigma = \langle (X-\mu)^2\rangle^{1/2}$ is the standard deviation, whereas in the case of a uniform distribution

$$p(x) = \left\{ \begin{array}{cc} \frac{1}{b-a} & a \le x \le b \\ 0 & x < a \text{ or } x > b \end{array} \right. , \tag{43}$$

Eq. (40) reads

$$S(p) = \log(b-a) = \log(2\sqrt{3}) + \log(\sigma)\,, \tag{44}$$

where here $\sigma = (b-a)/(2\sqrt{3})$. They are just linear increasing functions of the logarithm of the standard deviation.

Eq. (40) (and hence Eqs. (42) and (44)) is not invariant under a rescaling of the random variable: If $X_\alpha = \alpha X$ ($\alpha \neq 0$), its PDF is $p_\alpha(x) = p(x/\alpha)/|\alpha|$ and Eq. (40) leads to

$$S(p_\alpha) = S(p) + \log|\alpha|\,. \tag{45}$$

The extension of the joint entropy (10) to the continuous case is straightforward. The continuous joint entropy of two continuous random variables X, Y characterized by a joint probability density $p(x,y)$ is

$$S(p) = -\int_{-\infty}^{\infty}\int_{-\infty}^{\infty} p(x,y)\log p(x,y)\,dxdy\,. \tag{46}$$

The continuous marginal entropies $S(p^X)$, $S(p^Y)$, are similarly defined in terms of the marginal densities

$$p^X(x) = \int_{-\infty}^{\infty} p(x,y)\,dy\,, \quad p^Y(y) = \int_{-\infty}^{\infty} p(x,y)\,dx\,.$$

For independent variables, $p(x,y) = p^X(x)p^Y(y)$ and Eq. (13) remains then valid in terms of continuous entropies, *i.e.*, $S(p) = S(p^X) + S(p^Y)$. Moreover, Eq. (14), *i.e.*,

$$S(p) \le S(p^X) + S(p^Y)\,, \tag{47}$$

remains also valid in the continuous case, as the continuous mutual information

$$I(X;Y) = S(p||p^X p^Y) = S(p^X) + S(p^Y) - S(p)\,, \tag{48}$$

remains non-negative due to the positivity of the continuous relative entropy (see below).

The continuous extension of the conditional entropy (16) is similar, and Eqs. (17)–(19) remain then valid in terms of the continuous joint and marginal entropies:

$$S(Y|X) = S(p) - S(p^X) \leq S(p^Y)\,. \tag{49}$$

However, Eq. (21) no longer holds in general, as the continuous conditional entropy (49) can now be negative.

The relative entropy of two probability densities $p(x)$ and $q(x)$ is defined as

$$S(p||q) = -\int_{-\infty}^{\infty} p(x) \log \frac{q(x)}{p(x)}\, dx\,. \tag{50}$$

In contrast with Eq. (40), the relative entropy (50) *remains non-negative* in the continuous case: $S(p||q) \geq 0$, vanishing only if $p(x) = q(x)\ \forall\ x$ (except for a set of points of vanishing measure). The demonstration of positivity is the same as that of Eq. (32). Eq. (50) is in fact the limit for $\Delta x_i \to 0^+$ of the discrete relative entropy (29) for discrete distributions $p_i = p(x_i)\Delta x_i$, $q_i = q(x_i)\Delta x_i$.

For example, the relative entropy between *gaussian* distributions

$$p_i(x) = \frac{e^{-(x-\mu_i)^2/(2\sigma_i^2)}}{\sqrt{2\pi}\sigma_i}\,, \tag{51}$$

is given by (here we set log = ln)

$$S(p_1||p_2) = \frac{\sigma_1^2 - \sigma_2^2 + (\mu_1 - \mu_2)^2}{2\sigma_2^2} - \ln \frac{\sigma_1}{\sigma_2}\,, \tag{52}$$

which is verified to vanish if and only if $\mu_1 = \mu_2$ and $\sigma_1 = \sigma_2$, being otherwise positive (using again the bound $\ln x \leq x - 1$ in (52) leads to $S(p_1||p_2) \geq \frac{(\mu_1-\mu_2)^2+(\sigma_1-\sigma_2)^2}{2\sigma_2{}^2}$).

Another important property of (50) is the invariance under rescaling and under a change of variables. If $y(x)$ is some monotonous function of x and $\tilde{p}(y) = p(x)/|dy/dx|$, $\tilde{q}(y) = q(x)/|dy/dx|$ stand for the corresponding probability densities associated to the variable y (*i.e.*, to the random variable $Y = y(X)$), then

$$\begin{aligned} S(\tilde{p}||\tilde{q}) &= -\int_{-\infty}^{\infty} \tilde{p}(y) \log \frac{\tilde{q}(y)}{\tilde{p}(y)}\, dy \\ &= -\int_{-\infty}^{\infty} p(x) \log \frac{q(x)}{p(x)}\, dx = S(p||q)\,. \end{aligned} \tag{53}$$

Such invariance is not satisfied by the continuous entropy (40).

7 Fisher Information and Cramér-Rao Bound

Consider now the relative entropy from the displaced distribution $q(x) = p(x+\Delta x)$ to $p(x)$:

$$S(p||q) = -\int_{-\infty}^{\infty} p(x)\log\frac{p(x+\Delta x)}{p(x)}\,dx\,. \tag{54}$$

For a continuous $p(x)$ and a sufficiently small Δx, we may use the second order approximation (35) (setting again log = ln), which for a differentiable $p(x)$ leads to

$$S(p||q) \approx \frac{1}{2}\int_{-\infty}^{\infty}\frac{(p(x+\Delta x)-p(x))^2}{p(x)}\,dx \approx \frac{(\Delta x)^2}{2}I(p)\,, \tag{55}$$

$$I(p) = \int_{-\infty}^{\infty}\frac{[p'(x)]^2}{p(x)}\,dx\,, \tag{56}$$

where $p'(x) = dp/dx$ and Eq. (55) is exact up to $O((\Delta x)^2)$ ($I(p) = \frac{\partial^2 S(p||q)}{\partial(\Delta x)^2}|_{\Delta x=0}$).

The quantity (56) is called the *Fisher Information* [9, 10], and was developed much earlier than the Shannon entropy [9]. It is a measure of the uneveness of the distribution. The more localized or uneven is the density $p(x)$, the higher is $I(p)$, *i.e.*, the larger is the relative entropy between slightly displaced distributions. While the Shannon entropy (40) is a *global* measure, remaining unchanged under any reordering of the p_i or $p(x)$, $I(p)$ is a *local* measure, being essentially determined by the local variation as measured by $p'(x)$.

Eq. (56) may be rewritten in several different forms:

$$I(p) = 4\int_{-\infty}^{\infty}([p^{1/2}(x)]')^2\,dx = \int_{-\infty}^{\infty}p(x)([\ln p(x)]')^2\,dx = -\int_{-\infty}^{\infty}p(x)[\ln p(x)]''\,dx\,, \tag{57}$$

where the last expression assumes a vanishing $p'(x)$ for $x \to \pm\infty$. For instance, in the case of the gaussian distribution (51), the last expression leads immediately to

$$I(p) = \frac{1}{\sigma^2}\,,\quad (p(x) = \frac{\exp[-\frac{(x-\mu)^2}{2\sigma^2}]}{\sqrt{2\pi}\sigma})\,.$$

Interesting details and applications of $I(p)$ in the multidimensional case will be discussed in chapter 9 (where it is denoted as $F(\rho)$).

In general, for a distribution $p(x) = p_\theta(x)$ depending on a parameter θ, the relative entropy from $p_{\theta+\Delta\theta}$ to p_θ is given by

$$S(p_\theta||p_{\theta+\Delta\theta}) \approx \frac{1}{2}\int_{-\infty}^{\infty}\frac{(p_{\theta+\Delta\theta}(x)-p_\theta(x))^2}{p_\theta(x)}\,dx \approx \frac{(\Delta\theta)^2}{2}I_\theta(p)\,, \tag{58}$$

$$I_\theta(p) = \int_{-\infty}^{\infty}\frac{1}{p_\theta(x)}(\frac{\partial p_\theta(x)}{\partial\theta})^2\,dx = \int_{-\infty}^{\infty}p_\theta(x)(\frac{\partial\ln p_\theta(x)}{\partial\theta})^2\,dx\,, \tag{59}$$

where $I_\theta(p)$ is the Fisher information associated with the parameter θ. Eq. (56) corresponds to the particular case $p_\theta(x) = p(x+\theta)$ (or equivalently, $p(x-\theta)$), where $I_\theta(p) = I(p)$ is independent of the shift θ. For instance, for $\theta = \mu$ in the gaussian distribution (51), we obtain $I_\mu(p) = I(p) = 1/\sigma^2$, whereas for $\theta = \sigma^2$, $I_{\sigma^2}(p) = 1/(2\sigma^4)$.

A remarkable property of the Fisher Information is the *Cramér-Rao bound* [5, 10, 11, 12]: If $O(x)$ is a function of x whose average is the parameter θ, then

$$\langle (O-\theta)^2\rangle I_\theta(p) \geq 1\,, \tag{60}$$

where $\langle (O-\theta)^2\rangle = \int_{-\infty}^{\infty} p_\theta(x)(O(x)-\theta)^2\,dx = \langle O^2\rangle - \theta^2$ is the variance (mean square deviation) of O. Thus, $I_\theta(p)$ provides a lower bound to the variance: $\langle (O-\theta)^2\rangle \geq 1/I_\theta(p)$.

The proof of Eq. (60) is straightforward: Since $0 = \langle O\rangle - \theta = \int_{-\infty}^{\infty} p_\theta(x)(O(x)-\theta)dx\ \forall\ \theta$, then $0 = \int_{-\infty}^{\infty} \frac{\partial p_\theta(x)}{\partial\theta}(O(x)-\theta)dx - 1$, which leads to

$$\int_{-\infty}^{\infty} p_\theta(x)\frac{\partial \ln p_\theta(x)}{\partial\theta}(O(x)-\theta)dx = 1\,. \tag{61}$$

Applying in (61) the Cauchy-Schwartz inequality $|(f,g)|^2 \leq ||f||^2||g||^2$, where $(f,g) = \int_{-\infty}^{\infty} f(x)g(x)dx$ is the standard scalar product of two functions and $||f||^2 = (f,f)$, we obtain

$$[\int_{-\infty}^{\infty} p_\theta(x)(\frac{\partial \ln p_\theta(x)}{\partial\theta})^2\,dx][\int_{-\infty}^{\infty} p_\theta(x)(O(x)-\theta)^2\,dx] \geq 1\,,$$

which is in fact Eq. (60).

In particular, for $p_\theta(x) = p(x-\theta)$, Eq. (60) leads to

$$\langle (x-\mu)^2\rangle I(p) \geq 1\,, \tag{62}$$

where $\mu = \langle x\rangle = \int_{-\infty}^{\infty} p(x)x\,dx$. Eq. (62) is then equivalent to the well-known position-momentum uncertainty relation of quantum mechanics for a real wave function $\psi(x) = \sqrt{p(x)}$ vanishing for $|x| \to \infty$: Using Eq. (57),

$$I(p) = 4\int_{-\infty}^{\infty} [\psi'(x)]^2\,dx = \int_{-\infty}^{\infty} -4\psi(x)\psi''(x)dx = 4\langle P^2\rangle_\psi\,,$$

where $P = -i\frac{\partial}{\partial x}$ is the (dimensionless) momentum operator. Hence $I(p)$ is just proportional to the variance of P (as $\langle P\rangle_\psi = 0$), and Eq. (62) is equivalent to $\Delta X\Delta P \geq \frac{1}{2}$, where $\Delta X = \langle (x-\mu)^2\rangle_\psi^{1/2}$, $\Delta P = \langle P^2\rangle_\psi^{1/2}$ are the standard deviations ("uncertainties") of position and momentum.

The inequalities (60)–(62) are saturated for the gaussian distribution: For $\theta = \mu$, $O(x) = x$ and $\langle (x-\mu)^2\rangle = \sigma^2 = 1/I(p)$. The same holds for $\theta = \sigma^2$, where $O(x) = (x-\mu)^2$ and $\langle (O-\sigma^2)^2\rangle = 2\sigma^4 = 1/I_{\sigma^2}(p)$.

8 Entropy in Quantum Systems: The von Neumann Entropy

In quantum mechanics, the state of a system is characterized by a *density operator* ρ, also denoted as density matrix or statistical operator [4, 13, 14]. It is an hermitian operator whose eigenvalues are non-negative and form a probability distribution $\boldsymbol{p}$:

$$\rho = \sum_i p_i |i\rangle\langle i|, \quad p_i \geq 0, \quad \mathrm{Tr}\,\rho = \sum_i p_i = 1\,. \tag{63}$$

Here $|i\rangle$ denotes the normalized orthogonal eigenvectors of ρ ($\langle i|j\rangle = \delta_{ij}$) and p_i its eigenvalues, with $|i\rangle\langle i|$ the corresponding orthogonal projectors, such that $\rho|i\rangle = p_i|i\rangle$ and $\langle i|\rho|j\rangle = p_i\delta_{ij}$. The eigenvalue p_i represents the probability of finding the system in the pure quantum state $|i\rangle$. Accordingly, the density matrix ρ is said to be *pure* if $p_i = \delta_{ij}$ for some fixed j, such that $\rho = |j\rangle\langle j|$ and $\rho^2 = \rho$. In all other cases $\rho^2 \neq \rho$ and the state described by ρ is said to be *mixed*.

The complex vector space generated by all accessible states $|i\rangle$ is the Hilbert space of the system. Its dimension can be finite or infinite. The trace Tr of an operator O is defined as $\mathrm{Tr}\,O = \sum_i \langle i|O|i\rangle$, where the sum runs over all elements of any complete set of normalized orthogonal states which generate the Hilbert space. Any normalized pure quantum state $|\psi\rangle$ can then be written as $|\psi\rangle = \sum_i \alpha_i|i\rangle$, where $\alpha_i = \langle i|\psi\rangle$ and $\sum_i |\alpha_i|^2 = \langle\psi|\psi\rangle = 1$.

The von Neumann entropy of ρ is defined as the Shannon entropy of its eigenvalues:

$$S(\rho) = -\mathrm{Tr}\,\rho\log\rho = -\sum_i p_i \log p_i\,. \tag{64}$$

Eq. (64) then satisfies $S(\rho) \geq 0$, with $S(\rho) = 0$ if and only if ρ is pure ($\rho^2 = \rho$). For finite Hilbert space dimension n, $S(\rho)$ is maximum at the maximally mixed state $\rho_n = I_n/n$ (I_n denotes the identity operator), which has eigenvalues $p_i = 1/n\;\forall\, i$ and leads to $S(\rho_n) = \log n$.

The *quantum relative entropy* from a density operator $\tilde{\rho}$ to ρ is defined as

$$S(\rho||\tilde{\rho}) = -\mathrm{Tr}\,[\rho(\log\tilde{\rho} - \log\rho)]\,. \tag{65}$$

It continues to satisfy the fundamental property

$$S(\rho||\tilde{\rho}) \geq 0\,, \tag{66}$$

with $S(\rho||\tilde{\rho}) = 0$ if and only if $\rho = \tilde{\rho}$. Eq. (66) is denoted as Klein's inequality.

Proof: If $\rho = \sum_i p_i|i\rangle\langle i|$, $\tilde{\rho} = \sum_j \tilde{p}_j|\tilde{j}\rangle\langle\tilde{j}|$ and $\tilde{q}_i = \langle i|\tilde{\rho}|i\rangle = \sum_j |\langle i|\tilde{j}\rangle|^2\tilde{p}_j$, then

$$S(\rho||\tilde{\rho}) = -\sum_i p_i[\sum_j |\langle i|\tilde{j}\rangle|^2 \log\tilde{p}_j - \log p_i] \geq -\sum_i p_i[\log\tilde{q}_i - \log p_i] = S(\boldsymbol{p}||\tilde{\boldsymbol{q}}) \geq 0\,,$$

where $S(\boldsymbol{p}||\tilde{\boldsymbol{q}})$ is the "classical" relative entropy (29) associated with the distributions $\boldsymbol{p}$ and $\tilde{\boldsymbol{q}}$ ($\tilde{q}_i \geq 0$, $\sum_i \tilde{q}_i = 1$) and in the first step we have used the concavity property of the logarithm (see section 9), which implies $\log \tilde{q}_i \geq \sum_j |\langle i|\tilde{j}\rangle|^2 \log \tilde{p}_j$ (as $\sum_j |\langle i|\tilde{j}\rangle|^2 = 1$). Again, $S(\rho||\tilde{\rho}) = 0$ if and only if $\tilde{\boldsymbol{q}} = \boldsymbol{p}$ and $\tilde{q}_i = \tilde{p}_j$ for some $j(i)$ $\forall$ i (due to the strict concavity of the logarithm), which implies $|i\rangle\langle i| = |\tilde{j}\rangle\langle\tilde{j}|$ and hence $\tilde{\rho} = \rho$.

The state of a composite quantum system $A+B$ is described by a density operator ρ_{AB} acting in the tensor product space of states $\{|ij\rangle \equiv |i_A\rangle \otimes |j_B\rangle\}$, of dimension $n_A n_B$. The joint entropy of $A+B$ is then given by

$$S(A,B) = S(\rho_{AB}) = -\mathrm{Tr}\,\rho_{AB}\log\rho_{AB}\,. \tag{67}$$

The subsystems are described by the *reduced* density matrices $\rho_A = \mathrm{Tr}_B\,\rho_{AB}$, $\rho_B = \mathrm{Tr}_A\,\rho_{AB}$, where Tr_B, Tr_A denote the *partial trace, i.e.,* the trace over the states of B and A respectively [14]. The marginal entropies are then $S(A) = S(\rho_A)$ and $S(B) = S(\rho_B)$. The reduced states ρ_A, ρ_B play the role of the marginal distributions of the classical case but are nonetheless density operators. If A and B are *uncorrelated*, then $\rho_{AB} = \rho_A \otimes \rho_B$ and Eq. (67) leads to $S(\rho_{AB}) = S(\rho_A) + S(\rho_B)$, *i.e.,* $S(A,B) = S(A) + S(B)$, as in the classical case (13).

The quantum mutual information is similarly defined as

$$I(A;B) = S(\rho_A) + S(\rho_B) - S(\rho_{AB}) \tag{68}$$

$$= S(\rho_{AB}||\rho_A \otimes \rho_B)\,. \tag{69}$$

As in the classical case, it is again a *non-negative* quantity, as ensured by Eqs. (66)–(69): $I(A;B) \geq 0$, with $I(A;B) = 0$ if and only if $\rho_{AB} = \rho_A \otimes \rho_B$. The subadditivity property (14), $S(A,B) \leq S(A) + S(B)$, remains then valid also in the quantum case.

Finally, the quantum conditional entropy is defined as the quantum version of Eq. (17):

$$S(B|A) = S(A,B) - S(A)\,. \tag{70}$$

However, in contrast with the classical case (17), Eq. (70) can now be *negative* (even in the discrete case), since the eigenstates of ρ_{AB} are not necessarily product states and hence the eigenvalues of ρ_A cannot be expressed in general as partial sums of the eigenvalues of ρ_{AB}. Consequently, the marginal entropies $S(A)$ and $S(B)$ may be *larger* than $S(A,B)$. This remarkable possibility can actually happen only in *entangled* states, as discussed in more detail in chapters 5 and 10. Nevertheless, the positivity of the quantum mutual entropy (69) implies $S(B|A) \leq S(B)$ and $S(A|B) \leq S(A)$, as in the classical case.

9 Concavity

A fundamental yet straightforward property satisfied by the Shannon entropy is concavity [4, 13]. It directly follows from the (strict) concavity of the function $f(p) = -p \log_a p$ in the interval $[0, 1]$ ($f''(p) = -\frac{1}{p \ln a} < 0$), visualized in Fig. 1. This implies

$$f(qp_1 + (1-q)p_2) \geq qf(p_1) + (1-q)f(p_2), \tag{71}$$

for $q \in [0, 1]$ and $p_i \geq 0$, with equality holding just if $p_1 = p_2$ or $q(1-q) = 0$. This implies, in turn, $f(\sum_\alpha q_\alpha p_\alpha) \geq \sum_\alpha q_\alpha f(p_\alpha)$ if $\boldsymbol{q} = \{q_\alpha\}$ is a probability distribution ($q_\alpha \geq 0$, $\sum_\alpha q_\alpha = 1$) and the p_α's are arbitrary probabilities

Given now two discrete probability distributions $\boldsymbol{p}_1$, $\boldsymbol{p}_2$, we first note that the *convex combination*

$$\boldsymbol{p} = q\boldsymbol{p}_1 + (1-q)\boldsymbol{p}_2\,, \quad q \in [0, 1]$$

is another probability distribution ($p_i \geq 0$, $\sum_i p_i = 1$, where $p_i = qp_{1i} + (1-q)p_{2i}$). As a consequence of (71), the entropy of this convex combination is not less than the convex combination of the original entropies:

$$S(q\boldsymbol{p}_1 + (1-q)\boldsymbol{p}_2) \geq qS(\boldsymbol{p}_1) + (1-q)S(\boldsymbol{p}_2). \tag{72}$$

Similarly, we have

$$S(\sum_\alpha q_\alpha \boldsymbol{p}_\alpha) \geq \sum_\alpha q_\alpha S(\boldsymbol{p}_\alpha), \tag{73}$$

if $\boldsymbol{q} = \{q_\alpha\}$ and the $\boldsymbol{p}_\alpha$ are probability distributions. We then say that the entropy $S(\boldsymbol{p})$ is a concave function of $\boldsymbol{p}$. This result is actually an expected property of an information measure. For instance, in a binary experiment, if $\boldsymbol{p}_1 = (1, 0)$ and $\boldsymbol{p}_2 = (0, 1)$ are two minimum uncertainty distributions ($S(\boldsymbol{p}_i) = 0$, $i = 1, 2$), then $\boldsymbol{p} = q\boldsymbol{p}_1 + (1-q)\boldsymbol{p}_2 = (q, 1-q)$ has non-zero uncertainty if $q(1-q) > 0$, whence $S(\boldsymbol{p}) \geq q_1 S(\boldsymbol{p}_1) + q_2 S(\boldsymbol{p}_2) = 0$. Mixture of probabilities always enhances disorder and increases the lack of information. The same property holds in the continuous case for convex combinations of continuous distributions.

In particular, by suitable convex combinations of permutations of a given discrete distribution, we may always reach the uniform distribution $p_i = 1/n \; \forall \; i$, which should then have *maximum* entropy. For instance, if $\boldsymbol{p} = \boldsymbol{p}_1 = (p, 1-p)$ and $\boldsymbol{p}_2 = (1-p, p)$, then $\frac{1}{2}(\boldsymbol{p}_1 + \boldsymbol{p}_2) = (\frac{1}{2}, \frac{1}{2})$, whence

$$S(\tfrac{1}{2}, \tfrac{1}{2}) = S(\tfrac{1}{2}\boldsymbol{p}_1 + \tfrac{1}{2}\boldsymbol{p}_2) \geq \tfrac{1}{2}[S(\boldsymbol{p}_1) + S(\boldsymbol{p}_2)] = S(\boldsymbol{p}),$$

for *any* binary distribution $\boldsymbol{p}$, with equality valid just if $p=\frac{1}{2}$. A similar argument holds for any number of events $n>2$.

A much less obvious but still fundamental property associated with the Shannon entropy is the *joint convexity* of the relative entropy (29) [4], whose general validity (it is also valid in the quantum case) follows from a theorem by Lieb [15]:

$$S(q\boldsymbol{p}_1+(1-q)\boldsymbol{p}_2\,||\,q\boldsymbol{q}_1+(1-q)\boldsymbol{q}_2)\leq qS(\boldsymbol{p}_1||\boldsymbol{q}_1)+(1-q)S(\boldsymbol{p}_2||\boldsymbol{q}_2), \tag{74}$$

where $q\in[0,1]$ and $\boldsymbol{p}_i$, $\boldsymbol{q}_i$ are probability distributions. This is actually a very desirable property of a measure of distinguishability between probability distributions: Convex mixing of probability distributions should not increase the distinguishability between them. The corresponding relative entropy is, accordingly, not greater, and in general smaller, than the average of the original relative entropies.

Eqs. (74) and (33) imply that the conditional entropy *is a concave function of the joint distribution* $\boldsymbol{p}^{XY}$: If $\boldsymbol{p}^{XY}=q\boldsymbol{p}_1^{XY}+(1-q)\boldsymbol{p}_2^{XY}$, the marginals satisfy a similar relationship $\boldsymbol{p}^X=q\boldsymbol{p}_1^X+(1-q)\boldsymbol{p}_2^X$, and hence, for $q\in[0,1]$,

$$S(\boldsymbol{p}^{XY})-S(\boldsymbol{p}^X)\;\geq\;q[S(\boldsymbol{p}_1^{XY})-S(\boldsymbol{p}_1^X)]+(1-q)[S(\boldsymbol{p}_2^{XY})-S(\boldsymbol{p}_2^X)], \tag{75}$$

i.e., $S(Y|X)=S(X,Y)-S(X)\geq qS(Y_1|X_1)+(1-q)S(Y_2|X_2)$. Mixing of probability distributions enhances not only the total disorder but also the conditional disorder, making the knowledge of Y from X more uncertain: The final conditional entropy is not smaller, and in general greater, than the average of the original conditional entropies.

In the same way as probability distributions, in the quantum case the density operators also form a convex set: if ρ_α are density operators, the convex combination $\rho=\sum_\alpha q_\alpha\rho_\alpha$, where $q_\alpha\geq 0$, $\sum_\alpha q_\alpha=1$, is also a density operator, *i.e.*, it is also an operator with non-negative eigenvalues p_i satisfying $\sum_i p_i=1$. This holds, of course, even when the different ρ_α do not commute. Due to the concavity of $-p\log_a p$, the concavity of S remains also valid in the quantum case (see also chapter 5):

$$S(\sum_\alpha q_\alpha\rho_\alpha)\geq\sum_\alpha q_\alpha S(\rho_\alpha)\,,\quad q_\alpha\geq 0,\;\sum_\alpha q_\alpha=1\,. \tag{76}$$

Moreover, as stated above, the joint convexity of the relative entropy remains valid in the quantum case: Eqs. (74) holds also if $\boldsymbol{p}_i$ and $\boldsymbol{q}_i$ are replaced by density operators ρ_i and $\tilde{\rho}_i$, whereas Eq. (75) remains valid if $\boldsymbol{p}_i^{XY}$ and $\boldsymbol{p}_i^X$ are replaced, respectively, by the joint ρ_i^{AB} and reduced ρ_i^A density operators.

10 The Maxent Approach

So far we have assumed that the probability distribution $\boldsymbol{p}$ is known, and examined the properties of its Shannon entropy. Let us now briefly consider the inverse problem, namely, that of determining the distribution $\boldsymbol{p}$ from some set of data, with the aid of the Shannon entropy. A common problem problem is that where the available information is not sufficient to completely determine $\boldsymbol{p}$, such that several distributions compatible with the available data do exist. Of all such distributions, one may consider that which *maximizes* the entropy $S(\boldsymbol{p})$, *i.e.,* the lack of information, as the optimal choice. Such distribution is certainly the *least biased* distribution compatible with the data (and with the choice of (2) as a measure of lack of information), in the sense that all other feasible distributions contain some bias which has no a priori justification. For instance, if there is no data at all (apart from normalization), the previous maximum entropy approach just leads to the equally likely distribution $p_i = 1/n \; \forall \; n$, with any other distribution containing unjustified enhancement (or depletion) of the probability of certain events.

Let us now formalize the previous maximum entropy (maxent) scheme [16, 17]. We will assume that the available data consists of a set of m expectation values $\langle O_\alpha \rangle$ of random variables O_α, $\alpha = 1, \ldots, m$, having values $O_{\alpha i}$ when basic event i with probability p_i has occurred:

$$\langle O_\alpha \rangle = \sum_{i=1}^{n} p_i O_{\alpha i}, \quad \alpha = 1, \ldots, m\,, \tag{77}$$

where $m \leq n - 1$. Normalization can be considered as an additional constraint

$$\langle O_0 \rangle = \sum_{i=1}^{n} p_i = 1\,, \tag{78}$$

where $O_0 = 1$ is a trivial constant random variable ($O_{0i} = 1 \; \forall \; i$). Maximization of the entropy $S(\boldsymbol{p})$ subject to the previous $m+1$ *linear* (in $\boldsymbol{p}$) constraints is equivalent, after introducing $m+1$ Lagrange multipliers λ_α, $\alpha = 0, 1, \ldots, m$, to the unconstrained maximization of the quantity

$$\tilde{S}(\boldsymbol{p}) = S(\boldsymbol{p}) - \sum_{\alpha=0}^{m} \lambda_\alpha \langle O_\alpha \rangle. \tag{79}$$

This maximization is straightforward and leads to the distribution

$$p_i = \exp[-\tilde{\lambda}_0 - \sum_{\alpha=1}^{m} \lambda_\alpha O_{\alpha i}]\,, \tag{80}$$

where $\tilde{\lambda}_0 = \lambda_0 + 1$. These $m+1$ parameters are to be determined from Eqs. (77)–(78). The normalization constant $\tilde{\lambda}_0$ is simply given by

$$\tilde{\lambda}_0 = \ln Z, \quad Z = \sum_i \exp[-\sum_{\alpha=1}^{m} \lambda_\alpha O_{\alpha i}], \tag{81}$$

so that we may also rewrite Eq. (80) as

$$p_i \quad = \quad Z^{-1} \exp[-\sum_{\alpha=1}^{m} \lambda_\alpha O_{\alpha i}]. \tag{82}$$

Using (81), Eqs. (77) can be written as

$$-\frac{\partial \ln Z}{\partial \lambda_\alpha} = \langle O_\alpha \rangle, \quad \alpha = 1, \ldots, m. \tag{83}$$

Concavity of $S(\boldsymbol{p})$ ensures a unique maximum, and hence a unique solution of Eqs. (83), for any set of compatible $m \le n-1$ linear constraints.

Eqs. (80)–(83) determine the maxent distribution compatible with the available data. For instance, if just the expectation value $\mu = \langle X \rangle = \sum_i p_i x_i$ and the mean squared deviation $\sigma^2 = \langle (X - \langle X \rangle)^2 \rangle = \langle X^2 \rangle - \langle X \rangle^2$ of a discrete random variable X are known, such that $O_1 = X$ and $O_2 = X^2$, Eq. (82) becomes a discrete gaussian distribution:

$$p_i = Z^{-1} \exp[-\lambda (x_i - a)^2], \tag{84}$$

where $\lambda = \lambda_2$, $a = -\lambda_1/(2\lambda)$ and $Z = \sum_i \exp[-\lambda(x_i - a)^2]$, with λ, a to be determined from the constraints $\sum_i p_i x_i = \mu$, $\sum_i p_i x_i^2 = \sigma^2 + \mu^2$.

In the general continuous case, for constraints $\langle O_\alpha \rangle = \int_{-\infty}^{\infty} p(x) O_\alpha(x) dx$, Eq. (80) leads to the continuous maxent distribution

$$p(x) = Z^{-1} \exp[-\sum_{\alpha=1}^{m} \lambda_\alpha O_\alpha(x)], \tag{85}$$

where $Z = \int_{-\infty}^{\infty} \exp[-\sum_{\alpha=1}^{m} \lambda_\alpha O_\alpha(x)] dx$ (assumed finite) and the λ_α's are determined by the equations (83). For instance, for constraints $\mu = \langle X \rangle$ and $\sigma^2 = \langle X^2 \rangle - \langle X \rangle^2$ in the case of a continuous random variable X, Eq. (85) leads to the gaussian distribution (41), *i.e.*,

$$p(x) = \sqrt{\frac{\lambda}{\pi}} \exp[-\lambda (x - a)^2], \tag{86}$$

in the notation of Eq. (84), where here $a = \mu$, $\lambda = 1/(2\sigma^2)$ and $Z = \sqrt{\pi/\lambda}$. Hence, both the standard and the discrete gaussian distribution (84) have a very clear meaning: They maximize the Shannon entropy subject to the knowledge of the average and the variance of a random variable X.

Denoting with $\boldsymbol{p}_m$ the maximizing distribution (80) or (85), the ensuing maximum entropy is just

$$S(\boldsymbol{p}_m) = \tilde{\lambda}_0 + \sum_{\alpha=1}^{n} \lambda_\alpha \langle O_\alpha \rangle \,. \tag{87}$$

We then obtain, considering the maximum entropy (87) as a function of the m averages $\langle O_\alpha \rangle$ and using Eq. (81), the general complementary relationships

$$\frac{\partial \tilde{\lambda}_0}{\partial \lambda_\alpha} = -\langle O_\alpha \rangle, \tag{88}$$

$$\frac{\partial S}{\partial \langle O_\alpha \rangle} = \lambda_\alpha \,, \tag{89}$$

for $\alpha = 1, \ldots, m$, such that $d\tilde{\lambda}_0 = -\sum_{\alpha=1}^{m} \langle O_\alpha \rangle d\lambda_\alpha$, $dS = \sum_{\alpha=1}^{m} \lambda_\alpha d\langle O_\alpha \rangle$. The Lagrange parameters are, therefore, the variation rates of the maximum entropy when the corresponding expectation value is varied. We also obtain

$$\frac{\partial^2 \tilde{\lambda}_0}{\partial \lambda_\alpha \partial \lambda_\beta} = C_{\alpha\beta} \,, \tag{90}$$

$$\frac{\partial^2 S}{\partial \langle O_\alpha \rangle \partial \langle O_\beta \rangle} = -(C^{-1})_{\alpha\beta} \,. \tag{91}$$

where

$$C_{\alpha\beta} = \langle O_\alpha O_\beta \rangle - \langle O_\alpha \rangle \langle O_\beta \rangle \,, \tag{92}$$

represents the inferred covariance matrix associated with the random variables O_α. It is a symmetric *positive* matrix (its eigenvalues are all non-negative). Maximum entropy is then a *concave* function of the averages, while $\tilde{\lambda}_0$ a *convex* function of the Lagrange parameters.

All previous expressions remain valid in the quantum case, if the O_α's are now understood as hermitian operators O_α whose expectation values

$$\langle O_\alpha \rangle = \mathrm{Tr}\, \rho\, O_\alpha, \quad \alpha = 1, \ldots, m \tag{93}$$

are supposed to be known. The inferred maxent density operator ρ replacing the maxent distribution (82) is

$$\rho = Z^{-1} \exp[-\sum_\alpha \lambda_\alpha O_\alpha] \,, \tag{94}$$

where

$$Z = \mathrm{Tr} \exp[-\sum_\alpha \lambda_\alpha O_\alpha] \,.$$

Eqs. (88)–(91) remain valid (with $\tilde{\lambda}_0 = \ln Z$), the only exception being the expression (92) for $C_{\alpha\beta}$ in (91), which should be now replaced by

$$C_{\alpha\beta} = \langle \tilde{O}_\alpha O_\beta \rangle - \langle O_\alpha \rangle \langle O_\beta \rangle \,, \tag{95}$$

where $\tilde{O}_\alpha = \int_0^1 \rho^{-s} O_\alpha \rho^s ds$ (with $\langle \tilde{O}_\alpha O_\beta \rangle = \langle \tilde{O}_\beta O_\alpha \rangle$, such that the symmetry $C_{\alpha\beta} = C_{\beta\alpha}$ remains valid.)

Connection with statistical physics. The standard canonical and grand canonical distributions of statistical mechanics [13] can be derived in the present way, by considering respectively just the average energy $\langle E \rangle$ (usually in a system with fixed particle number), and the average energy $\langle E \rangle$ plus the average particle number $\langle N \rangle$, as the only constraints in addition to normalization [16, 17]. The associated Lagrange parameters are then $\lambda_E = \beta = 1/k_B T$, where T is the absolute temperature and k_B is Boltzmann's constant (such that $k_B T$ has units of energy) and $\lambda_N = -\beta\mu$, with μ the chemical potential. The thermodynamic entropy is just

$$\mathcal{S} = -k_B \sum_i p_i \ln p_i \,. \tag{96}$$

For instance, the celebrated Boltzmann-Gibbs exponential distribution,

$$p_i = Z^{-1} \exp[-\beta E_i] \,, \tag{97}$$

where E_i is the energy of state i and $Z = \sum_i \exp[-\beta E_i]$ the *partition function*, with $\beta = 1/k_B T$, can be obtained from the maximization of (96) with an average energy constraint. The connection between the average energy $\langle E \rangle$ and β is determined by Eq. (83), *i.e.*,

$$-\frac{\partial \ln Z}{\partial \beta} = \langle E \rangle, \tag{98}$$

in the present situation. The associated entropy (87) is

$$S = \ln Z + \beta \langle E \rangle, \tag{99}$$

and the thermodynamic entropy (96) is just $k_B S$. Eq. (89) becomes

$$\frac{\partial S}{\partial \langle E \rangle} = \beta, \tag{100}$$

i.e., $d\mathcal{S} = dE/T$, whereas Eqs. (91) lead here to

$$\frac{\partial^2 \ln Z}{\partial \beta^2} = \langle E^2 \rangle - \langle E \rangle^2, \quad \frac{\partial^2 S}{\partial \langle E \rangle^2} = \frac{-1}{\langle E^2 \rangle - \langle E \rangle^2} \,. \tag{101}$$

Since $\langle E^2 \rangle - \langle E \rangle^2 \geq 0$, Eqs. (98)–(101) imply that the average energy is a decreasing function of β (*i.e.*, an increasing function of temperature) while the thermodynamic entropy is a concave function of energy.

11 Uniqueness Theorem and Generalized Entropies

We close this chapter by briefly describing the assumptions which lead to the Shannon entropy (2) (see also chapter 4). Given a discrete probability distribution $\boldsymbol{p} = (p_1, \ldots, p_n)$, the Shannon entropy is the *unique* quantity S satisfying the following axioms (Khinchin axioms [3, 18]):

1) S should be a continuous function of the probabilities p_i only.

2) S should be maximum for the uniform distribution $p_i = 1/n\ \forall\ i$:
$S(p_1, \ldots, p_n) \leq S(\frac{1}{n}, \ldots, \frac{1}{n})$.

3) S should remain unchanged if an event with zero probability is added:
$S(p_1, \ldots, p_n) = S(p_1, \ldots, p_n, 0)$.

4) $S(X, Y) = S(X) + S(Y|X)$, *i.e.*, $S(\boldsymbol{p}^{XY}) = S(\boldsymbol{p}^X) + \sum_i p_i^X S(\boldsymbol{p}^{Y/i})$, where $\boldsymbol{p}_i^X = \sum_j p_{ij}^{XY}$ is the marginal distribution and $\boldsymbol{p}^{Y/i}$ the distribution of conditional probabilities $p_{j/i}^Y = p_{ij}^{XY}/p_i^X$. This states that the joint entropy should be the entropy of X plus the average of the conditional entropies $S(Y/X = x_i)$.

This result can be verified by considering first the case of $n = n_X n_Y$ equally likely events, corresponding to independent and uniform random variables X, Y (such that $p_{ij}^{XY} = 1/n$, $p_i^X = 1/n_X$, $p_j^Y = 1/n_Y$). Axiom 1 implies that the entropy of a uniform distribution is a certain function of n, *i.e.*, $s(n)$. Axiom 4 then leads in this case to $s(n_X n_Y) = s(n_X) + s(n_Y)$ $\forall\ n_X, n_Y$, which implies $s(n) = k \ln n = \log_a n$, with $\ln a = 1/k$. Axioms 2 and 3 then imply that $s(n)$ should be an increasing function of n ($s(n+1) \geq s(n)$), which leads to $k > 0$, *i.e.*, $a > 1$.

Consider next the case of n equally likely events, but with $p_i^X = n_i/n$ an arbitrary probability distribution and Y taking one of n_i values with the same probability if $X = x_i$, such that $p_{j/i}^Y = 1/n_i$ if y_j is one of these values. Axiom 4 leads then to $s(n) = S(X) + \sum_i p_i^X s(n_i)$, whence $S(X) = \sum_i p_i^X (s(n) - s(n_i)) = -\sum_i p_i^X \log_a p_i^x$.

Other sets of axioms, such as those of the original Shannon uniqueness theorem [1], also specify Eq. (2) as the unique information measure. In particular, the fundamental additivity property was originally stated as [13]

$$S(p_1, \ldots, p_m, p_{m+1}, \ldots, p_n) = S(q_A, q_B) + q_A S(\frac{p_1}{q_A}, \ldots, \frac{p_m}{q_A}) + q_B S(\frac{p_{m+1}}{q_B}, \ldots, \frac{p_n}{q_B}), \quad (102)$$

where $q_A = p_1 + \ldots + p_m$, $q_B = p_{m+1} + \ldots + p_n = 1 - q_A$. Eq. (102) can be regarded, mathematically, as a particular case of the previous fourth postulate where X takes two values (with $\boldsymbol{p}^X = (q_A, q_B)$ and $\boldsymbol{p}^{Y/A} = (\frac{p_1}{q_A}, \ldots, \frac{p_m}{q_A})$, $\boldsymbol{p}^{Y/B} = (\frac{p_{m+1}}{q_B}, \ldots, \frac{p_n}{q_B})$) and indicates that the average information acquired in two stages (stage 1 is the choice between A and B and stage 2 the choice within A or B) must be added.

When X and Y are independent variables, axiom 4 implies *additivity*, *i.e.*, $S(X,Y) = S(X)+S(Y)$ (Eq. (13)). It is important to notice, however, that just imposition of additivity ($S(X,Y) = S(X) + S(Y)$ for X, Y independent) *does not select* the Shannon entropy as the *unique* candidate. For instance, the *Rényi entropy* [18, 19]

$$S_q^R(\boldsymbol{p}) = \frac{1}{1-q} \ln \sum_i p_i^q \,, \tag{103}$$

which also satisfies axioms 1–3 for $q > 0$ and approaches the Shannon entropy (2) for $q \to 1$ ($p^q \approx p + (q-1)p \ln p$ in this limit), satisfies as well $S(X,Y) = S(X)+S(Y)$ if X and Y are independent, as easily checked ($\sum_{i,j}(p_i p_j)^q = \sum_i p_i^q \sum_j p_j^q$). However, it does not comply with Khinchin axiom 4 in the general case.

When the axiom 4 is removed, many other functions of the probabilities satisfying axioms 1–3 (and hence possessing an *entropic* character) become feasible. These functions are still quite reasonable disorder measures and can be convenient and useful in a wide variety of scenarios, as described in other chapters of this book (particularly in chapters 3, 4 and 5), providing a more detailed characterization of the probability distribution and its information content. In addition to the Rényi entropy, a most famous generalization is the *Tsallis entropy* [20, 21] (see chapter 3),

$$S_q(\boldsymbol{p}) = \frac{1 - \sum_i p_i^q}{q-1}\,, \quad q > 0\,, \tag{104}$$

which satisfies again axioms 1–3 for $q > 0$ and reduces to the Shannon entropy for $q \to 1$. It does not satisfy Axiom 4, being instead sub (super) additive for $q > 1$ ($q < 1$) and satisfying

$$S_q(X,Y) = S_q(X) + S_q(Y) + (1-q)S_q(X)S_q(Y)\,, \tag{105}$$

when X and Y are independent. In the general case, $S_q(\boldsymbol{p}^{XY}) = S_q(\boldsymbol{p}^X) + \sum_i (p_i^X)^q S_q(\boldsymbol{p}^{Y/i})$. Uniqueness axioms leading to Eq. (104) also exist, which essentially modify previous axiom 4 [21]. See chapter 3 of this book for more details.

Notice that Eqs. (103) and (104) are obviously related:

$$S_q^R(\boldsymbol{p}) = \frac{\ln[1 + (1-q)S_q(\boldsymbol{p})]}{1-q}\,, \tag{106}$$

implying that the Rényi entropy S_q^R is just a strictly increasing function of the Tsallis entropy S_q.

Other generalizations of the Shannon entropy in terms of arbitrary concave functions and their applications are described in other chapters of this book (see in particular chapters 4 and 5).

Conflict of interest: None.

Acknowledgments

The authors acknowledge financial support from CIC of Argentina (RDR and AMK) and the brazilian scientific agencies CNPq, FAPERJ and CAPES (EMFC).

References

[1] C.E. Shannon, A Mathematical Theory of Communication. Bell System Tech. J. **27**, pp. 379–423 & 623-656, 1948.

[2] M. Tribus, Thermodynamics and Thermostatics, Van Nostrand, NY, 1961.

[3] A.I. Khinchin, Uspekhi matem. Nauk **8**, 3, 1953 (R.A. Silverman, M.D. Friedman, Mathematical Foundations of Information Theory, Dover, NY, 1957).

[4] A. Wehrl, General properties of entropy, Rev. Mod. Phys. **50**, 221, 1978.

[5] T.M. Cover, J.A. Thomas, Elements of Information Theory, J. Wiley and sons, NY, 1991.

[6] S. Kullback, R.A. Leibler, On information and sufficiency, Ann. Math. Stat. **22**, pp. 79–86, 1951.

[7] C. Rao, IMS-Lect. Notes **10**, 217, 1987.

[8] D.M. Endres, J. E. Schindlin, A new metric for probability distributions, IEEE Trans. Inf. Theory **49**, pp. 1858–1860, 2003.

[9] R.A. Fisher, Theory of statistical estimation, Proc. Cam. Phil. Soc. 22, pp. 700–725, 1925.

[10] B.R. Frieden, *Science from Fisher Information: A unification*, Cambridge University Press, Cambridge, 2004.

[11] H. Cramér, Mathematical Methods of Statistics, Princeton University Press, Princeton, 1946.

[12] C. Rao, Information and the attainable in the estimation of statistical parameters, B. of the Calcutta Mathematical Society **37**, pp. 81–89, 1945.

[13] R. Balian, From Microphysics to Macrophysics, Springer-Verlag, Berlin, 1991.

[14] M.A. Nielsen, I.L. Chuang, *Quantum Computation and Quantum Information*, Cambridge University Press, Cambridge, UK, 2000.

[15] E. H. Lieb, Convex trace functions and the Wigner-Yanase-Dyson conjecture, Ad. Math. **11**, pp. 267–288, 1973.

[16] E.T. Jaynes, Information Theory and Statistical Mechanics, Phys. Rev. **106**, pp. 620–630, 1957.

[17] E.T. Jaynes, Information Theory and Statistical Mechanics, Phys. Rev. **108**, pp. 171-190, 1957.

[18] C. Beck, F. Schlögl, Thermodynamic of Chaotic Systems, Cambridge Univ. Press, UK, 1993.

[19] A. Renyi, Probability Theory, North Holland, Amsterdam (1970).

[20] C. Tsallis, Possible Generalizations of Boltzmann-Gibbs Statistics, J. Stat. Phys. **52**, 479, 1988.

[21] C. Tsallis, Introduction to Nonextensive Statistical Mechanics, Springer, NY, 2009.

Send Orders for Reprints to reprints@benthamscience.net
Concepts and Recent Advances in Generalized Information Measures and Statistics, 2013, 57-80

CHAPTER 3

The Nonadditive Entropy S_q: A Door Open to the Nonuniversality of the Mathematical Expression of the Clausius Thermodynamic Entropy in Terms of the Probabilities of the Microscopic Configurations

Constantino Tsallis*

Centro Brasileiro de Pesquisas Fisicas and
National Institute of Science and Technology for Complex Systems
Rua Xavier Sigaud 150, 22290-180 Rio de Janeiro-RJ, Brazil
and
Santa Fe Institute, 1399 Hyde Park Road, Santa Fe, NM 87501, USA

Abstract

Clausius introduced, in the 1860s, a thermodynamical quantity which he named *entropy* S. This thermodynamically crucial quantity was proposed to be *extensive*, i.e., in contemporary terms, $S(N) \propto N$ in the thermodynamic limit $N \to \infty$. A decade later, Boltzmann proposed a functional form for this quantity which connects S with the occurrence probabilities of the microscopic configurations (referred to as *complexions* at that time) of the system. This functional is, if written in modern words referring to a system with W possible discrete states, $S_{BG} = -k_B \sum_{i=1}^{W} p_i \ln p_i$, with $\sum_{i=1}^{W} p_i = 1$, k_B being nowadays called the Boltzmann constant (BG stands for *Boltzmann-Gibbs*, to also acknowledge the fact that Gibbs provided a wider sense for W). The BG entropy is *additive*, meaning that, if A and B are two probabilistically independent systems, then $S_{BG}(A+B) = S_{BG}(A) + S_{BG}(B)$. These two words, *extensive* and *additive*, were practically treated by physicists, for over more than one century, as almost synonyms, and S_{BG} was considered to be the unique form that S could take. In other words, the functional S_{BG} was considered to be universal. It has become increasingly clear today that it is *not* so, and that those two words are *not* synonyms, but happen to coincide whenever we are dealing with paradigmatic Hamiltonians involving *short-range* interactions between their elements, presenting no strong frustration and other "pathologies". Consistently, it is today allowed to think that the entropic functional connecting S with the microscopic world transparently appears to be *nonuniversal*, but is rather dictated by the nature of possible strong correlations between the elements of the system. These facts constitute the basis of a generalization of the BG entropy and statistical

* Corresponding author. E-mail: tsallis@cbpf.br, tsallis@santafe.edu

Andres M. Kowalski, Raul D. Rossignoli and Evaldo M. F. Curado (Eds)

mechanics, introduced in 1988, and frequently referred to as nonadditive entropy S_q and nonextensive statistical mechanics, respectively. We briefly review herein these points, and exhibit recent as well as typical applications of these concepts in natural, artificial, and social systems, as shown through theoretical, experimental, observational and computational predictions and verifications.

Keywords: Nonadditive entropy; Nonextensive statistical mechanics; Complex systems; Central Limit theorems.

1 Introduction

In 1865, Rudolf Clausius introduced in thermodynamics a quantity S that, in analogy with the word *energy*, he named *entropy*. The basic motivation was to introduce an integrating factor in the differential transfer of heat δQ, so that it would become an *exact differential.* The relation is now referred to as the Clausius equality for quasi-stationary process, namely $dS = \frac{\delta Q}{T}$, T being the absolute temperature. Less than a decade later, Ludwig Boltzmann introduced a quantity which was defined in terms of the microscopic configurations of the system. This quantity gradually became identified as being a microscopic expression for the Clausius entropy S. We refer hereafter to this quantity as S_{BG}, where BG stands for *Boltzmann-Gibbs*, thus acknowledging the great contribution of Gibbs to the same concept. This functional form constitutes one of the cornerstones of the celebrated Boltzmann-Gibbs (BG) statistical mechanics. For a discrete set of probabilities $\{p_i\}$, it is given by

$$S_{BG} = -k \sum_{i=1}^{W} p_i \ln p_i = k \sum_{i=1}^{W} p_i \ln \frac{1}{p_i} \quad \Big(\sum_{i=1}^{W} p_i = 1\Big) , \tag{1}$$

where W is the total number of possibilities, and k a positive conventional constant chosen once for ever (the most frequent choices are $k = k_B$, or $k = 1$). If all probabilities are equal, we obtain the celebrated expression

$$S_{BG} = k \ln W . \tag{2}$$

If the variable which characterizes the microscopic state of the system is a D-dimensional continuous one $\mathbf{x}$, this expression is taken to be

$$S_{BG} = -k \int d\mathbf{x}\, p(\mathbf{x}) \ln p(\mathbf{x}) = k \int d\mathbf{x}\, p(\mathbf{x}) \ln \frac{1}{p(\mathbf{x})} \quad \Big(\int d\mathbf{x}\, p(\mathbf{x}) = 1\Big) , \tag{3}$$

and, if we are dealing with a quantum system, it is taken to be

$$S_{BG} = -k\,Tr\rho \ln \rho = k\,Tr\rho \ln \frac{1}{\rho} \quad \Big(Tr\rho = 1\Big), \tag{4}$$

where ρ is the density matrix. We shall indistinctively use here one or the other of these forms, depending on the particular point we are addressing.

If our system is a dynamical one with an unique stationary state (a very frequent case), this state (referred to as *thermal equilibrium*, when we are dealing, in one way or another, with a macroscopically large amount of particles) is the one which, under appropriate constraints, maximizes S_{BG}. For example, for $D = 1$, if we happen to know $\langle x \rangle \equiv \int dx\, x\, p(x)$, the maximizing distribution is given by

$$p(x) = \frac{e^{-\beta x}}{\int dx\, e^{-\beta x}} \quad (\beta > 0)\,, \tag{5}$$

where β is the corresponding Lagrange parameter.

If $\langle x \rangle = 0$, and we happen to know $\langle x^2 \rangle$, the maximizing distribution is given by

$$p(x) = \frac{e^{-\beta x^2}}{\int dx\, e^{-\beta x^2}} \quad (\beta > 0)\,. \tag{6}$$

This Gaussian form is known to be consistent with the Maxwellian distribution of velocities for classical statistical mechanics, with the solution of the standard Fokker-Planck equation in the presence of a linear drift, and also with the Central Limit Theorem (CLT). The latter basically states that if we consider the sum $S_N = \sum_{i=1}^{N} X_i$ of N *independent* (or nearly independent in some sense) random variables $\{X_i\}$, each of them having a *finite* variance, this sum converges for $N \to \infty$, after appropriate centering and rescaling, to a Gaussian. This most important theorem can be proved in a variety of manners and under slightly different hypothesis. One of those standard proofs uses the Fourier transform, which we shall address later on.

It is well known that the BG entropy is the one to be used for a wide and important class of physical systems, basically those whose (nonlinear) dynamics is strongly chaotic (*positive* maximal Lyapunov exponent for classical systems), hence mixing, hence ergodic. What can be done with the others? In particular with those that might be weakly chaotic, with *vanishing* maximal Lyapunov exponent? It has been proposed in 1988 [1] that the current statistical mechanical methods can be extended for an even wider class of physical systems just by generalizing S_{BG}. This generalization is reviewed in Section II. Before addressing it, let us make a crucial remark.

Let us consider a system whose total number of microscopic posibilities is given by $W^{(1)}$, hence, using Eq. (1), we have that

$$S_{BG}^{(1)} = -k \sum_{i_1=1}^{W^{(1)}} p_{i_1}^{(1)} \ln p_{i_1}^{(1)} \quad \Big(\sum_{i_1=1}^{W^{(1)}} p_{i_1}^{(1)} = 1\Big), \tag{7}$$

Let us assume next that we have N such systems and that they are *probabilistically independent*, i.e., $p_{i_1,i_2,...,i_N}^{(N)} = p_{i_1}^{(1)} p_{i_2}^{(2)} ... p_{i_N}^{(N)}$, we straightforwardly prove that

$$S_{BG}(N) \equiv -k \sum_{i_1,i_2,...i_N}^{W(N)} p_{i_1,i_2,...,i_N}^{(N)} \ln p_{i_1,i_2,...,i_N}^{(N)} \quad \Big(\sum_{i_1,i_2,...,i_N}^{W(N)} p_{i_1,i_2,...,i_N}^{(N)} = 1\Big), \tag{8}$$

is given by

$$S_{BG}(N) = S_{BG}^{(1)} + S_{BG}^{(2)} + ... + S_{BG}^{(N)}, \tag{9}$$

where

$$W(N) = W^{(1)} W^{(2)} ... W^{(N)}. \tag{10}$$

It follows from Eq. (9) that S_{BG} is *additive* [1].

If the N independent subsystems are equal we have that

$$W(N) = [W^{(1)}]^N, \tag{11}$$

and

$$S_{BG}(N) = N S_{BG}(1) \propto N. \tag{12}$$

Consequently the BG entropy of such a system is *extensive* since it satisfies the thermodynamical requirement that

$$0 < \lim_{N\to\infty} \frac{S_{BG}(N)}{N} < \infty, \quad i.e., \quad S(N) \propto N \quad (N \to \infty). \tag{13}$$

What happens if the N subsystems (or elements) are not strictly independent but only nearly so? In other words, what happens if, for example, the system satisfies the following relation?

$$W(N) \sim \mu^N \quad (N \to \infty;\ \mu > 1). \tag{14}$$

This is the case, at infinite temperature, of the Ising model with first-neighboring ferromagnetic interactions in any Bravais lattice, of the Lennard-Jones gas, and of a plethora

[1]We are following the definition in [8] for entropic additivity, namely an entropy S is *additive* if, for two probabilistically independent systems A and B, $S(A+B) = S(A) + S(B)$. Otherwise it is *nonadditive*.

of other important systems. For all systems satisfying Eq. (14), S_{BG} is extensive since definition (13) is satisfied.

But what happens if we have very *strong* generic correlations? More specifically, if the following relation is satisfied?

$$W(N) \sim N^{\tau} \quad (N \to \infty;\ \tau > 0)\,. \tag{15}$$

We immediately verify that $S_{BG}(N) = k \ln W(N) \propto \ln N$, hence it is *nonextensive* for any system satisfying (15). In other words, it violates classical thermodynamics. It is to avoid such disagreable situation that we are led to generalize the BG entropic functional.

2 Generalizing Boltzmann-Gibbs entropy and statistical mechanics

The generalization we are focusing here is based on the following entropy [1]:

$$S_q = k \frac{1 - \sum_{i=1}^{W} p_i^q}{q-1} = -k \sum_{i=1}^{W} p_i^q \ln_q p_i = k \sum_{i=1}^{W} p_i \ln_q \frac{1}{p_i} \quad \Big(\sum_{i=1}^{W} p_i = 1 \Big)\,, \tag{16}$$

$$S_q = -k \int d\mathbf{x}\, [p(\mathbf{x})]^q \ln_q p(\mathbf{x}) = k \int d\mathbf{x}\, p(\mathbf{x}) \ln_q \frac{1}{p(\mathbf{x})} \quad \Big(\int d\mathbf{x}\, p(\mathbf{x}) = 1 \Big)\,, \tag{17}$$

and

$$S_q = -k\, Tr\rho^q \ln_q \rho = k\, Tr\rho \ln_q \frac{1}{\rho} \quad \Big(Tr\rho = 1 \Big)\,, \tag{18}$$

where $q \in R$, and

$$\ln_q z \equiv \frac{z^{1-q} - 1}{1-q} \quad (\ln_1 z = \ln z)\,. \tag{19}$$

Eqs. (16), (17) and (18) respectively recover Eqs. (1), (3) and (2) in the $q \to 1$ instance. This theory is currently referred to as *nonextensive statistical mechanics* [1, 2, 3, 4, 5, 6, 7]. The entropy S_q satisfies the following property: if A and B are two probabilistically *independent* systems, then

$$\frac{S_q(A+B)}{k} = \frac{S_q(A)}{k} + \frac{S_q(B)}{k} + (1-q) \frac{S_q(A)}{k} \frac{S_q(B)}{k}\,. \tag{20}$$

In other words, S_q is *additive* [8] for $q = 1$, and *nonadditive* for $q \neq 1$.

Furthermore, if the probabilities are all equal, we have that

$$S_q = k \ln_q W\,. \tag{21}$$

In remarkable contrast with $S_{BG}(N)$, $S_q(N) \propto N$ for $q = 1 - 1/\tau$ for all systems satisfying (15). *For such complex systems, the additive entropy S_{BG} is nonextensive, whereas the nonadditive entropy S_q is extensive for a special (system-dependent) value of q, and therefore thermodynamically admissible!* Mathematical and physical examples can be seen in [9, 10, 11, 12]. We shall hereafter note q_{ent} the value of q such that $S_q(N) \propto N$. So, if the system satisfies relation (14), we have $q_{ent} = 1$, whereas if it satisfies (15) we have $q_{ent} = 1 - \frac{1}{\tau}$. It is convenient to emphasize at this point that particularly complex systems might exist (and probably do exist) for which no value q exists such that S_q is extensive.

In the $D = 1$ continuous case, if an appropriate first moment of x is different from zero (see details in [3]), the extremization of S_q yields

$$p(x) = \frac{e_q^{-\beta x}}{\int dx\, e_q^{-\beta x}} \quad (q < 2;\ \beta > 0)\,, \tag{22}$$

where e_q^z is the inverse function of $\ln_q z$, more precisely

$$e_q^z \equiv [1 + (1-q)z]_+^{\frac{1}{1-q}} \quad (e_1^z = e^z)\,, \tag{23}$$

$[...]_+$ being equal to its argument when this is positive, and zero otherwise [2]. Clearly Eq. (22) recovers Eq. (5) for the $q = 1$ particular case.

If the appropriate first moment vanishes and we happen to know the appropriate second moment, the extremization of S_q yields, for the continuous case, the so-called q-Gaussian distribution

$$p(x) = \frac{e_q^{-\beta x^2}}{\int dx\, e_q^{-\beta x^2}} \quad (q < 3;\ \beta > 0)\,, \tag{24}$$

The q-Gaussian distributions [3] have a finite support for $q < 1$, and an infinite one for $q \geq 1$; for $q > 1$ they asymptotically decay as power-laws (more precisely like $x^{-2/(q-1)}$); they have a finite variance for $q < 5/3$, and an infinite one for $q \geq 5/3$; they are normalizable only for $q < 3$.

Around 2000 [16], q-Gaussians have been conjectured (see details in [17]) to be attractors in the CLT sense whenever the N random variables that are being summed are strongly

---

[2]To be more explicit, e_q^z equals e^z $(\forall z)$ if $q = 1$, vanishes for $z \leq -1/(1-q)$ and equals $[1 + (1-q)z]^{\frac{1}{1-q}}$ for $z > -1/(1-q)$ if $q < 1$, and equals $[1 + (1-q)z]^{\frac{1}{1-q}}$ for all $z < 1/(q-1)$ (value at which it blows up to infinity) if $q > 1$.

[3]Since long known in plasma physics under the name *suprathermal* or *κ distributions* [13] if $q > 1$, and equal to the Student's t-distributions [14] for special values of $q > 1$. They are also occasionally referred to as *generalized Lorentzians* [15].

correlated in a specific manner. The conjecture was recently proved in the presence of q-independent variables [18, 19, 20, 21, 22]. The proof presented in [20] is based on a q-generalization of the Fourier transform, denoted as q-Fourier transform, and the theorem is currently referred to as the q-CLT. The validity of this proof has been recently challenged by Hilhorst [23]. His criticism is constructed on the inexistence of inverse q-Fourier transform for $q > 1$, which he illustrates with counterexamples. The inverse, as used in the [20] paper, indeed does not exist in general, which essentially makes the proof in [20] a proof of existence, but not of uniqueness. The q-generalization of the inverse Fourier transform appears then to be a quite subtle mathematical problem. It has nevertheless been solved recently [24], and further considerations are coming [25] related to q-moments [26]. Work is in progress attempting to transform the existence proof in [20] into a uniqueness one. In the meanwhile, several other forms [27, 28] of closely related q-generalized CLT's have already been published which do *not* use the inverse q-Fourier transform.

Probabilistic models have been formulated [29, 30] which, in the $N \to \infty$ limit, yield q-Gaussians. These models are scale-invariant, which might suggest that q-independence implies scale-invariance, but this is an open problem at the present time. However, definitively, scale-invariance does *not* imply q-independence. Indeed, (strictly or asymptotically) scale-invariant probabilistic models are known [31, 32] which do *not* yield q-Gaussians, but other limiting distributions instead [33]. In addition to all these, q-Gaussians are exact solutions of *nonlinear homogeneous* [34, 35, 36, 37] as well as of *linear inhomogenous* Fokker-Planck equations [38]. In fact, these various cases have been recently unified as particular cases of *nonlinear inhomogenneous* Fokker-Planck equations [39].

In Section 3 we remind some basic properties, and in Section 4 we briefly review various predictions, verifications and applications of q-exponentials and q-Gaussians through analytical, experimental, observational and computational methods in natural, artificial and social systems. However, before presenting them, let us make a few epistemological remarks.

Physics differs from mathematics. Within the traditional methodology of theoretical physics, mathematical tools are quite frequently (and fruitfully!) used *before* their complete mathematical understanding is achieved on rigorous grounds. In Henri Poincaré's words: *Deviner avant de démontrer! Ai-je besoin de rappeler que c'est ainsi que se sont faites toutes les découvertes importantes?* [4].

[4] *Guessing before proving! Need I remind you that it is so that all important discoveries have been made?,*

For example, up to now there is no theorem determining the precise necessary and sufficient conditions for a classical or quantum many-body Hamiltonian to be legitimately admissible in order to know with mathematical certainty that its thermostatistical equilibrium state is given by the celebrated BG weight. Nevertheless, physics has exhibited amazingly interesting results and progress by so doing in a relatively *cavalier* manner!

As a second example, we are not aware of any mathematical proof (rigorous by definition!) that the expansions based on the entire set of Feynman diagrams do converge in general to the exact answer. The same can be said for the successive truncations within the BBGKY hierarchical scheme [40]. In fact these two latter illustrations most probably have some deep relation with the so-called *moment problem* in theory of probabilities. This problem focuses on the necessary and sufficient conditions for a probability distribution to be completely and uniquely determined by the complete set of its momenta, even when they are all *finite* (see [26, 41] for cases where not all momenta are finite). Hundreds (maybe thousands) of useful developments in theoretical physics assume (sometimes explicitly, but more often just implicitly) that the full set of momenta perfectly determines the distribution. This question is so subtle that, in mathematics, three different moment problems are normally discussed separately for various classes of support of the probability distribution, namely for the $[0,1]$ support (*Hausdorff* moment problem), the $[0,\infty)$ support (*Stieltjes* moment problem), and the $(-\infty,\infty)$ support (*Hamburger* moment problem). General necessary and sufficient conditions are still ellusive [42, 43, 44, 45].

However, the frequent lack of well established mathematical rigorous has never been a serious obstacle to the advancement of theoretical and experimental physics. We may once again quote Galileo: *Nelle prove naturali non si deve ricercare l'esattezza geometrica* [5].

3 Some basic properties of S_q

We remind below some of the most basic properties of the S_q entropy, besides the nonadditivity property shown in Eq. (20). These properties are valid for all values of q, otherwise explicitly stated. For a more detailed list of the properties the reader can look at reference [6].

Henri Poincaré, *La valeur de la science*, in Anton Bovier, *Statistical Mechanics of Disordered Systems*, (2006) page 218.

[5] *In the natural proofs we should not search for geometrical exactness.*

- *Continuous dependence on the probabilites*: The S_q function depends continuously on the microscopic probabilities.

- *Nul event*: The addition of a new event with zero probability doesn't change the S_q value for $q > 0$. For $q < 0$, by definition, the sum in Eq. (16) runs only for states whose probability is positive.

- *Certainty*: If one of the microscopic states has probability equal to one, the S_q entropy equals to zero.

- *Non-negativity*: $S_q(\{p_i\}) \geq 0$ for any set of probabilities $\{p_i\}$.

- *Invariance with respect to the permutation of states*: S_q given by Eq. (16) is invariant with respect to the permutation of all the states or, given by Eq. (18), is invariant with respect to unitary transformations.

- *Concavity*: The entropy S_q satisfies the property $S_q(\{p_i''\}) \geq \lambda S_q(\{p_i'\})+(1-\lambda)S_q(\{p_i\})$ for $p_i'' = \lambda p_i' + (1-\lambda)p_i$, $\forall i$ $(0 < \lambda < 1)$, i.e., it is concave with respect to the probabilities for $q > 0$. For $q < 0$ it is a convex function with respect to the probabilities.

- *Extremum*: S_q has its maximum value at equiprobability for $q > 0$. A simple proof can be seen by noticing that the function $s(p) = (p - p^q)/(q-1)$ is a concave function and that S_q can be written like $\sum_i^W s(p_i)$. As $s(p)$ is a concave function, it satisfies $s(\frac{1}{W}\sum_i^W p_i) = s(\frac{1}{W}) \geq \frac{1}{W}\sum_i^W s(p_i)$ for any configuration $\{p_i\}$. For equiprobability we have $s(\frac{1}{W}\sum_i^W \frac{1}{W}) = \frac{1}{W}\sum_i^W s(1/W) = s(1/W) \geq \frac{1}{W}\sum_i^W s(p_i)$, for any configuration $\{p_i\}$. The last inequality implies that $S_q(\{\frac{1}{W}\}) = \sum_i^W s(1/W) = Ws(1/W) \geq \sum_i^W s(p_i) = S(\{p_i\})$ for any configuration $\{p_i\}$ other than equiprobability, showing that equiprobability extremizes S_q [119]. The same reasoning can be used to show that, for $q < 0$, S_q has a minimum at equiprobability as now the function $s(p) = (p - p^q)/(q-1)$ is a convex one.

- *Shannon additivity*: The Shannon additivity is slightly modified. For a partition of the W microscopic configurations into two subsets A and B, containing respectively W_A and W_B configurations satisfying $W_A + W_B = W$, with respective probabilities $\{p_1, \cdots, p_{W_A}\}$ and $\{p_{W_A+1}, \cdots, p_W\}$, the well-known Shannon additivity turns out to be generalized as $S_q(p_1, \cdots, p_W) = S_q(p_A, p_B) + p_A^q\, S_q(p_1/p_A, \cdots, p_{W_A}/p_A) +$

$p_B^q\, S_q(p_{W_A+1}/p_B, \cdots, p_W/p_B)$, where, in the above expression, $p_A = \sum_{i=1}^{W_A} p_i$ and $p_B = \sum_{j=W_A+1}^{W} p_j$.

- *Lesche stability*: The entropy S_q is Lesche-stable (see details in [6]).

4 Applications

Let us list here some of the many verifications and applications of the present theory (i.e., of the nonadditive entropy S_q, and of its associated nonextensive statistical mechanics) that are available in the literature (see [46] for full bibliography). (i) The velocity distribution of (cells of) *Hydra viridissima* follows a q-Gaussian probability distribution function (PDF) with $q = 3/2$ [47]; (ii) The velocity distribution of (cells of) *Dictyostelium discoideum* follows a q-Gaussian PDF with $q = 5/3$ in the vegetative state and with $q = 2$ in the starved state [48]; (iii) The velocity distribution in defect turbulence [49]; (iv) The velocity distribution of cold atoms in a dissipative optical lattice [50]; (v) The velocity distribution during silo drainage [51, 52]; (vi) The velocity distribution in a driven-dissipative 2D dusty plasma, with $q = 1.08 \pm 0.01$ and $q = 1.05 \pm 0.01$ at temperatures of $30000\,K$ and $61000\,K$ respectively [53]; (vii) The spatial (Monte Carlo) distributions of a trapped $^{136}Ba^+$ ion cooled by various classical buffer gases at $300\,K$ [54]; (viii) The distributions of price returns and stock volumes at stock exchange, as well as the volatility smile [55, 56, 57, 58]; (ix) Biological evolution [59]; (x) The distributions of returns in the Ehrenfest's dog-flea model [60, 61]; (xi) The distributions of returns in the coherent noise model [62]; (xii) The distributions of returns of the avalanche sizes in the self-organized critical Olami-Feder-Christensen model, as well as in real earthquakes [63]; (xiii) The distributions of angles in the HMF model [64]; (xiv) Turbulence in electron plasma [65]; (xv) The relaxation in various paradigmatic spin-glass substances through neutron spin echo experiments [66]; (xvi) Various properties directly related with the time dependence of the width of the ozone layer around the Earth [67]; (xvii) Various properties for conservative and dissipative nonlinear dynamical systems [68, 69, 70, 71, 72, 73, 74, 75, 76, 77, 78, 79, 80, 81, 82, 83, 84, 85]; (xviii) The degree distribution of (asymptotically) scale-free networks [86, 87, 88, 89]; (xix) Tissue radiation response [90]; (xx) Overdamped motion of interacting particles [91]; (xxi) Rotational population in molecular spectra in plasmas [92]; (xxii) High energy physics [93, 94, 95, 96, 97, 98, 99, 100, 101, 102, 103, 104, 105, 106]; (xxiii) Astrophysics [107, 108, 109]; (xxiv) Analysis of the magnetic field in the solar wind plasma using data from Voyager 1 and Voyager 2 [110, 111, 112]; (xxv) Nonlinear generalizations of quantum and relativistic equations [113]. The systematic study of metastable or long-living states in long-range versions of magnetic models such as the Ising [114] and Heisenberg [115] ones, or in hydrogen-like atoms [116, 117, 118], might provide further applications.

5 Final remarks

The BG entropy and its associated *exponential* distribution for thermal equilibrium have been extended, during the last two decades, in the sense of thermodynamics and statistical mechanics, into other forms. Stationary states have been discussed which correpond to the *q-exponential* form [1], *logarithmic* form [119, 120], *stretched-exponential* form [121], as well as other, more general, forms [122, 123].

Let us make a comment on the results obtained in [123]. By using group entropies, general forms of entropies are found which correspond to various Dirichlet *zeta* functions. The simplest of these functions is the Riemann *zeta* function, which turns out to correspond to the entropy S_q. This constitutes a rather striking and deep connection of nonextensive statistical mechanics with the theory of numbers. As a peculiarity it is also comes out in [123] that the $q \to 1$ limit can not be straightforwardly implemented, thus making that the entropy S_{BG} remains outside the class with real $q \neq 1$ (a remarkable mathematical fact which also occurs in [122]!).

While by all means interesting, the connection with the Riemann ζ function established in [123] is not a full surprise. Indeed, in 1993, the possibility of some kind of such connection was raised (see Fig. 1) in a very particular context, namely the q-partition function of a single quantum harmonic oscillator. Let us review this point. If we choose as zero energy the energy of the fundamental state, the spectrum of a single harmonic oscillator is given by

$$E_n = \hbar\omega n \quad (n = 0, 1, 2, ...) , \tag{25}$$

hence the canonical q-partition function is given by

$$Z_q(\beta) = \sum_{n=0} e_q^{-\beta\,\hbar\omega n} . \tag{26}$$

For $q > 1$, this expression becomes

$$
\begin{aligned}
Z_q(\beta) &= \sum_{n=0}^{\infty} \frac{1}{[1+(q-1)\beta\,\hbar\omega n]^{\frac{1}{q-1}}} \\
&= \frac{1}{\left[(q-1)\beta\hbar\omega\right]^{\frac{1}{q-1}}} \sum_{n=0}^{\infty} \frac{1}{\left[\frac{1}{(q-1)\beta\,\hbar\omega}+n\right]^{\frac{1}{q-1}}} \\
&= \frac{1}{\left[(q-1)\beta\hbar\omega\right]^{\frac{1}{q-1}}} \zeta\Big(\frac{1}{q-1}, \frac{1}{(q-1)\beta\,\hbar\omega}\Big), \qquad (27)
\end{aligned}
$$

where $\zeta(s,Q)$ is the Hurwitz zeta function. The Riemann zeta function is given by

$$
\zeta(s) = \zeta(s,1) = \sum_{n=1}^{\infty} \frac{1}{n^s} = \prod_{p\ prime} \frac{1}{1-p^{-s}} = \frac{1}{1-2^{-s}} \cdot \frac{1}{1-3^{-s}} \cdot \frac{1}{1-5^{-s}} \cdot \frac{1}{1-7^{-s}} \cdot \frac{1}{1-11^{-s}} \cdots, \qquad (28)
$$

which establishes a connection with the prime numbers, thus with the theory of numbers. We notice also that here the $q \to 1$ limit is pathological, like in [122, 123]. Clearly, the partition functions Z_q of systems different from the harmonic oscillator are not necessarily going to exhibit any relationship with the Riemman zeta function. It is nevertheless a kind of amusing coincidence that this function already appeared before in connection with q-statistics.

As a final comment let us state that it appears that BG entropy and statistics are sufficient but not necessary for thermodynamics. In other words, thermodynamics might well be more powerful than the role attributed to it by BG statistical mechanics. A question which then arises naturally is: What is the most general form of entropy which can be consistent with thermodynamics, more precisely with the zeroth, first, second, and third principles? What are the most general superstatistical forms [124, 125, 126] which would correspond to this general entropy and statistics? By following in depth along the lines of [122, 123, 127], it might be possible to clarify this important — though hard — question.

Conflict of interest: None.

Acknowledgments

I acknowledge useful remarks by T. Bountis, E.M.F. Curado, R. Hanel, H.J. Hilhorst, M. Jauregui, F.D. Nobre, A. Rodriguez, G. Ruiz, P. Tempesta, S. Thurner and S. Umarov, as well as partial financial support by Faperj and CNPq (Brazilian agencies).

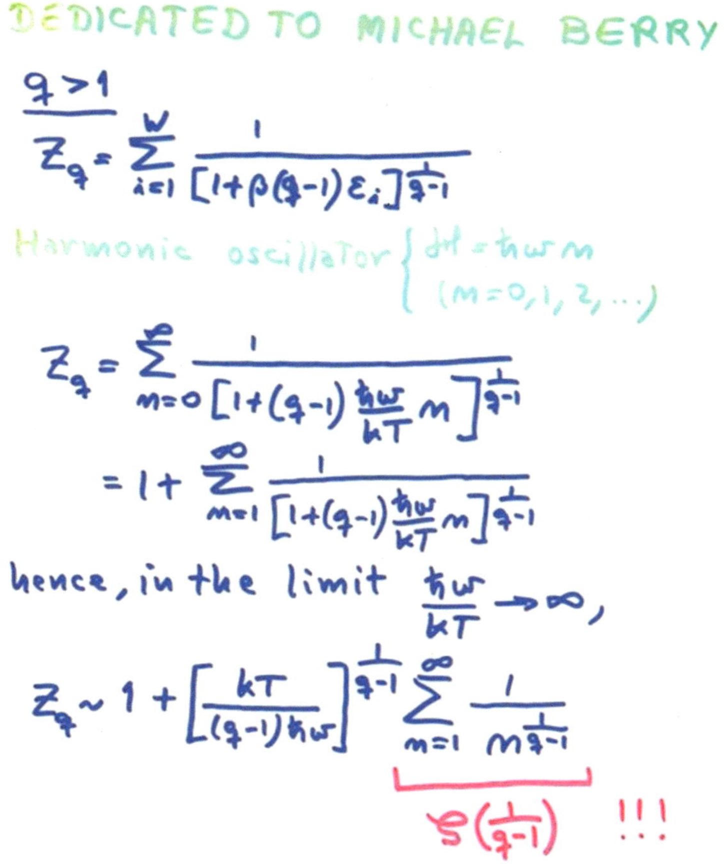

Figure 1: First occasion when it became apparent that there was a connection between S_q (hence q-statistics) and the Riemann ζ function. I projected this slide during my lecture at the Workshop held, in 2-6 August 1993, in Mar del Plata, Argentina. The day before Michael Berry, in his own (wonderful) lecture, had emphasized that it would be very exciting if one could find fundamental connections of the prime numbers with physics. It is precisely such connection what the slide shows.

References

[1] C. Tsallis, *Possible generalization of Boltzmann-Gibbs statistics*, J. Stat. Phys. **52**, 479-487 (1988).

[2] E.M.F. Curado and C. Tsallis, *Generalized statistical mechanics: connection with thermodynamics*, J. Phys. A **24**, L69-72 (1991); Corrigenda: **24**, 3187 (1991) and **25**, 1019 (1992).

[3] C. Tsallis, R.S. Mendes and A.R. Plastino, *The role of constraints within generalized nonextensive statistics*, Physica A **261**, 534-554 (1998).

[4] M. Gell-Mann and C. Tsallis, eds., *Nonextensive Entropy - Interdisciplinary Applications*, eds. M. Gell-Mann and C. Tsallis (Oxford University Press, New York, 2004).

[5] J.P. Boon and C. Tsallis, eds., *Nonextensive Statistical Mechanics: New Trends, New Perspectives*, Europhysics News **36** (6) (2005).

[6] C. Tsallis, *Introduction to Nonextensive Statistical Mechanics - Approaching a Complex World* (Springer, New York, 2009).

[7] C. Tsallis, *Entropy*, in *Encyclopedia of Complexity and Systems Science*, ed. R.A. Meyers (Springer, Berlin, 2009), 11 volumes [ISBN: 978-0-387-75888-6].

[8] O. Penrose, *Foundations of Statistical Mechanics: A Deductive Treatment* (Pergamon, Oxford, 1970), p. 167.

[9] C. Tsallis, *Nonextensive statistical mechanics: Construction and physical interpretation*, in *Nonextensive Entropy - Interdisciplinary Applications*, eds. M. Gell-Mann and C. Tsallis (Oxford University Press, New York, 2004).

[10] C. Tsallis, M. Gell-Mann and Y. Sato, *Asymptotically scale-invariant occupancy of phase space makes the entropy S_q extensive*, Proc. Natl. Acad. Sc. USA **102**, 15377 (2005).

[11] F. Caruso and C. Tsallis, *Nonadditive entropy reconciles the area law in quantum systems with classical thermodynamics*, Phys. Rev. E **78**, 021101 (2008).

[12] A. Saguia and M.S. Sarandy, *Nonadditive entropy for random quantum spin-S chains*, Phys. Lett. A **374**, 3384 (2010).

[13] V.M. Vasyliunas, J. Geophys. Res., **73**, 2839 (1968); L.A. Rios and R.M.O. Galvao, Phys. Plasmas **17** (4), 042116 (2010).

[14] A.M.C. Souza and C. Tsallis, *Student's t- and r-distributions: Unified derivation from an entropic variational principle*, Physica A, **236**, 52 (1997).

[15] R.A. Treumann, *Generalized-Lorentziann path integrals*, Phys. Rev. E **57**, 5150 (1998).

[16] M. Bologna, C. Tsallis and P. Grigolini, *Anomalous diffusion associated with nonlinear fractional derivative Fokker-Planck-like equation: Exact time-dependent solutions*, Phys. Rev. E **62**, 2213 (2000).

[17] C. Tsallis, *Nonextensive statistical mechanics, anomalous diffusion and central limit theorems*, Milan J. Math. **73**, 145-176 (2005).

[18] C. Tsallis and S.M.D. Queiros, *Nonextensive statistical mechanics and central limit theorems I - Convolution of independent random variables and q-product*, in *Complexity, Metastability and Nonextensivity*, eds. S. Abe, H.J. Herrmann, P. Quarati, A. Rapisarda and C. Tsallis, American Institute of Physics Conference Proceedings **965**, 8 (New York, 2007).

[19] S.M.D. Queiros and C. Tsallis, *Nonextensive statistical mechanics and central limit theorems II - Convolution of q-independent random variables*, in *Complexity, Metastability and Nonextensivity*, eds. S. Abe, H.J. Herrmann, P. Quarati, A. Rapisarda and C. Tsallis, American Institute of Physics Conference Proceedings **965**, 21 (New York, 2007).

[20] S. Umarov, C. Tsallis and S. Steinberg [2008], *On a q-central limit theorem consistent with nonextensive statistical mechanics*, Milan J. Math. **76**, 307-328.

[21] S. Umarov, C. Tsallis, M. Gell-Mann and S. Steinberg, *Generalization of symmetric α-stable Lévy distributions for $q > 1$*, J. Math. Phys. **51**, 033502 (2010).

[22] K.P. Nelson and S. Umarov, *Nonlinear statistical coupling*, Physica A **389**, 2157 (2010).

[23] H.J. Hilhorst, *Central limit theorems for correlated variables: some critical remarks*, Braz. J. Phys. **39**, 371-379 (2009), and *Note on a q-modified central limit theorem*, J. Stat. Mech., P10023 (2010).

[24] M. Jauregui and C. Tsallis, *q-generalization of the inverse Fourier transform*, Phys. Lett. A **375**, 2085 (2011). See also M. Jauregui and C. Tsallis, *New representations of π and Dirac delta using the nonextensive-statistical-mechanics q-exponential function*, J. Math. Phys. **51**, 063304 (2010).

[25] M. Jauregui, C. Tsallis and E.M.F. Curado, *q-moments remove the degeneracy associated with the inversion of the q-Fourier transform*, J. Stat. Mech. P10016 (2011).

[26] C. Tsallis, A.R. Plastino and R.F. Alvarez-Estrada, *Escort mean values and the characterization of power-law-decaying probability densities*, J. Math. Phys **50**, 043303 (2009).

[27] M.G. Hahn, X.X. Jiang and S. Umarov, *On q-Gaussians and exchangeability*, J. Phys. A **43** (16), 165208 (2010).

[28] C. Vignat and A. Plastino, *Central limit theorem and deformed exponentials*, J. Phys. A **40**, F969 (2007). See also C. Vignat and A. Plastino, *Geometry of the central limit theorem in the nonextensive case*, Phys. Lett. A **373**, 1713 (2009).

[29] A. Rodriguez, V. Schwammle and C. Tsallis, *Strictly and asymptotically scale-invariant probabilistic models of N correlated binary random variables having* q–*Gaussians as $N \to \infty$ limiting distributions*, JSTAT P09006 (2008).

[30] R. Hanel, S. Thurner and C. Tsallis, *Limit distributions of scale-invariant probabilistic models of correlated random variables with the q-Gaussian as an explicit example*, Eur. Phys. J. B **72**, 263 (2009).

[31] L.G. Moyano, C. Tsallis and M. Gell-Mann, *Numerical indications of a q-generalised central limit theorem*, Europhys. Lett. **73**, 813 (2006).

[32] W.J. Thistleton, J.A. Marsh, K.P. Nelson and C. Tsallis, *q-Gaussian approximants mimic non-extensive statistical-mechanical expectation for many-body probabilistic model with long-range correlations*, Cent. Eur. J. Phys. **7**, 387 (2009).

[33] H.J. Hilhorst and G. Schehr, *A note on q-Gaussians and non-Gaussians in statistical mechanics*, J. Stat. Mech. (2007) P06003.

[34] A.R. Plastino and A. Plastino, *Non-extensive statistical mechanics and generalized Fokker-Planck equation*, Physica A **222**, 347 (1995).

[35] C. Tsallis and D.J. Bukman, *Anomalous diffusion in the presence of external forces: exact time-dependent solutions and their thermostatistical basis*, Phys. Rev. E **54**, R2197 (1996).

[36] L. Borland, *Ito-Langevin equations within generalized thermostatistics*, Phys. Lett. A **245**, 67 (1998).

[37] M.A. Fuentes and M.O. Caceres, *Computing the non-linear anomalous diffusion equation from first principles*, Phys. Lett. A **372**, 1236 (2008). To compare with the present paper, it must be done $q \to 2 - q$.

[38] C. Anteneodo and C. Tsallis, *Multiplicative noise: A mechanism leading to nonextensive statistical mechanics*, J. Math. Phys. **44**, 5194 (2003).

[39] A.M. Mariz and C. Tsallis, *Unified long-memory mesoscopic mechanism consistent with nonextensive statistical mechanics*, Phys. Lett. A **376**, 3088-3091(2012).

[40] R. Balescu, *Equilibrium and Nonequilibrium Statistical Mechanics*, (John Wiley and Sons, 1975, New York).

[41] A. Rodriguez and C. Tsallis, *A generalization of the cumulant expansion. Application to a scale-invariant probabilistic model*, J. Math. Phys. **51**, 073301 (2010).

[42] J.A. Shohat and J.D. Tamarkin, *The Problem of Moments*, Amer. Math. Soc. Mathematical Surveys, vol II (American Mathematical Society, New York, 1943).

[43] N.I. Akhiezer, *The classical moment problem and some related questions in analysis*, translated by N. Kemmer (Hafner Publishing Co., New York, 1965).

[44] T.S. Chihara, *An Introduction to Orthogonal Polynomials* (Gordon and Breach, 1978).

[45] A.G. Bakan, *Codimension of polynomial subspace in $L_2(\mathcal{R}, d\mu)$ for discrete indeterminate measure μ*, Proc. Amer. Math. Soc. **130**, 3545-3553 (2002) (electronic) .

[46] http://tsallis.cat.cbpf.br/biblio.htm

[47] A. Upadhyaya, J.-P. Rieu, J.A. Glazier and Y. Sawada, *Anomalous diffusion and non-Gaussian velocity distribution of Hydra cells in cellular aggregates*, Physica A **293**, 549 (2001). See also S. Thurner, N. Wick, R. Hanel, R. Sedivy and L.A. Huber, *Anomalous diffusion on dynamical networks: A model for epithelial cell migration*, Physica A **320**, 475 (2003).

[48] A.M. Reynolds, *Can spontaneous cell movements be modelled as Lévy walks?*, Physica A **389**, 273 (2010).

[49] K.E. Daniels, C. Beck and E. Bodenschatz, *Defect turbulence and generalized statistical mechanics*, Physica D **193**, 208 (2004).

[50] P. Douglas, S. Bergamini and F. Renzoni, *Tunable Tsallis distributions in dissipative optical lattices*, Phys. Rev. Lett. **96**, 110601 (2006); G.B. Bagci and U. Tirnakli, *Self-organization in dissipative optical lattices*, Chaos **19**, 033113 (2009).

[51] R. Arevalo, A. Garcimartin and D. Maza, *Anomalous diffusion in silo drainage*, Eur. Phys. J. E **23**, 191 (2007).

[52] R. Arevalo, A. Garcimartin and D. Maza, *A non-standard statistical approach to the silo discharge*, in *Complex Systems - New Trends and Expectations*, eds. H.S. Wio, M.A. Rodriguez and L. Pesquera, Eur. Phys. J.-Special Topics **143** (2007).

[53] B. Liu and J. Goree, *Superdiffusion and non-Gaussian statistics in a driven-dissipative 2D dusty plasma*, Phys. Rev. Lett. **100**, 055003 (2008).

[54] R.G. DeVoe, *Power-law distributions for a trapped ion interacting with a classical buffer gas*, Phys. Rev. Lett. **102**, 063001 (2009).

[55] L. Borland, *Closed form option pricing formulas based on a non-Gaussian stock price model with statistical feedback*, Phys. Rev. Lett. **89**, 098701 (2002).

[56] L. Borland, *A theory of non-gaussian option pricing*, Quant. Finance **2**, 415 (2002).

[57] R. Osorio, L. Borland and C. Tsallis, *Distributions of high-frequency stock-market observables*, in *Nonextensive Entropy - Interdisciplinary Applications*, eds. M. Gell-Mann and C. Tsallis (Oxford University Press, New York, 2004).

[58] S.M.D. Queiros, *On non-Gaussianity and dependence in financial in time series: A nonextensive approach*, Quant. Finance **5**, 475 (2005).

[59] F.A. Tamarit, S.A. Cannas and C. Tsallis, *Sensitivity to initial conditions in the Bak-Sneppen model of biological evolution*, Eur. Phys. J. B **1**, 545 (1998).

[60] B. Bakar and U. Tirnakli, *Analysis of self-organized criticality in Ehrenfest's dog-flea model*, Phys. Rev. E **79**, 040103(R) (2009).

[61] B. Bakar and U. Tirnakli, *Return distributions in dog-flea model revisited*, Physica A **389**, 3382-3386 (2010).

[62] A. Celikoglu, U. Tirnakli and S.M.D. Queiros, *Analysis of return distributions in the coherent noise model*, Phys. Rev. E **82**, 021124.

[63] F. Caruso, A. Pluchino, V. Latora, S. Vinciguerra and A. Rapisarda, *Analysis of self-organized criticality in the Olami-Feder-Christensen model and in real earthquakes*, Phys. Rev. E **75**, 055101(R) (2007).

[64] L.G. Moyano and C. Anteneodo, *Diffusive anomalies in a long-range Hamiltonian system*, Phys. Rev. E **74**, 021118 (2006).

[65] C. Anteneodo and C. Tsallis, *Two-dimensional turbulence in pure-electron plasma: A nonextensive thermostatistical description*, J. Molecular Liquids **71**, 255 (1997).

[66] R.M. Pickup, R. Cywinski, C. Pappas, B. Farago and P. Fouquet, *Generalized spin glass relaxation*, Phys. Rev. Lett. **102**, 097202 (2009).

[67] G.L. Ferri, M.F. Reynoso Savio and A. Plastino , *Tsallis q-triplet and the ozone layer*, Physica A **389**, 1829 (2010).

[68] C. Tsallis, A.R. Plastino and W.-M. Zheng, *Power-law sensitivity to initial conditions - New entropic representation*, Chaos, Solitons and Fractals **8**, 885 (1997).

[69] F. Baldovin and A. Robledo [2002a], "Sensitivity to initial conditions at bifurcations in one-dimensional nonlinear maps: Rigorous nonextensive solutions", Europhys. Lett. **60**, 518.

[70] F. Baldovin and A. Robledo [2002b], "Universal renormalization-group dynamics at the onset of chaos in logistic maps and nonextensive statistical mechanics", Phys. Rev. E **66**, R045104.

[71] A. Robledo, *Incidence of nonextensive thermodynamics in temporal scaling at Feigenbaum points*, Physica A **370**, 449 (2006).

[72] G. Miritello, A. Pluchino and A. Rapisarda, *Central limit behavior in the Kuramoto model at the 'edge of chaos'*, Physica A **388**, 4818 (2009).

[73] O. Afsar and U. Tirnakli, *Probability densities for the sums of iterates of the sine-circle map in the vicinity of the quasi-periodic edge of chaos*, Phys. Rev. E **82**, 046210 (2010).

[74] M. Leo, R.A. Leo and P. Tempesta, *Thermostatistics in the neighborhood of the π-mode solution for the Fermi-Pasta-Ulam β system: From weak to strong chaos*, J. Stat. Mech. P04021 (2010).

[75] F. Baldovin and A. Robledo, *Nonextensive Pesin identity. Exact renormalization group analytical results for the dynamics at the edge of chaos of the logistic map*, Phys. Rev. E **69**, 045202(R) (2004).

[76] E. Mayoral and A. Robledo, *Tsallis' q index and Mori's q phase transitions at edge of chaos*, Phys. Rev. E **72**, 026209 (2005).

[77] C. Anteneodo and C. Tsallis, *Breakdown of the exponential sensitivity to the initial conditions: Role of the range of the interaction*, Phys. Rev. Lett. **80**, 5313 (1998).

[78] A. Pluchino, A. Rapisarda and C. Tsallis, *Nonergodicity and central limit behavior in long-range Hamiltonians*, Europhys. Lett. **80**, 26002 (2007).

[79] A. Pluchino, A. Rapisarda and C. Tsallis, *A closer look at the indications of q-generalized Central Limit Theorem behavior in quasi-stationary states of the HMF model*, Physica A **387**, 3121 (2008).

[80] M.L. Lyra and C. Tsallis, *Nonextensivity and multifractality in low-dimensional dissipative systems*, Phys. Rev. Lett. **80**, 53 (1998).

[81] E.P. Borges, C. Tsallis, G.F.J. Ananos and P.M.C. Oliveira, *Nonequilibrium probabilistic dynamics at the logistic map edge of chaos*, Phys. Rev. Lett. **89**, 254103 (2002).

[82] G.F.J. Ananos and C. Tsallis, *Ensemble averages and nonextensivity at the edge of chaos of one-dimensional maps*, Phys. Rev. Lett. **93**, 020601 (2004).

[83] U. Tirnakli, C. Beck and C. Tsallis, *Central limit behavior of deterministic dynamical systems*, Phys. Rev. E **75**, 040106(R) (2007).

[84] U. Tirnakli, C. Tsallis and C. Beck, *A closer look on the time-average attractor at the edge of chaos of the logistic map*, Phys. Rev. E **79**, 056209 (2009).

[85] C. Tsallis and U. Tirnakli, *Nonadditive entropy and nonextensive statistical mechanics - Some central concepts and recent applications*, J. Phys. Conf. Series **201**, 012001 (2010).

[86] D.J.B. Soares, C. Tsallis, A.M. Mariz and L.R. da Silva, *Preferential attachment growth model and nonextensive statistical mechanics*, Europhys. Lett. **70**, 70 (2005).

[87] S. Thurner and C. Tsallis, *Nonextensive aspects of self-organized scale-free gas-like networks*, Europhys. Lett. **72**, 197-203 (2005).

[88] D.R. White, N. Kejzar, C. Tsallis, D. Farmer and S. White, *A generative model for feedback networks*, Phys. Rev. E **73**, 016119 (2006).

[89] S. Thurner, F. Kyriakopoulos and C. Tsallis, *Unified model for network dynamics exhibiting nonextensive statistics*, Phys. Rev. E **76**, 036111 (2007).

[90] O. Sotolongo-Grau, D. Rodriguez-Perez, J.C. Antoranz and O. Sotolongo-Costa, *Tissue radiation response with maximum Tsallis entropy*, Phys. Rev. Lett. **105**, 158105 (2010).

[91] J. S. Andrade Jr., G.F.T. da Silva, A.A. Moreira, F.D. Nobre and E.M.F. Curado, *Thermostatistics of overdamped motion of interacting particles*, Phys. Rev. Lett. **105**, 260601 (2010).

[92] J.L. Reis Jr., J. Amorim and A. Dal Pino Jr., *Occupancy of rotational population in molecular spectra based on nonextensive statistics*, Phys. Rev. E **83**, 017401 (2011) (4 pages).

[93] G. Kaniadakis, A. Lavagno and P. Quarati, *Generalized statistics and solar neutrinos*, Phys. Lett. B **369**, 308 (1996).

[94] I. Bediaga, E.M.F. Curado and J. Miranda, *A nonextensive thermodynamical equilibrium approach in* $e^+e^- \to$ *hadrons*, Physica A **286**, 156 (2000).

[95] C. Beck, *Non-extensive statistical mechanics and particle spectra in elementary interactions*, Physica A **286**, 164 (2000).

[96] C. Tsallis, J.C. Anjos and E.P. Borges, *Fluxes of cosmic rays: A delicately balanced stationary state*, Phys. Lett. A **310**, 372 (2003).

[97] C. Beck, *Generalized statistical mechanics of cosmic rays*, Physica A **331**, 173 (2003).

[98] G. Wilk and Z. Wlodarczyk, *Power laws in elementary and heavy-ion collisions - A story of fluctuations and nonextensivity?*, Eur. Phys. J. A **40**, 299 (2009).

[99] T.S. Biro, G. Purcsel and K. Urmossy, *Non-extensive approach to quark matter*, in *Statistical Power-Law Tails in High Energy Phenomena*, Eur. Phys. J. A **40**, 325 (2009).

[100] V. Khachatryan *et al.* (CMS Collaboration), *Transverse-momentum and pseudorapidity distributions of charged hadrons in pp collisions at* $\sqrt{s} = 0.9$ *and* $2.36\ TeV$, J. High Energy Phys. **02**, 041 (2010).

[101] V. Khachatryan *et al.* (CMS Collaboration), *Transverse-momentum and pseudorapidity distributions of charged hadrons in pp collisions at* $\sqrt{s} = 7\ TeV$, Phys. Rev. Lett. **105**, 022002 (2010).

[102] D. d'Enterria, R. Engel, T. Pierog, S. Ostapchenko and K. Werner, *Constraints from the first LHC data on hadronic event generators for ultra-high energy cosmic-ray physics*, Astroparticle Phys. **35** (2), 98-113 (2011).

[103] Adare *et al.* (PHENIX Collaboration), *Measurement of neutral mesons in* $p+p$ *collisions at* $\sqrt{s} = 200\ GeV$ *and scaling properties of hadron production*, 1005.3674 [hep-ex] (2010).

[104] M. Shao, L. Yi, Z.B. Tang, H.F. Chen, C. Li and Z.B. Xu, *Examination of the species and beam energy dependence of particle spectra using Tsallis statistics*, J. Phys. G **37** (8), 085104 (2010); C.Y. Wong and G. Wilk, *Tsallis fits to* pT *spectra and relativistic hard scattering in pp collisions at LHC*, Phys. Rev. D **87**, 114007 (2013); L. Marques,

E. Andrade-II and A. Deppman, *Nonextensivity of hadronic systems*, Phys. Rev. D **87**, 114022 (2013).

[105] T. Wibig, *The non-extensivity parameter of a thermodynamical model of hadronic interactions at LHC energies*, J. Phys. G: Nucl. Part. Phys. **37**, 115009 (2010).

[106] W.M. Alberico, A. Lavagno and P. Quarati, *Non-extensive statistics, fluctuations and correlations in high energy nuclear collisions*, Eur. Phys. J C **12**, 499 (2000).

[107] J.C. Carvalho, R. Silva, J.D. do Nascimento and J.R. de Medeiros, *Power law statistics and stellar rotational velocities in the Pleiades*, Europhys. Lett. **84**, 59001 (2008).

[108] A. Esquivel and A. Lazarian, *Tsallis statistics as a tool for studying interstellar turbulence*, Astrophys. J. **710**, 125-132 (2010).

[109] M.A. Moret, V. de Senna, G.F. Zebende and P. Vaveliuk, *X-ray binary systems and nonextensivity*, Physica A **389**, 854-858 (2010).

[110] L.F. Burlaga and A.F.-Vinas, *Triangle for the entropic index q of non-extensive statistical mechanics observed by Voyager 1 in the distant heliosphere*, Physica A **356**, 375 (2005).

[111] L.F. Burlaga, A.F. Vinas, N.F. Ness and M.H. Acuna, *Tsallis statistics of the magnetic field in the heliosheath*, Astrophys. J. **644**, L83 (2006).

[112] L.F. Burlaga and N.F. Ness, *Compressible "turbulence" observed in the heliosheath by Voyager 2*, Astrophys. J. **703**, 311 (2009).

[113] F.D. Nobre, M.A. Rego-Monteiro and C. Tsallis, *Nonlinear generalizations of relativistic and quantum equations with a common type of solution*, Phys. Rev. Lett. **106**, 140601 (2011).

[114] F.D. Nobre and C. Tsallis, *Infinite-range Ising ferromagnet: thermodynamic limit within generalized statistical mechanics*, Physica A **213**, 337 (1995); Erratum: **216**, 369 (1995).

[115] A.O. Caride, C. Tsallis and S. I. Zanette, *Criticality of the anisotropic quantum Heisenberg model on a self-dual hierarchical lattice*, Phys. Rev. Lett. **51**, 145 (1983); **51**, 616 (1983).

[116] L.S. Lucena, L.R. da Silva and C. Tsallis, *Departure from Boltzmann-Gibbs statistics makes the hydrogen-atom specific heat a computable quantity*, Phys. Rev. E **51**, 6247 (1995).

[117] N.M. Oliveira-Neto, E.M.F. Curado, F.D. Nobre and M.A. Rego-Monteiro, *Approach to equilibrium of the hydrogen atom at low temperature*, Physica A **374**, 251-262 (2007).

[118] N.M. Oliveira-Neto, E.M.F. Curado, F.D. Nobre and M.A. Rego-Monteiro, *A simple model to describe the low-temperature behaviour of some atoms and molecules: An application to the hydrogen atom*, J. Phys. B - Atomic, Molecular and Optical Physics **40**, 1975-1989 (2007).

[119] E.M.F. Curado, *General aspects of the thermodynamical formalism*, in *Nonextensive Statistical Mechanics and Thermodynamics*, eds. S.R.A. Salinas and C. Tsallis, Braz. J. Phys. **29**, 36 (1999).

[120] E.M.F. Curado and F.D. Nobre, *On the stability of analytic entropic forms*, Physica A **335**, 94 (2004).

[121] C. Anteneodo and A.R. Plastino, *Maximum entropy approach to stretched exponential probability distributions*, J. Phys. A **32**, 1089 (1999).

[122] R. Hanel and S. Thurner, *A comprehensive classification of complex statistical systems and an axiomatic derivation of their entropy and distribution functions*, Europhys. Lett. **93**, 20006 (2011).

[123] P. Tempesta, *Group entropies, correlation laws, and zeta functions*, Phys. Rev. E **84**, 021121 (2011).

[124] C. Beck and E.G.D. Cohen, *Superstatistics*, Physica A **322**, 267 (2003).

[125] C. Tsallis and A.M.C. Souza, *Constructing a statistical mechanics for Beck-Cohen superstatistics*, Phys. Rev. E **67**, 026106 (2003).

[126] R. Hanel, S. Thurner and M. Gell-Mann, *Generalized entropies and the transformation group of superstatistics*, Proc. Nat. Acad. Sci. (USA) 1103539108 (2011).

[127] R. Hanel and S. Thurner, *When do generalised entropies apply? How phase space volume determines entropy*, EPL **96**, 50003 (2011).

Send Orders for Reprints to reprints@benthamscience.net
Concepts and Recent Advances in Generalized Information Measures and Statistics, 2013, 81-99

CHAPTER 4

What do Generalized Entropies Look Like? An Axiomatic Approach for Complex, Non-Ergodic Systems

Stefan Thurner [a,b,*] **and Rudolf Hanel** [b]

[a] Section for Science of Complex Systems, Medical University of Vienna
Spitalgasse 23, A-1090, Austria

[b] Santa Fe Institute, 1399 Hyde Park Road, Santa Fe, NM 87501, USA

Abstract

Shannon and Khinchin showed that assuming four information theoretic axioms the entropy must be of Boltzmann-Gibbs type, $S = -\sum_i p_i \log p_i$. Here we note that in physical systems one of these axioms may be violated. For non-ergodic systems the so called separation axiom (Shannon-Khinchin axiom 4) will in general not be valid. We show that when this axiom is violated the entropy takes a more general form, $S_{c,d} \propto \sum_i^W \Gamma(d+1, 1-c\log p_i)$, where c and d are scaling exponents and $\Gamma(a,b)$ is the incomplete gamma function. The exponents (c,d) define equivalence classes for all interacting and non-interacting systems and unambiguously characterize any statistical system in its thermodynamic limit. The proof is possible because of two newly discovered scaling laws which any entropic form has to fulfill, if the first three Shannon-Khinchin axioms hold. (c,d) can be used to define equivalence classes of statistical systems. A series of known entropies can be classified in terms of these equivalence classes. We show that the corresponding distribution functions are special forms of Lambert-$\mathcal{W}$ exponentials containing – as special cases – Boltzmann, stretched exponential and Tsallis distributions (power-laws). In the derivation we assume trace form entropies, $S = \sum_i g(p_i)$, with g some function, however more general entropic forms can be classified along the same scaling analysis.

Keywords: Generalized Entropies; Nonextensive Statistical Mechanics; Complex Systems; Distribution Functions.

1 Introduction

Theorem number 2 in the seminal 1948 paper, *The Mathematical Theory of Communication* [3], by Claude Shannon, proves the existence of the one and only form of entropy, given

* Corresponding author. E-mail: stefan.thurner@meduniwien.ac.at

that three fundamental requirements hold. A few years later A.I. Khinchin remarked in his *Mathematical Foundations of Information Theory* [4]: "However, Shannon's treatment is not always sufficiently complete and mathematically correct so that, besides having to free the theory from practical details, in many instances I have amplified and changed both the statement of definitions and the statement of proofs of theorems." Khinchin adds a fourth axiom. The three fundamental requirements of Shannon, in the 'amplified' version of Khinchin, are known as the Shannon-Khinchin (SK) axioms. These axioms list the requirements needed for an entropy to be a reasonable measure of the 'uncertainty' about a finite probabilistic system. Khinchin further suggests to also use entropy as a measure of the information *gained* about a system when making an 'experiment', *i.e.*, by observing a realization of the probabilistic system.

Khinchin's first axiom states that for a system with W potential outcomes (states) each of which is given by a probability $p_i \geq 0$, with $\sum_{i=1}^{W} p_i = 1$, the entropy $S(p_1, \cdots, p_W)$ as a measure of uncertainty about the system must take its maximum for the equi-distribution $p_i = 1/W$, for all i.

Khinchin's second axiom (missing in [3]) states that any entropy should remain invariant under adding zero-probability states to the system, *i.e.*, $S(p_1, \cdots, p_W) = S(p_1, \cdots, p_W, 0)$.

Khinchin's third axiom (separability axiom) finally makes a statement of the composition of two finite probabilistic systems A and B. If the systems are independent of each other, entropy should be additive, meaning that the entropy of the combined system $A+B$ should be the sum of the individual systems, $S(A+B) = S(A) + S(B)$. If the two systems are dependent on each other, the entropy of the combined system, *i.e.*, the information given by the realization of the two finite schemes A and B, $S(A+B)$, is equal to the information gained by a realization of system A, $S(A)$, plus the mathematical expectation of information gained by a realization of system B, after the realization of system A, $S(A+B) = S(A) + S|_A(B)$.

Khinchin's fourth axiom is the requirement that entropy is a continuous function of all its arguments p_i and does not depend on anything else.

Given these axioms, the *Uniqueness theorem* [4] states that the one and only possible entropy is $S(p_1, \cdots, p_W) = -k \sum_{i=1}^{W} p_i \log p_i$, where k is an arbitrary positive constant. The result is of course the same as Shannon's. We call the combination of 4 axioms the Shannon-Khinchin (SK) axioms.

From information theory now to physics, where systems may exist that violate the separability axiom. This might especially be the case for non-ergodic, complex systems exhibiting long-range and strong interactions. Such complex systems may show extremely rich behavior in contrast to simple ones, such as gases. There exists some hope that it should be possible to understand such systems also on a thermodynamical basis, meaning that a few measurable quantities would be sufficient to understand their macroscopic phenomena. If this would be possible, through an equivalent to the second law of thermodynamics, some appropriate entropy would enter as a fundamental concept relating the number of microstates in the system to its macroscopic properties. Guided by this hope, a series of so called generalized entropies have been suggested over the past decades, see [5, 6, 7, 8, 9, 10] and Table 1. These entropies have been designed for different purposes and have not been related to a fundamental origin. Here we ask how generalized entropies can look like if they fulfill three of the Shannon-Khinchin axioms, but explicitly violate the separability axiom. We do this axiomatically as first presented in [1]. By doing so we can relate a large class of generalized entropies to a single root.

The reason why this axiom is violated in some physical, biological or social systems is *broken ergodicity*, *i.e.*, that not all regions in phase space are visited and many micro states are effectively 'forbidden'. Entropy relates the number of micro states of a system to an *extensive* quantity, which plays the fundamental role in the systems thermodynamical description. Extensive means that if two initially isolated, *i.e.*, sufficiently separated systems, A and B, with W_A and W_B the respective numbers of states, are brought together, the entropy of the combined system $A+B$ is $S(W_{A+B}) = S(W_A) + S(W_B)$. W_{A+B} is the number of states in the combined system $A+B$. This is not to be confused with *additivity* which is the property that $S(W_A W_B) = S(W_A) + S(W_B)$. Both, extensivity and additivity coincide if number of states in the combined system is $W_{A+B} = W_A W_B$. Clearly, for a non-interacting system Boltzmann-Gibbs-Shannon entropy, $S_{\mathrm{BG}}[p] = -\sum_i^W p_i \ln p_i$, is extensive *and* additive. By 'non-interacting' (short-range, ergodic, sufficiently mixing, Markovian, ...) systems we mean $W_{A+B} = W_A W_B$. For interacting statistical systems the latter is in general not true; phase space is only partly visited and $W_{A+B} < W_A W_B$. In this case, an additive entropy such as Boltzmann-Gibbs can no longer be extensive and vice versa. To ensure extensivity of entropy, an entropic form should be found for the particular interacting statistical systems at hand. These entropic forms are called *generalized entropies* and usually assume trace form [5, 6, 7, 8, 9, 10]

$$S_g[p] = \sum_{i=1}^{W} g(p_i) \quad , \tag{1}$$

W being the number of states. Obviously not all generalized entropic forms are of this type. Rényi entropy e.g., is of the form $G(\sum_i^W g(p_i))$, with G a monotonic function. We use trace forms Eq. (1) for simplicity. Rényi forms can be studied in exactly the same way (as is shown later), however at more technical cost.

Let us revisit the Shannon-Khinchin axioms in the light of generalized entropies of trace form Eq. (1). Specifically axioms SK1-SK3 (now re-ordered) have implications on the functional form of g

- SK1: The requirement that S depends continuously on p implies that g is a continuous function.
- SK2: The requirement that the entropy is maximal for the equi-distribution $p_i = 1/W$ (for all i) implies that g is a concave function.
- SK3: The requirement that adding a zero-probability state to a system, $W + 1$ with $p_{W+1} = 0$, does not change the entropy, implies that $g(0) = 0$.
- SK4 (separability axiom): The entropy of a system – composed of sub-systems A and B – equals the entropy of A plus the expectation value of the entropy of B, conditional on A. Note that this also corresponds exactly to Markovian processes.

As mentioned, if SK1 to SK4 hold, the only possible entropy is the Boltzmann-Gibbs-Shannon entropy. We are now going to derive the extensive entropy when the separability axiom SK4 is violated. Obviously this entropy will be more general and should contain BG entropy as a special case.

We now assume that axioms SK1, SK2, SK3 hold, *i.e.*, we restrict ourselves to trace form entropies with g continuous, concave and $g(0) = 0$. These systems we call *admissible* systems.

This generalized entropy for (large) admissible statistical systems (SK1-SK3 hold) is derived from two hitherto unexplored fundamental scaling laws of extensive entropies. Both scaling laws are characterized by exponents c and d, respectively, which allow to uniquely define equivalence classes of entropies, meaning that two entropies are equivalent in the

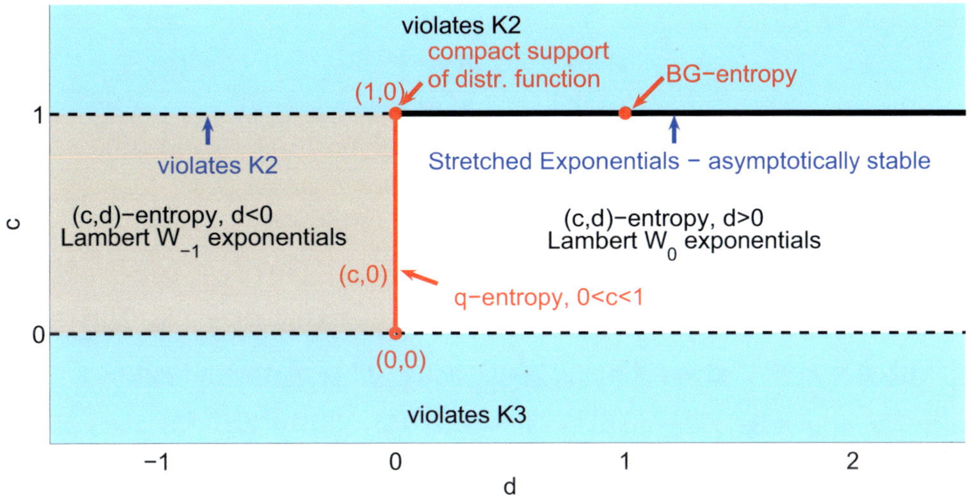

Figure 1: Equivalence classes of entropies parametrized in the (c,d)-plane, with their associated distribution functions. BG entropy corresponds to $(1,1)$, Tsallis entropy to $(c,0)$, and entropies for stretched exponentials to $(1,d>0)$. All entropies leading to distribution functions with compact support, belong to equivalence class $(1,0)$. An example are S_q entropies with $q>1$ (using the maximum entropy principle with usual expectation values in the constraints [10, 14]).

thermodynamic limit if their exponents (c,d) coincide. Each admissible system belongs to one of these equivalence classes (c,d), [1].

In terms of the exponents (c,d) we show in the following that all generalized entropies have the form $S_{c,d} \propto \sum_i^W \Gamma(d+1, 1-c\log p_i)$, with $\Gamma(a,b)=\int_b^\infty dt\, t^{a-1}\exp(-t)$ the incomplete Gamma-function.

Admissible systems when combined with a maximum entropy principle show remarkably simple mathematical properties as recently demonstrated in [11].

2 The scaling laws of entropies

We discuss two – a primary and a secondary – scaling properties of generalized entropies of trace form and assume the validity of the first 3 KS axioms. For equi-distribution $p_i=\frac{1}{W}$ (for all i) obviously, $\sum_{i=1}^W g(p_i) = Wg(\frac{1}{W})$.

The *primary scaling law* is found from the relation

$$\frac{S_g(\lambda W)}{S_g(W)} = \lambda \frac{g(\frac{1}{\lambda W})}{g(\frac{1}{W})} \quad . \tag{2}$$

We define a scaling function

$$f(z) \equiv \lim_{x\to 0} \frac{g(zx)}{g(x)} \qquad (0 < z < 1) \quad . \tag{3}$$

This function f for systems satisfying SK1-SK3, but violating SK4, can only be a power $f(z) = z^c$, with $0 < c \leq 1$, given f being continuous. This mathematical fact is shown as a theorem in the appendix (Theorem 1). Inserting Eq. (3) in Eq. (2) gives the primary asymptotic scaling law for entropies

$$\lim_{W\to\infty} \frac{S_g(\lambda W)}{S_g(W)} = \lambda^{1-c} \quad . \tag{4}$$

To identify the *secondary scaling law* start from the observation that

$$\lim_{W\to\infty} \frac{S(\lambda W)}{S(W)} \lambda^{c-1} = 1 \quad . \tag{5}$$

and substitute λ in Eq. (5) by $\lambda \to W^a$. We define $h_c(a)$

$$h_c(a) \equiv \lim_{W\to\infty} \frac{S(W^{1+a})}{S(W)} W^{a(c-1)} = \lim_{x\to 0} \frac{g(x^{1+a})}{x^{ac} g(x)} \quad , \tag{6}$$

with $x \equiv 1/W$. It can be proved as a mathematical fact (Theorem 2 in the appendix) that

$$h_c(a) = (1+a)^d \qquad (d \text{ constant}) \quad . \tag{7}$$

Remarkably, h_c does not explicitly depend on c. $h_c(a)$ is an asymptotic property which is *independent* of property Eq. (4). Note that if $c = 1$, concavity of g implies $d \geq 0$. In principle this scheme can be iterated to higher orders, see appendix.

3 Derivation of entropy

We now ask which families of entropies, *i.e.*, functions $g_{c,d}$ fulfill the primary and the secondary scaling law. A particularly simple choice is

$$g_{c,d,r}(x) = rA^{-d} e^A \, \Gamma(1+d\,,\, A - c\ln x) - rcx \quad , \tag{8}$$

where $A = \frac{cdr}{1-(1-c)r}$ and r is an arbitrary constant $r > 0$ (see below). For all choices of r the function $g_{c,d,r}$ is a representative of the class (c, d). This allows to choose r as a suitable function of c and d. For example choose $r = (1 - c + cd)^{-1}$, so that $A = 1$, and

$$S_{c,d}[p] = \frac{e \sum_i^W \Gamma\left(1 + d\,,\, 1 - c \ln p_i\right)}{1 - c + cd} - \frac{c}{1 - c + cd} \quad . \tag{9}$$

It can be easily verified that $S_{c,d}$ has the correct asymptotic properties, see Theorem 3 in the appendix.

3.1 Special cases of entropic equivalence classes

Let us look at some specific equivalence classes (c, d)

- Boltzmann-Gibbs entropy belongs to the $(c, d) = (1, 1)$ class. One gets from Eq. (8)

$$S_{1,1}[p] = \sum_i g_{1,1}(p_i) = -\sum_i p_i \ln p_i + 1 \quad . \tag{10}$$

- Tsallis entropy belongs to the $(c, d) = (c, 0)$ class. From Eq. (8) and the choice $r = 1/(1 - c)$ (see below) we get

$$S_{c,0}[p] = \textstyle\sum_i g_{c,0}(p_i) = \frac{1-\sum_i p_i^c}{c-1} + 1\,. \tag{11}$$

  Note, that although the *pointwise* limit $c \to 1$ of Tsallis entropy yields BG entropy, the asymptotic properties $(c, 0)$ do *not* change continuously to $(1, 1)$ in this limit! In other words the thermodynamic limit and the limit $c \to 1$ do not commute.

- The entropy related to stretched exponentials [6] belongs to the $(c, d) = (1, d)$ classes, see Table 1. As a specific example we compute the $(c, d) = (1, 2)$ case,

$$S_{1,2}[p] = 2\left(1 - \sum_i p_i \ln p_i\right) + \frac{1}{2}\sum_i p_i \left(\ln p_i\right)^2 , \tag{12}$$

  leading to a superposition of two entropy terms, the asymptotic behavior being dominated by the second.

- All entropies associated to distributions with compact support belong to $(c, d) = (1, 0)$. Clearly, all of these have the same (trivial) asymptotic behavior.

Other entropies which are special cases of our scheme are found in Table 1.

Inversely, for any given entropy we are now in the remarkable position to characterize *all* large SK1-SK3 systems by a pair of two exponents (c, d), *i.e.*, their scaling functions f and h_c. See Fig. 1. For example, for $g_{\rm BG}(x) = -x\ln(x)$ we have $f(z) = z$, *i.e.*, $c = 1$, and $h_c(a) = 1 + a$, *i.e.*, $d = 1$. $S_{\rm BG}$ therefore belongs to the universality class $(c, d) = (1, 1)$. For $g_q(x) = (x - x^q)/(1 - q)$ (Tsallis entropy) and $0 < q < 1$ one finds $f(z) = z^q$, *i.e.*, $c = q$ and $h_c(a) = 1$, *i.e.*, $d = 0$, and Tsallis entropy, S_q, belongs to the universality class $(c, d) = (q, 0)$. Other examples are listed in Table 1.

The universality classes (c, d) are equivalence classes with the equivalence relation given by: $g_\alpha \equiv g_\beta \Leftrightarrow c_\alpha = c_\beta$ and $d_\alpha = d_\beta$. This relation partitions the space of all admissible g into equivalence classes completely specified by the pair (c, d).

4 Distribution functions

Distribution functions associated with our Γ-entropy, Eq. (9), can be derived from so-called generalized logarithms of the entropy. Under the maximum entropy principle (given ordinary constraints) the inverse functions of these logarithms, $\mathcal{E} = \Lambda^{-1}$, are the distribution functions, $p(\epsilon) = \mathcal{E}_{c,d,r}(-\epsilon)$. Following [10, 14] the generalized logarithm Λ is found in closed form

$$\Lambda_{c,d,r}(x) = r - r\,x^{c-1}\left[1 - \frac{1-(1-c)r}{rd}\ln x\right]^d , \tag{13}$$

and its inverse function is

$$\mathcal{E}_{c,d,r}(x) = e^{-\frac{d}{1-c}\left[\mathcal{W}_k\left(B(1-x/r)^{\frac{1}{d}}\right) - \mathcal{W}_k(B)\right]} , \tag{14}$$

with the constant $B \equiv \frac{(1-c)r}{1-(1-c)r}\exp\left(\frac{(1-c)r}{1-(1-c)r}\right)$. The function $\mathcal{W}_k$ is the k'th branch of the Lambert-$\mathcal{W}$ function [13] which – as a solution to the equation $x = \mathcal{W}(x)\exp(\mathcal{W}(x))$ – has only two real solutions $\mathcal{W}_k$, the branch $k = 0$ and branch $k = -1$. Branch $k = 0$ covers the classes for $d \geq 0$, branch $k = -1$ those for $d < 0$.

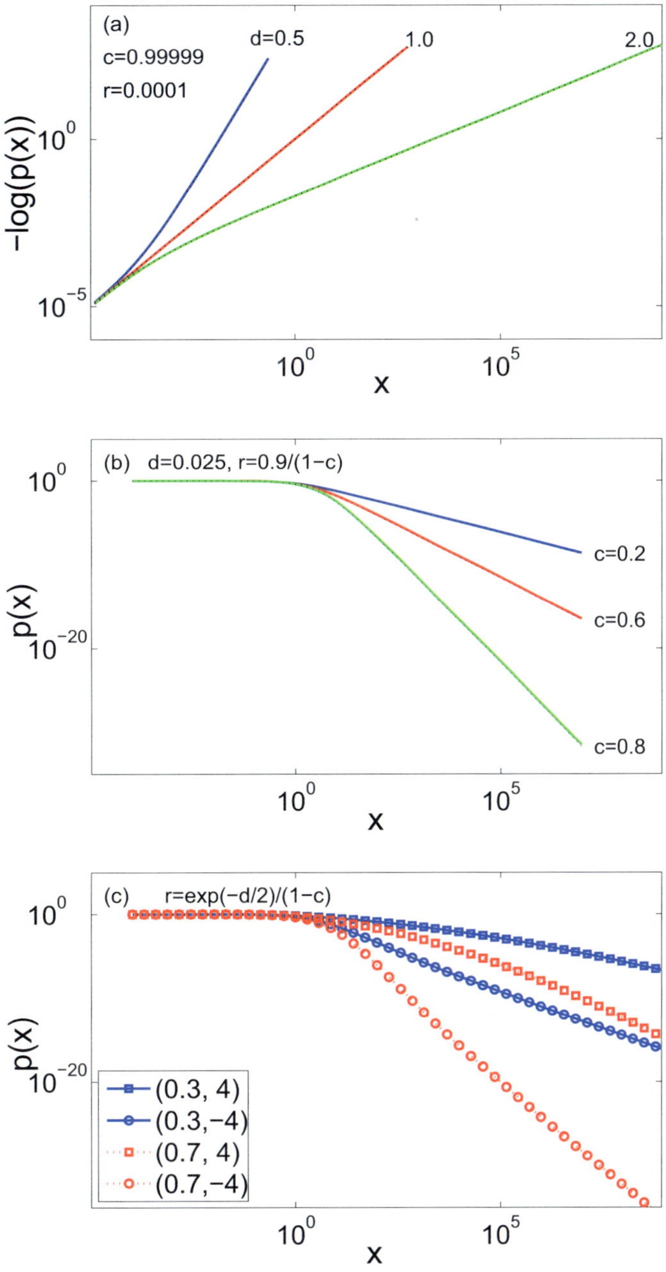

Figure 2: Distribution functions based on the Lambert exponential, $p(x) = \mathcal{E}_{c,d,r}(-x)$, for various (c, d) values. (a) Stretched exponential limit $c \to 1$. It includes the Boltzmann distribution for $d = 1$. (b) $d \to 0$ limit – *i.e.*, the q-exponential limit. In (a) and (b) the black dashed lines represent the stretched exponential ($c = 1$) or q-exponential ($d = 0$) limit functions. (c) General case for distribution functions for several values of (c, d) different from the limits $c \sim 1$ or $d \sim 0$. They must not be confused with power-laws.

4.1 Special cases of distribution functions

It is easy to verify that the class $(c,d)=(1,1)$ leads to Boltzmann distributions, and the class $(c,d)=(c,0)$ yields power-laws, or more precisely, Tsallis distributions, *i.e.*, q-exponentials. All classes associated with $(c,d)=(1,d)$, for $d>0$ are associated with stretched exponential distributions. Expanding the $k=0$ branch of the Lambert-$\mathcal{W}$ function $\mathcal{W}_0(x)\sim x-x^2+\ldots$ for $1\gg|x|$, the limit $c\to 1$ is seen to be a stretched exponential

$$\lim_{c\to 1}\mathcal{E}_{c,d,r}(x)=e^{-dr\left[\left(1-\frac{x}{r}\right)^{\frac{1}{d}}-1\right]} \quad . \tag{15}$$

Clearly, r does not effect its asymptotic properties (tail of the distribution), but can be used to incorporate finite size properties of the distribution function for small x. Examples of all of the above distribution functions are shown in Fig. 2.

4.2 Finite size effects and the parameter r

In Eq. (9) we chose $r=(1-c+cd)^{-1}$. This is not the most general case and the only limitations on r – if one requires the generalized logarithms to have the usual properties $\Lambda(1)=0$ and $\Lambda'(1)=1$ to hold – are

$$\begin{aligned} d>0: &\quad r<\tfrac{1}{1-c} \quad , \\ d=0: &\quad r=\tfrac{1}{1-c} \quad , \\ d<0: &\quad r>\tfrac{1}{1-c} \quad . \end{aligned} \tag{16}$$

Every choice of r gives a representative of the equivalence class (c,d), *i.e.*, r has no effect on the thermodynamic limit and therefore can be used to encode potential finite-size characteristics of a system at hand. In case of no such effects, practical choices are $r=(1-c+cd)^{-1}$ for $d>0$, and $r=\exp(-d)/(1-c)$ for $d<0$.

5 A physical system as an example

From axiomatic derivations no physical relevance can be inferred. To demonstrate the applicability of the proposed classification scheme to concrete physical systems, consider over-damped interacting particles moving in a narrow channel,

$$\mu\vec{v}_i=\sum_{j\neq i}\vec{J}(\vec{r}_i-r_j)+\vec{F}(\vec{r}_i)+\eta(\vec{r}_i,t) \quad , \tag{17}$$

Table 1: Several trace form entropies $S = \sum_i^W g(p_i)$, for which SK1-SK3 hold. They are special cases of the entropy given in Eq. (8) and their asymptotic behavior is uniquely determined by c and d. It can be seen immediately that $S_{q>1}$, S_b and S_E are asymptotically identical; so are $S_{q<1}$ and S_κ, as well as S_η and S_γ.

entropy		c	d	reference
$S_{c,d} = er\sum_i \Gamma(d+1, 1-c\ln p_i) - cr$	$(r = (1-c+cd)^{-1})$	c	d	
$S_{BG} = \sum_i p_i \ln(1/p_i)$		1	1	[4]
$S_{q<1}(p) = \frac{1-\sum p_i^q}{q-1}$	$(q < 1)$	$c = q < 1$	0	[5]
$S_\kappa(p) = -\sum_i p_i \frac{p_i^\kappa - p_i^{-\kappa}}{2\kappa}$	$(0 < \kappa \leq 1)$	$c = 1-\kappa$	0	[7]
$S_{q>1}(p) = \frac{1-\sum p_i^q}{q-1}$	$(q > 1)$	1	0	[5]
$S_b(p) = \sum_i(1-e^{-bp_i}) + e^-b - 1$	$(b > 0)$	1	0	[8]
$S_E(p) = \sum_i p_i(1-e^{\frac{p_i-1}{p_i}})$		1	0	[9]
$S_\eta(p) = \sum_i \Gamma(\frac{\eta+1}{\eta}, -\ln p_i) - p_i\Gamma(\frac{\eta+1}{\eta})$	$(\eta > 0)$	1	$d = \frac{1}{\eta}$	[6]
$S_\gamma(p) = \sum_i p_i \ln^{1/\gamma}(1/p_i)$		1	$d = 1/\gamma$	[12], footnote 11, page 60
$S_\beta(p) = \sum_i p_i^\beta \ln(1/p_i)$		$c = \beta$	1	[13]

where v_i is velocity of ith particle, μ viscosity, F an external force, $\vec{J}(\vec{r})$ a linear repulsive particle-particle interaction, η uncorrelated thermal noise with $\langle\eta\rangle = 0$ and $\langle\eta^2\rangle = \frac{kT}{\mu}$, with λ a characteristic length of the pair interaction. For details see [20], where the non-linear Fokker-Planck equation for the spatial distribution $\rho(x)$ of the particles was solved. It was shown that the corresponding entropy is a superposition of BG and Tsallis entropy with $q = 2$. Note the similarity to our example in Eq. (12) where a similar superposition emerges naturally. At high temperatures the BG contribution dominates *i.e.*, particles diffuse normally, while at zero temperature the system is governed by Tsallis entropy with a non-Gaussian parabolic diffusion profile of compact support. The concrete stationary distribution functions given in [20] can be used to demonstrate that the system is found either in the asymptotic equivalence class $(c, d) = (1, 1)$ (BG entropy) or in $(c, d) = (1, 0)$ (compact support entropies) depending on the temperature of the heat bath. Expressing the integral by a discrete sum the entropy given in [20] can be used to demonstrate the existence of the two sets of classes. This shows that the classification is applicable for concrete physical systems and suggests further that the equivalence class (c, d) may even depend on macro variables of the system such as the temperature of the heat bath.

6 Rényi-type entropies

Rényi entropy is obtained by relaxing SK4 to the pure (unconditional) additivity condition. Following the same scaling idea for Rényi-type entropies, $S = G(\sum_{i=1}^{W} g(p_i))$, with G and g some functions, one gets

$$\lim_{W\to\infty} \frac{S(\lambda W)}{S(W)} = \lim_{s\to\infty} \frac{G\left(\lambda f_g(\lambda^{-1})s\right)}{G(s)} \quad , \tag{18}$$

where $f_g(z) = \lim_{x\to 0} g(zx)/g(x)$. The expression $f_G(s) \equiv \lim_s G(sy)/G(s)$, provides the starting point for deeper analysis which now gets more involved. In particular, for Rényi entropy with $G(x) \equiv \ln(x)/(1-\alpha)$ and $g(x) \equiv x^\alpha$, the asymptotic properties yield the class $(c,d) = (1,1)$, (BG entropy) meaning that Rényi entropy is additive. However, in contrast to the trace form entropies used above, Rényi entropy can be shown to be *not* Lesche stable, as was observed before [15, 16, 17, 18, 19]. All of the $S = \sum_i^W g(p_i)$ entropies can be shown to be Lesche stable, see Theorem 4 in the appendix.

7 Discussion

We studied the scaling laws of trace form entropies which are constrained by the first three Shannon-Khinchin axioms in the thermodynamic limit. In analogy to critical exponents these laws are characterized by two scaling exponents (c,d), which define equivalence relations on entropic forms. We showed that a single entropic form – parametrized by the two exponents – covers all *admissible* systems (Shannon-Khinchin axioms 1-3 hold, 4 is violated). In other words every statistical system has its pair of unique exponents in the large size limit, its entropy is then given by $S_{c,d} \sim \sum_i^W \Gamma\left(1+d\,,\,1-c\ln p_i\right)$.

The exponents for BG systems are $(c,d) = (1,1)$, systems characterized by stretched exponentials belong to the class $(c,d) = (1,d)$, and Tsallis systems have $(c,d) = (q,0)$. In the context of a maximum entropy principle, the associated distribution functions of *all* systems (c,d) are shown to belong to a class of exponentials involving Lambert-$\mathcal{W}$ functions, given in Eq. (14). There are no other options for tails in distribution functions other than these.

The equivalence classes characterized by the exponents (c,d) form *basins of asymptotic equivalence*. In general these basins characterize interacting statistical (non-additive) systems. There exists an analogy between these basins of asymptotic *equivalence* and the *basin of attraction* of weakly interacting, uncorrelated systems subject to the law of large numbers, *i.e.*, the central limit theorem. Any system within a given equivalence class may show individual characteristics as long as it is small. Systems belonging to the same class will start behaving similarly as they become larger and in the thermodynamic limit they become identical. Distribution functions converge to those uniquely determined by (c, d).

Our framework shows that for non-interacting systems $c = 1$. Setting $\lambda = W_B$ in Eq. (2) and Eq. (3), immediately implies $S(W_A W_B)/S(W_A) \sim W_B^{1-c}$. This means that if for such a system it would be true that $c \neq 1$, then adding only a few independent states to a system would explosively change its entropy and extensivity would be strongly violated. A further interesting feature of admissible systems is that they all are what has been called *Lesche stable.* As a practical note Lesche stability corresponds one-to-one to the continuity of the scaling function f and can now be checked by a trivial verification of this property (Eq. (3)). The proof is given in Theorem 4 in the appendix.

Finally, we remark that the classification scheme for generalized entropic forms of type $S = \sum_i g(p_i)$ can be extended to entropies of *e.g.*, Rényi type, *i.e.*, $S = G(\sum_i g(p_i))$. We demonstrated that generalized entropies can be applied to actual physical systems. We hypothesize that many complex statistical systems are indeed admissible systems of equivalence classes (c, d), with $0 < c < 1$.

8 Appendix

Theorem 1: Let g be a continuous, concave function on $[0, 1]$ with $g(0) = 0$ and let $f(z) = \lim_{x \to 0^+} g(zx)/g(x)$ be continuous, then f is of the form $f(z) = z^c$ with $c \in (0, 1]$.

Proof. Note that $f(ab) = \lim_{x\to 0} g(abx)/g(x) =$ $\lim_{x\to 0}(g(abx)/g(bx))(g(bx)/g(x)) = f(a)f(b)$. All pathological solutions are excluded by the requirement that f is continuous. So $f(ab) = f(a)f(b)$ implies that $f(z) = z^c$ is the only possible solution of this equation. Further, since $g(0) = 0$, also $\lim_{x\to 0} g(0x)/g(x) = 0$, and it follows that $f(0) = 0$. This necessarily implies that $c > 0$. $f(z) = z^c$ also has to be concave since $g(zx)/g(x)$ is concave in z for arbitrarily small, fixed $x > 0$. Therefore $c \leq 1$. □

Theorem 2: Let g be like in Theorem 1 and let $f(z) = z^c$ then h_c in Eq. (6) is a constant of the form $h_c(a) = (1 + a)^d$ for some constant d.

Proof. We can determine $h_c(a)$ again by a similar trick as we have used for f.

$$\begin{aligned} h_c(a) &= \lim_{x\to 0} \frac{g(x^{a+1})}{x^{ac} g(x)} \\ &= \frac{g\left((x^b)^{\left(\frac{a+1}{b}-1\right)+1}\right)}{(x^b)^{\left(\frac{a+1}{b}-1\right)c} g(x^b)} \frac{g(x^b)}{x^{(b-1)c} g(x)} \\ &= h_c\left(\frac{a+1}{b}-1\right) h_c(b-1) \quad , \end{aligned} \tag{19}$$

for some constant b. By a simple transformation of variables, $a = bb' - 1$, one gets $h_c(bb' - 1) = h_c(b-1)h_c(b'-1)$. Setting $H(x) = h_c(x-1)$ one again gets $H(bb') = H(b)H(b')$. So $H(x) = x^d$ for some constant d and consequently $h_c(a)$ is of the form $(1+a)^d$. □

Theorem 3: The entropy based on $g_{c,d,r}$, Eq. (8), has the desired asymptotic properties.

Proof. Let g be like in Theorem 1, *i.e.*, let $f(z) = z^c$ with $0 < c \leq 1$, then

$$\lim_{x\to 0^+} \frac{g'(x)}{\frac{1}{x} g(x)} = c\,. \tag{20}$$

Consider

$$\lim_{x\to 0^+} \frac{\frac{g(x)-g(zx)}{(1-z)x}}{\frac{1}{x}g(x)} = \frac{1}{1-z}\left(\frac{g(x)-g(zx)}{g(x)}\right) = \frac{z^c-1}{z-1}\,. \tag{21}$$

Taking the limit $z \to 1$ on both sides completes the first part of the proof. Further, two functions g_A and g_B generate equivalent entropic forms if $\lim_{x\to 0^+} g_A(x)/g_B(x) = \phi$ and $0 < \phi < \infty$. This clearly is true since

$$\begin{aligned} \lim_{x\to 0^+} \frac{g_A(zx)}{g_A(x)} &= \frac{g_A(zx)}{g_B(zx)} \frac{g_B(x)}{g_A(x)} \frac{g_B(zx)}{g_B(x)} \\ &= \phi\phi^{-1} \frac{g_B(zx)}{g_B(x)} \\ &= \lim_{x\to 0^+} \frac{g_B(zx)}{g_B(x)}\,. \end{aligned} \tag{22}$$

By an analogous argument the same result can be obtained for the second asymptotic property, Eq. (6). A simple lemma is that given $g_B(x) = a g_A(bx)$, for some suitable constants a and b, then g_B and g_A are equivalent.

A lemma, following from Eq. (20) is that

$$\lim_{x\to 0^+} \frac{g_A(x)}{g_B(x)} = \lim_{x\to 0^+} \frac{g_A'(x)}{g_B'(x)}\,, \tag{23}$$

which is just the rule of L'Hospital shown to hold for the considered families of functions g. This is true since, either $\lim_{x\to 0^+} g_A(x)/g_B(x) = \phi$ with $0 < \phi < \infty$ and $c_A = c_B$, *i.e.*, g_A and g_B are equivalent, or g_A and g_B are inequivalent, *i.e.*, $c_A \neq c_B$ but $\phi = 0$ or $\phi \to \infty$.

So if one can find a function $g_{\rm test}$, having the desired asymptotic exponents c and d, it suffices to show that $0 < -\lim_{x\to 0^+} \Lambda_{c,d,r}(x)/g'_{\rm test}(x) < \infty$, where $\Lambda_{c,d,r}$ is the generalized logarithm Eq. (13) associated with the generalized entropy Eq. (9). The test function $g_{\rm test}(x) = x^c \log(1/x)^d$ is of class (c,d), as can be verified easily. Unfortunately $g_{\rm test}$ can not be used to define the generalized entropy due to several technicalities. In particular $g_{\rm test}$ lacks concavity around $x \sim 1$ for a considerable range of (c,d) values, which then makes it impossible to define proper generalized logarithms and generalized exponential functions on the entire interval $x \in [0,1]$. However, we only need the asymptotic properties of $g_{\rm test}$ and for $x \sim 0$ the function $g_{\rm test}$ does not violate concavity or any other required condition. The first derivative is $g'_{\rm test}(x) = x^{c-1}\log(1/x)^{d-1}(c\log(1/x) - d)$. With this we finally get

$$\lim_{x\to 0+} \frac{\Lambda_{c,d,r}(x)}{g'_{\rm test}(x)} = \frac{r - D^{-d}\left(\frac{x}{z}\right)^{c-1}\left[\log\left(\frac{z}{x}\right)\right]^d}{x^{c-1}\log(\frac{1}{x})^{d-1}(c\log\frac{1}{x} - d)} = -\frac{z^{1-c}}{cD^d} \,. \tag{24}$$

Since $0 < \frac{z^{1-c}}{cD^d} < \infty$ this proves that the Gamma-entropy $g_{c,d,r}$, Eq. (10), represents the equivalence classes (c,d). □

Lesche stability of trace form entropies

The Lesche stability criterion is a uniform-equi-continuity property of functionals $S[p]$ on families of probability functions $\{p^{(W)}\}_{W=1}^{\infty}$ where $p^{(W)} = \{p_i^W\}_{i=1}^W$. The criterion is phrased as follows:

Let $p^{(W)}$ and $q^{(W)}$ be probabilities on W states. An entropic form S is Lesche stable if for all $\epsilon > 0$ and all W there is a $\delta > 0$ such that

$$||p^{(W)} - q^{(W)}||_1 < \delta \Rightarrow |S[p^{(W)}] - S[q^{(W)}]| < \epsilon \hat{S}(W) \quad , \tag{25}$$

where $\hat{S}(W)$ is again the maximal possible entropy for W states.

We characterize Lesche stability on the class of our generalized entropic forms in terms of continuity of f in

Theorem 4: Let $p_i \geq 0$ be a probability and W the number of states i. Let g be a concave, continuous function on $[0,1]$, continuously differentiable on the semi-open interval $(0,1]$ and $g(0) = 0$. The entropic form $S_g[p] = \sum_{i=1}^W g(p_i)$ is Lesche stable iff the function $f(z) = \lim_{x\to 0} g(zx)/g(x)$ is continuous on $z \in [0,1]$.

Proof. SK2 states that maximal entropy is given by $\hat{S}_g(W) = Wg(1/W)$. We identify the worst case scenario for $|S_g[p] - S_g[q]|$, where p and q are probabilities on the W states. For this maximize $G[p,q] = |S_g[p] - S_g[q]| - \alpha(\sum_i p_i - 1) - \beta(\sum_i q_i - 1) - \gamma(\sum_i |p_i - q_i| - \delta)$, where α, β and γ are Lagrange multipliers. Without loss of generality assume that $S_g[p] > S_g[q]$. Thus condition $\partial G/\partial p_i = 0$ gives $g'(p_i) + \gamma\,\mathrm{sign}(p_i - q_i) - \alpha = 0$, where g' is the derivative of g and sign is the sign function. Similarly, $\partial G/\partial q_i = 0$ leads to $g'(q_i) + \gamma\,\mathrm{sign}(p_i - q_i) + \beta = 0$. From this we see that both p and q can only possess two values p_+, p_- and q_+ and q_-, where one can assume (without loss of generality) that $p_+ > q_-$ and $q_+ > p_-$. We can now assume that for w indices i $p_+ = p_i > q_i = q_-$ and for $W - w$ indices j $p_- = p_j < q_j = q_+$ where w may range from 1 to $W - 1$. This leads to seven equations

$$\begin{array}{ll} wp_+ + (W-w)p_- = 1 & , \quad g'(p_+) + \gamma - \alpha = 0 \\ wq_- + (W-w)q_+ = 1 & , \quad g'(p_-) - \gamma - \alpha = 0 \\ w(p_+ - q_-) - & , \quad g'(p_+) + \gamma + \beta = 0 \\ -(W-w)(p_- - q_+) = \delta & , \quad g'(p_+) - \gamma + \beta = 0 \end{array} \tag{26}$$

which allow to express p_-, q_-, and q_+ in terms of p_+

$$\begin{aligned} p_- &= \frac{1 - wp_+}{W - w} \\ q_- &= p_+ - \frac{\delta}{2w} \\ q_+ &= \frac{1 - wp_+}{W - w} + \frac{\delta}{2(W-w)} \quad . \end{aligned} \tag{27}$$

Further, we get the equation

$$g'(p_+) - g'(p_-) + g'(q_+) - g'(q_-) = 0 \quad . \tag{28}$$

Since g is concave g' is monotonically decreasing and $g'(p_+) - g'(q_-) > 0$ and $g'(q_+) - g'(p_-) > 0$. Thus Eq. (28) has no solution, meaning that there is no extremum with $p_\pm$ and $q_\pm$ in $(0,1)$, and extrema are at the boundaries. The possibilities are $p_+ = 1$ or $p_- = 0$, then $q_+ = 1$ and $q_- = 0$. Only $p_+ = 1$ or $p_- = 0$ are compatible with the assumption that $S[p] > S[q]$; $p_+ = 1$ is only a special case of $p_- = 0$ with $n = 1$. Since $g(0) = 0$ this immediately leads to the inequality

$$\frac{|S_g[p] - S_g[q]|}{S_{\max}} \leq (1-\phi)\frac{g\left(\frac{\delta}{2(1-\phi)W}\right)}{g(\frac{1}{W})}$$

$$+ \quad \phi \left| \frac{g\left(\frac{1}{\phi W}\right)}{g(\frac{1}{W})} - \frac{g\left(\frac{1-\frac{\delta}{2}}{\phi W}\right)}{g(\frac{1}{W})} \right| , \tag{29}$$

where $\phi = w/W$ is chosen such that the right hand side of the equation is maximal. Obviously, for any finite W the right hand side can always be made as small as needed by choosing $\delta > 0$ small enough. Now take the limit $W \to \infty$. If f is continuous

$$\begin{aligned} \frac{|S_g[p] - S_g[q]|}{S_{\max}} &\leq \\ &\leq (1-\phi)\left(\frac{\delta}{2(1-\phi)}\right)^c + \phi\left|\left(\frac{1}{\phi}\right)^c - \left(\frac{1-\frac{\delta}{2}}{\phi}\right)^c\right| \\ &\leq (1-\phi)^{1-c}\delta^c + \phi^{1-c}\left|1 - \left(1-\frac{\delta}{2}\right)^c\right| \\ &\leq \delta^c + \left|1 - \left(1 - c\frac{\delta}{2}\right)\right| \leq \delta^c + \delta \quad . \end{aligned} \tag{30}$$

Lesche-stability of S_g follows since the right hand side of Eq. (30) can be made smaller than any given $\epsilon > 0$ by choosing $\delta > 0$ small enough. This completes the first direction of the proof.

If, on the other hand, S_g is not Lesche-stable then there exists an $\epsilon > 0$, such that $|S_g[p] - S_g[q]|/S_{\max} \geq \epsilon$, $\forall N$, implying

$$(1-\phi)f\left(\frac{\delta}{2(1-\phi)}\right) + \phi\left|f\left(\frac{1}{\phi}\right) - f\left(\frac{1-\frac{\delta}{2}}{\phi}\right)\right| \geq \epsilon, \tag{31}$$

$\forall \delta > 0$. This again means that either $f(z)$ is discontinuous at $z = 1/\phi$, or $\lim_{z\to 0} f(z) > 0$. Since $g(0) = 0$ implies that $f(0) = 0$, $f(z)$ has to be discontinuous at $z = 0$. □

Remark on higher order scaling exponents:

In principle the scheme of finding scaling exponents can be iterated to higher orders, *i.e.*, to find a sequence of exponents $(c, d, d_2, d_3, \ldots)$. To see this define $\log^{(n+1)}(x) = \log(\log^{(n)}(x))$, with $\log^{(1)}(x) = \log(x)$ and $\exp^{(n+1)}(x) = \exp(\exp^{(n)}(x))$, with $\exp^{(1)}(x) = \exp(x)$, then up to a precision M

$$g(x) \sim x^c \prod_{m=1}^{M} \log^{(m)}\left(\frac{1}{x}\right) \quad , \tag{32}$$

for small x. The n'th scaling exponent d_n can be determined by

$$\lim_{x\to 0} \frac{g\left(\lambda_n(x)x\right)}{\lambda_n(x)^c g(x)} \prod_{m=1}^{n-1} \left(\frac{\log^{(m)}\left(\frac{1}{x\lambda_n(x)}\right)}{\log^{(m)}\left(\frac{1}{x}\right)}\right)^{-d_m} = (1+a_n)^{d_n}, \tag{33}$$

where the generalized scaling factor reads

$$\lambda_n(x)^{-1} = x \exp^{(n)}\left((1 + a_n) \log^{(n)}\left(\frac{1}{x}\right)\right) \quad . \tag{34}$$

Note that this is a nested scheme of sequentially incorporating more information with higher levels, *i.e.*, if $g \in (c, d, d_2, \ldots, d_{n+1})$, then also $g \in (c, d, d_2, \ldots, g_n)$. Using higher order asymptotic exponents makes it difficult to find a concave representative $g_{c,d,d_2,\ldots,d_M}(x)$ and g quickly becomes experimentally inaccessible. For macroscopic systems there may exist an upper limit M such that measuring d_m for $m > M$ becomes senseless. For example, consider a system with $N \sim 10^{23}$ particles and $\Omega \propto e^N$ states. Then $x \propto 1/\Omega$ and it does not make any sense to measure d_6 or higher because $\log^{(6)}(x) \sim -1.13$ is already negative, and $\log^{(7)}(x)$ is no longer defined on the principal branch of the logarithm.

Conflict of interest: None.

Disclosure: Part of information included in this chapter has been previously published in EPL (Europhysics Letters) 93:20006, (2011) and EPL (Europhysics Letters) 96:50003, (2011).

Acknowledgments

None.

References

[1] Hanel R., Thurner S., *Europhys Lett* 93:20006 (2011).

[2] Hanel R., Thurner S., *Europhys Lett* 96:50003, (2011).

[3] Shannon C. E., *The Bell System Technical Journal* 27:379,623 (1948).

[4] Khinchin A.I., *Mathematical foundations of information theory* (Dover Publ., New York), (1957).

[5] Tsallis C., *J Stat Phys* 52:479 (1988).

[6] Anteneodo C., Plastino A.R., *J Phys A: Math Gen* 32:1089, (1999).

[7] Kaniadakis G., *Phys Rev* E 66:056125, (2002).

[8] Curado E.M.F., Nobre F.D., *Physica A* 335:94, (2004).

[9] Tsekouras G.A., Tsallis C., *Phys Rev* E 71:046144, (2005).

[10] Hanel R., Thurner S., *Physica A* 380:109, (2007).

[11] Hanel R., Thurner S., Gell-Mann M., *PNAS* 108:6390-6394, (2011).

[12] Tsallis C., Introduction to Nonextensive Statistical Mechanics (Springer, New York) (2009).

[13] Shafee F., *IMA J Appl Math* 72:785, (2007).

[14] Hanel R., Thurner S., *Braz J Phys* 39:413, (2009).

[15] Lesche B., *J Stat Phys* 27:419, (1982).

[16] Abe S., *Phys Rev E* 66:046134, (2002).

[17] Jizba P., Arimitsu T., *Phys Rev E* 69:026128, (2004).

[18] Kaniadakis G., Scarfone A.M., *Physica A* 340:102, (2004).

[19] Hanel R., Thurner S., Tsallis C., *Europhys Lett* 85:20005, (2009).

[20] Andrade J.S., da Silva G.F.T., Moreira A.A., Nobre F.D., Curado E.M.F., Phys Rev Lett 105:260601 (2010).

CHAPTER 5

Majorization and Generalized Entropies

Norma Canosa [a,b] and Raúl D. Rossignoli [a,c,*]

[a] Departamento de Física-IFLP, Universidad Nacional de La Plata, C.C.67, La Plata (1900), Argentina

[b] CONICET, Argentina

[c] Comisión de Investigaciones Científicas (CIC), Argentina

Abstract

We review the concept of majorization and its relation with generalized information measures. Majorization theory provides an elegant framework for comparing two probability distributions, leading to a rigorous concept of disorder which is more stringent than that based on the Shannon entropy. Nevertheless, it is shown that it can be fully captured through general entropic inequalities based on generalized entropic forms. A brief review of generalized entropies is also provided. As illustration, we discuss the majorization properties of generalized thermal distributions derived from generalized entropies, and identify rigorous mixing parameters. We also describe majorization in quantum systems. We discuss in particular its capability for providing a disorder based criterion for the detection of quantum entanglement, which is stronger than that based on the von Neumann entropy and leads to a generalized entropic separability criterion.
Keywords: Generalized Entropies; Majorization; Generalized Thermal Distributions; Mixing Parameters; Quantum Mixed States; Quantum Entanglement; Separability.

1 Introduction

The theory of majorization [1–4] is an elegant mathematical framework for comparing probability distributions. It provides the basis for a rigorous concept of disorder or "mixedness" which is more fundamental than that implied by the Shannon entropy [5] or by any single choice of entropy. The basic idea is that a probability distribution, or in quantum mechanics a *density matrix*, can be said to be *more mixed or "spread out"* than another distribution only when it is *majorized* by the latter. This implies a higher entropy of the majorized distribution, although the converse relation, *i.e.,* that a higher entropy implies majorization,

* Corresponding author. E-mail: rossigno@fisica.unlp.edu.ar

Andres M. Kowalski, Raul D. Rossignoli and Evaldo M. F. Curado (Eds)

is not necessarily true. The stringent concept of disorder derived from majorization has consequently found many applications not only in mathematics, but also in a wide variety of fields such as economy, computer science and more recently quantum information [6–16].

At the same time, there has been in the last decades an enormous interest on possible generalizations of the Shannon entropy [2,17–32], motivated in part by the need of more general statistical descriptions of physical systems and also more general characterizations of the information content associated with a probability distribution. The Rényi and Tsallis entropies (see section 10 in chapter 2) and the ensuing generalized thermodynamics [20], are main examples of these extensions. In this chapter, we will link the ideas of majorization with those of generalized entropic measures, following the arguments of refs. [11–13,28,29]. We show that the possibility of capturing the rigorous concept of disorder provided by majorization by means of purely entropic arguments is realized when a sufficiently general yet simple class of entropic forms are considered.

This chapter is organized as follows. In Section 2, *Majorization*, we first review, in an accessible level, the concept of majorization and its main fundamental properties. Section 3, *Majorization and Shannon entropy*, describes its relation with the standard entropy.

Section 4, *Generalized entropies*, describes trace form generalizations of the Shannon entropy, based on arbitrary concave functions [2,22–32], together with their main features and properties. Several important particular cases are examined. Section 5, *Majorization and generalized entropies*, discusses the connection of these general entropic forms with majorization. A complete characterization of majorization in terms of generalized entropic inequalities [12] is provided.

Section 6, *Generalized maxent distributions*, describes the probability distributions obtained from the maximization of these generalized entropic forms subject to standard average constraints [18–26]. These distributions include as particular cases the well-known exponential distribution derived from the Shannon entropy [33, 34] and the generalized power-law distributions obtained from the Tsallis entropy [18, 19, 20]. Logarithmic and generalized exponential distributions are also discussed.

Section 7, *Majorization and generalized maxent distributions*, examines the rigorous majorization properties satisfied by these distributions [13, 36, 37]. We show that the corresponding effective temperature is always a proper *mixing parameter*. Section 8, *Mixing parameters in generalized distributions*, discusses the concept of mixing parameters [36, 37] in even more general distributions. Rigorous conditions, which ensure majorization as some parameter characterizing the distribution is increased, are provided. The majorization properties of power-law distributions are also analyzed.

Section 9, *Majorization and generalized entropies in the quantum case*, discusses the extension of the concept of majorization and entropy to the quantum case. Previous results remain essentially valid although some fundamental differences, connected with quantum entanglement in composite quantum systems, do arise.

Section 10, *Quantum entanglement and majorization: The generalized entropic criterion for separability*, discusses these last aspects. Quantum entanglement is a fundamental feature of correlations in composite quantum systems, which has no classical analogue [6]. In recent years it has attracted an enormous interest due to the key role it plays in the new field of quantum information, where it is considered a *resource*: It provides the basis which allows completely new forms of information transmission (quantum teleportation [35]) and processing (quantum computation [6]). In this section it is shown how majorization can be used to derive a generalized entropic criterion for detecting quantum entanglement [9, 11, 12, 28, 29]. This criterion extends previous entropic criteria based on the von Neumann [38] and Tsallis [39, 40] entropies, and exhausts the possibilities of a simple disorder based approach. An illustrative example is discussed in detail. Conclusions are finally drawn in section 11.

2 Majorization

Let us consider two sets of real numbers $\boldsymbol{p} = (p_1, \ldots, p_n)$ and $\boldsymbol{p}' = (p'_1, \ldots, p'_n)$. We will denote as $\boldsymbol{p}^s$ the corresponding set of numbers sorted in *decreasing* order, such that $p^s_i \geq p^s_j$ if $i < j$, with $p^s_1 = \text{Max}(p_1, \ldots, p_n)$ and $p^s_n = \text{Min}(p_1, \ldots, p_n)$. The set $\boldsymbol{p}$ is said to be *majorized* by $\boldsymbol{p}'$, which is denoted as $\boldsymbol{p} \prec \boldsymbol{p}'$, if [1, 2, 3, 4]

$$
\begin{aligned}
p^s_1 &\leq p'^s_1 \,, \\
p^s_1 + p^s_2 &\leq p'^s + p'^s_2 \,, \\
&\ldots \\
p^s_1 + \ldots + p^s_{n-1} &\leq p'^s_1 + \ldots + p'^s_{n-1} \,, \\
p^s_1 + \ldots + p^s_n &= p'^s_1 + \ldots + p'^s_n \,.
\end{aligned}
\tag{2.1}
$$

We may succinctly express the previous conditions as

$$\boldsymbol{p} \prec \boldsymbol{p}' \text{ iff } \sum_{j=1}^{i} p_j^s \leq \sum_{j=1}^{i} p'^s_j, \;\; i = 1, \ldots, n-1, \;\; \sum_{i=1}^{n} p_i = \sum_{i=1}^{n} p'_i \,. \tag{2.2}$$

If the numbers are sorted in increasing order ($p_i^s \leq p_j^s$ if $i < j$), we should just change $\leq$ by $\geq$ in the previous expressions to obtain a set of equivalent conditions. If the last condition in (2.1)–(2.2) is softened by replacing $=$ with $\leq$, $\boldsymbol{p}$ is said to be submajorized by $\boldsymbol{p}'$.

In the rest of this article we will restrict ourselves to *probability distributions*

$$\boldsymbol{p} \equiv (p_1, \ldots, p_n)\,, \;\; p_i \geq 0\,, \;\; i = 1, \ldots, n\,, \;\; \sum_{i=1}^{n} p_i = 1\,, \tag{2.3}$$

associated with a complete set of n disjoint events i, in which case the last condition in (2.1)–(2.2) is directly fulfilled. The meaning of Eqs. (2.1) is then apparent: If $\boldsymbol{p} \prec \boldsymbol{p}'$, the distribution $\boldsymbol{p}$ is more even than $\boldsymbol{p}'$, *i.e.*, it is more "spread out" or "random" (Fig. 1). In economy, if p_i and p'_i stand for the fraction of the total income received by a subgroup i in two different countries or regions, $\boldsymbol{p} \prec \boldsymbol{p}'$ would indicate a more even distribution of incomes in $\boldsymbol{p}$ than in $\boldsymbol{p}'$.

The *uniform distribution* $p_i = 1/n \; \forall i$, which represents *minimum knowledge* in the case of a probability distribution, is then majorized by *any* other distribution, while the *delta distribution* $p_i = \delta_{ij}$ for some j (*i.e.*, $p_i = 1$ if $i = j$ and 0 otherwise), which represents *maximum knowledge*, majorizes *any* other distribution:

$$(\frac{1}{n}, \ldots, \frac{1}{n}) \prec \boldsymbol{p} \prec (0, \ldots, 1, \ldots, 0) \;\; \forall\, \boldsymbol{p} = (p_1, \ldots, p_n)\,. \tag{2.4}$$

Majorization can also be applied to distributions of unequal length ($n \neq n'$) by completing with zeros that with the smaller number of components. For instance, we may write $(\frac{1}{n+1}, \ldots, \frac{1}{n+1}) \prec (\frac{1}{n}, \ldots, \frac{1}{n})$ for the uniform distributions of $n+1$ and n components respectively, which means $(\frac{1}{n+1}, \ldots, \frac{1}{n+1}) \prec (\frac{1}{n}, \ldots, \frac{1}{n}, 0)$.

The relation $\boldsymbol{p} \prec \boldsymbol{p}'$ is obviously reflexive and transitive. Nonetheless, it provides just a *partial* order relation between *sorted* probability distributions: It may happen that $\boldsymbol{p} \nprec \boldsymbol{p}'$ and $\boldsymbol{p}' \nprec \boldsymbol{p}$, in which case the distributions are incomparable according to the majorization criterion. If $\boldsymbol{p} \prec \boldsymbol{p}'$ and $\boldsymbol{p}' \prec \boldsymbol{p}$, then $\boldsymbol{p}^s$ and $\boldsymbol{p}'^s$ are identical (except for the number of zeros if lengths n and n' differ), although this does not imply, of course, $\boldsymbol{p} = \boldsymbol{p}'$ (unsorted distributions).

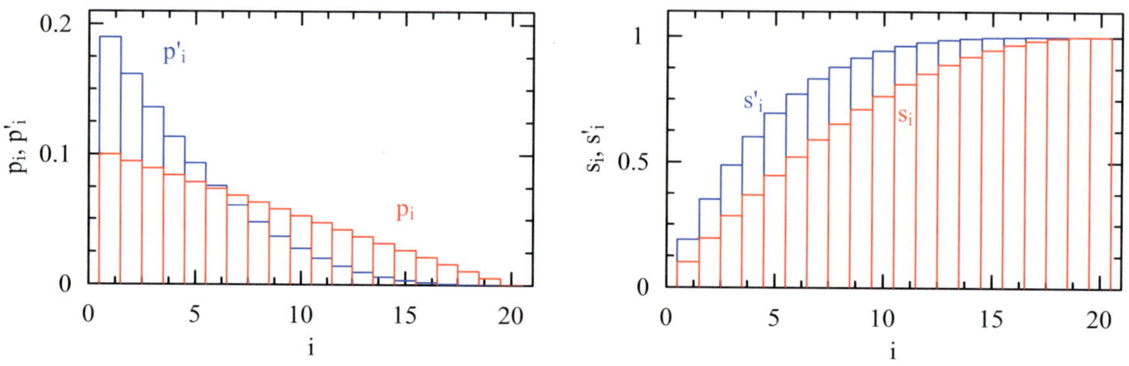

Figure 1: Example of a (sorted) probability distribution $\boldsymbol{p} = (p_1, \ldots, p_n)$ majorized by a distribution $\boldsymbol{p}'$. The right panel depicts the partial sums $s_i = \sum_{j=1}^{i} p_i$, $s'_i = \sum_{j=1}^{i} p'_i$.

The following theorem may help to understand the deep significance of the concept of majorization: For distributions of the same length, it states that $\boldsymbol{p} \prec \boldsymbol{p}'$ if and only if $\boldsymbol{p}$ can be obtained from $\boldsymbol{p}'$ by a suitable convex combination of permutations P_α of $\boldsymbol{p}'$ [4]:

$$\boldsymbol{p} \prec \boldsymbol{p}' \text{ iff } \boldsymbol{p} = \sum_{\alpha} q_\alpha P_\alpha(\boldsymbol{p}')\,, \tag{2.5}$$

where $q_\alpha \geq 0$, $\sum_\alpha q_\alpha = 1$. For instance, if $\boldsymbol{p} = (\frac{1}{2}, \frac{1}{2})$ and $\boldsymbol{p}' = (p_1, 1 - p_1)$, with $p_1 \in [0, 1]$, then $\boldsymbol{p} = \frac{1}{2}\boldsymbol{p}' + \frac{1}{2}P(\boldsymbol{p}')$, where $P(\boldsymbol{p}') = (1 - p_1, p_1)$. Permuting the elements at random always leads to a decrease of information, which is reflected in a majorization relation between the final and the initial distribution. Eq. (2.5) states, in addition, that this is always the case: If $\boldsymbol{p} \prec \boldsymbol{p}'$ then $\boldsymbol{p}$ can be obtained from $\boldsymbol{p}'$ by averaging permutations of $\boldsymbol{p}'$. The relation $\boldsymbol{p} \prec \boldsymbol{p}'$ is accordingly also denoted as $\boldsymbol{p}$ being *more mixed* than $\boldsymbol{p}'$.

In close relation with this theorem, we may also say that [4] $\boldsymbol{p} \prec \boldsymbol{p}'$ if and only if $\boldsymbol{p} = M\boldsymbol{p}'$, where M is a $n \times n$ doubly stochastic matrix ($M_{ij} \geq 0$, $\sum_i M_{ij} = \sum_j M_{ij} = 1$). This is due to the fact that a matrix is doubly stochastic if and only if it can be written as a convex combination of permutation matrices [4].

A basic important occurrence of a majorization relation is that taking place between a joint probability distribution $\boldsymbol{p}^{AB} = \{p_{ij}\}$ associated with a composite system $A + B$ and its marginals $p_i^A = \sum_j p_{ij}$, $p_j^B = \sum_i p_{ij}$: The joint distribution is always more mixed than its marginals [11]:

$$\boldsymbol{p}^{AB} \prec \boldsymbol{p}^A, \;\; \boldsymbol{p}^{AB} \prec \boldsymbol{p}^B\,. \tag{2.6}$$

This reflects the basic fact that the spreading of the marginal cannot be greater than that of the whole distribution (of course, this holds for *classical* composite systems, *i.e.*, those described by a joint probability distribution; for quantum systems this property is not necessarily valid, as discussed in the last section).

3 Majorization and Shannon entropy

The Shannon entropy of a distribution $\boldsymbol{p}$ is defined as [5]

$$S(\boldsymbol{p}) = -\sum_{i=1}^{n} p_i \ln p_i \,, \tag{3.1}$$

and is a measure of the lack of information associated with the distribution. As is well known, it vanishes for a delta type distribution $p_i = \delta_{ij}$ and is maximum for the uniform distribution $p_i = \frac{1}{n}\ \forall\ i$. It can then be expected that if $\boldsymbol{p} \prec \boldsymbol{p}'$, $\boldsymbol{p}$ has also a larger entropy than $\boldsymbol{p}'$, since it is more spread out. This is indeed the case [4]: It can be demonstrated (see the following section) that

$$\boldsymbol{p} \prec \boldsymbol{p}' \Rightarrow S(\boldsymbol{p}) \geq S(\boldsymbol{p}') \,, \tag{3.2}$$

with equality holding (when $\boldsymbol{p} \prec \boldsymbol{p}'$) only if $\boldsymbol{p}^s = \boldsymbol{p}'^s$ (except for the number of zeros). Hence, the fundamental property of S being maximum for the uniform distribution is just a direct consequence of Eqs. (3.2) and (2.4).

The important point is, however, that *the converse relation does not hold in general*:

$$S(\boldsymbol{p}) \geq S(\boldsymbol{p}') \not\Rightarrow \boldsymbol{p} \prec \boldsymbol{p}' \,, \tag{3.3}$$

In other words, the concept of disorder implied by majorization is more stringent than that based on the Shannon entropy: When $S(\boldsymbol{p}) \geq S(\boldsymbol{p}')$, it may also happen that there is no majorization relation between $\boldsymbol{p}$ and $\boldsymbol{p}'$. The trivial exception are of course binary distributions $\boldsymbol{p}^s = (p_1^s, 1 - p_1^s)$, where there is a single free parameter p_1^s and hence $\boldsymbol{p} \prec \boldsymbol{p}'$ iff $p_1^s \leq p'^s_1$, *i.e.*, iff $S(\boldsymbol{p}) \geq S(\boldsymbol{p}')$, as $S(\boldsymbol{p})$ is here just a decreasing function of the largest eigenvalue p_1^s.

A very simple example of (3.3) is just

$$\boldsymbol{p} = (0.4, 0.4, 0.2), \quad \boldsymbol{p}' = (0.5, 0.25, 0.25) \,, \tag{3.4}$$

which leads to $S(\boldsymbol{p}) \approx 1.055 > S(\boldsymbol{p}') \approx 1.04$, yet $p_1 + p_2 = 0.8 > p_1' + p_2' = 0.75$ (and $p_1 = 0.4 < p_1' = 0.5$), indicating that $\boldsymbol{p} \not\prec \boldsymbol{p}'$ despite the fact that $S(\boldsymbol{p}) > S(\boldsymbol{p}')$.

Thus, we may be missing some fundamental information about the relative evenness of the distributions $\boldsymbol{p}$ and $\boldsymbol{p}'$ if we just restrict ourselves to the information $S(\boldsymbol{p}) \geq S(\boldsymbol{p}')$. For instance, in the case of joint distributions, it is well known that

$$S(\boldsymbol{p}^{AB}) \geq S(\boldsymbol{p}^{A}), \;\; S(\boldsymbol{p}^{AB}) \geq S(\boldsymbol{p}^{B})\,, \tag{3.5}$$

i.e., that the entropy of the joint distribution is not less than that of its marginals. This fact can be derived from the positivity of the (classical) conditional Shannon entropy [2], which can be written as $S(B|A) = S(\boldsymbol{p}^{AB}) - S(\boldsymbol{p}^{A})$. However, we now see from (2.6) that there is a deeper relation between the evenness of $\boldsymbol{p}^{AB}$ and that of its marginals, Eq. (3.5) being just one if its consequences.

The natural question that now arises is if it is possible to capture the concept of majorization just by entropic means (entailing no need to sort the distribution). We will show that this is in fact feasible [12], by introducing simple generalized entropic forms.

4 Generalized entropies

Given a general probability distribution $\boldsymbol{p} \equiv (p_1, \ldots, p_n)$, we will consider the entropic forms [2,22–26]

$$S_f(\boldsymbol{p}) = \sum_{i=1}^{n} f(p_i)\,, \tag{4.1}$$

where $f(p)$ is a smooth *concave* function defined for $p \in [0,1]$ satisfying $f(0) = f(1) = 0$ (Fig. 2). By smooth and concave we mean here $f(p)$ continuous for $p \in [0,1]$ and $f'(p)$ strictly decreasing for $p \in (0,1)$, which ensures strict concavity:

$$f(qp_1 + (1-q)p_2) \geq qf(p_1) + (1-q)f(p_2), \;\; q \in [0,1]\,,$$

with equality holding if and only if $p_1 = p_2$ or $q(1-q) = 0$.

Eq. (4.1) is the most simple generalization of the Shannon entropy and comprises well known extensions: i) The Shannon entropy (3.1) is of course recovered for

$$f(p) = -p \ln p\,, \tag{4.2}$$

where the continuous limit $f(0) = \lim_{p\to 0^+} -p\ln p = 0$ is understood.

ii) The so called *linear* entropy

$$S_L(\boldsymbol{p}) = 1 - \sum_{i=1}^{n} p_i^2\,, \tag{4.3}$$

which is related to (4.2) by the linear approximation $-\ln p \approx 1-p$ (valid for p close to 1) but is otherwise a quadratic function of $\boldsymbol{p}$, corresponds to

$$f(p) = p - p^2\,, \tag{4.4}$$

iii) The *Tsallis* entropy [18–20]

$$S_q(\boldsymbol{p}) = \frac{1-\sum_{i=1}^{n} p_i^q}{q-1}\,, \quad q > 0\,, \tag{4.5}$$

which reduces to (4.3) for $q = 2$ and to (3.1) for $q \to 1$, corresponds to

$$f(p) = \frac{p - p^q}{q-1}\,, \quad q > 0\,. \tag{4.6}$$

All functions (4.2), (4.4) and (4.6) satisfy $f(0) = f(1) = 0$, with $f''(p) < 0$ for $p \in (0,1)$, so that the previous entropies are just particular instances of the general form (4.1). A positive prefactor $k > 0$ can be attached to the previous entropies, which can be chosen such that $S_f(\boldsymbol{p}) = 1$ for some reference distribution (such as the binary random distribution $\boldsymbol{p} = (\frac{1}{2}, \frac{1}{2})$).

Another example of a suitable entropic function f is [25, 41]

$$f(p) = \frac{\Gamma(1/q+1, -\ln p)}{\Gamma(1/q+1)} - p\,, \tag{4.7}$$

where $q > 0$ and $\Gamma(z,x) = \int_x^\infty e^{-t} t^{z-1} dt$ is the incomplete gamma function, with $\Gamma(z) = \Gamma(z,0)$. The Shannon entropy is here also recovered for $q \to 1$, as $\Gamma(2,x) = e^{-x}(1+x)$. Still, a further example of (1) is [24, 26, 30]

$$f(p) = \frac{1}{q}[p - \frac{e^{qp}-1}{e^q - 1}]\,, \tag{4.8}$$

which is again concave for any $q \in \Re$, satisfies $f(0) = f(1) = 0$ and approaches $\frac{1}{2}p(1-p)$ (*i.e.*, half the linear entropy (4.3)) for $q \to 0$. It satisfies $f_{-q}(p) = f_q(1-p)$.

With the exception of *additivity,* the basic properties satisfied by the Shannon entropy are also satisfied by an arbitrary $S_f(\boldsymbol{p})$ of this form:

1) $S_f(\boldsymbol{p}) \geq 0$, with $S_f(\boldsymbol{p}) = 0$ if and only if $p_i = \delta_{ij}$ for some $j \in \{1, \ldots, n\}$.

This is apparent, as strict concavity and the border conditions imply $f(p) > 0$ for $p \in (0,1)$ (Fig. 2), with $f(p) = 0$ only if $p = 0$ or $p = 1$.

2) $S_f(\boldsymbol{p})$ is a *concave* function of $\boldsymbol{p}$ [2, 25]:

$$S_f(q\boldsymbol{p} + (1-q)\boldsymbol{p}') \geq qS_f(\boldsymbol{p}) + (1-q)S_f(\boldsymbol{p}'),\;\; q \in [0,1]\,. \tag{4.9}$$

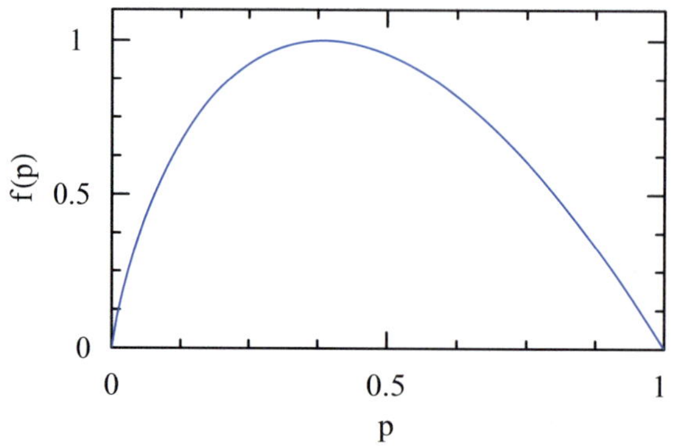

Figure 2: The concave functions $f(p)$ defining the generalized entropies (4.1).

with equality holding if and only if $q(1-q) = 0$ or $\boldsymbol{p} = \boldsymbol{p}'$ (assuming the same length). Here $q\boldsymbol{p} + (1-q)\boldsymbol{p}'$ denotes the convex combination of the distributions $\boldsymbol{p}$ and $\boldsymbol{p}'$, of elements $qp_i + (1-q)p'_i$, which is also a probability distribution (distributions form a convex set). Eq. (4.9) follows immediately from the concavity of f. Eq. (4.9) also implies the generalized inequality

$$S_f(\sum_\alpha q_\alpha \boldsymbol{p}^\alpha) \geq \sum_\alpha q_\alpha S_f(\boldsymbol{p}^\alpha)\,, \tag{4.10}$$

where $\boldsymbol{p}^\alpha$ and $\boldsymbol{q}$ are probability distributions ($q_\alpha \geq 0$, $\sum_\alpha q_\alpha = 1$).

3) The maximum of $S_f(\boldsymbol{p})$ is attained at the *uniform distribution* $p_i = 1/n\ \forall\ i$.

This can be shown by different methods, but is an immediate consequence of concavity: Given an arbitrary normalized distribution $\boldsymbol{p} = (p_1, \ldots, p_n)$, the uniform distribution $\boldsymbol{p}^n = (1/n, \ldots, 1/n)$ can be obtained from $\boldsymbol{p}$ by summing all n cyclic permutations P_α of $\boldsymbol{p}$, *i.e.*, as the convex combination $\boldsymbol{p}^n = \frac{1}{n}\sum_{\alpha=1}^n P_\alpha(\boldsymbol{p})$ (where $P_1(\boldsymbol{p}) = \boldsymbol{p}$, $P_2(\boldsymbol{p}) = (p_2, p_3, \ldots, p_1)$, *etc.*). Hence, by concavity,

$$S_f(\boldsymbol{p}^n) \geq \sum_{\alpha=1}^n \frac{1}{n} S_f(P_\alpha(\boldsymbol{p})) = S(\boldsymbol{p})\,, \tag{4.11}$$

since $S_f(P_\alpha(\boldsymbol{p})) = S_f(\boldsymbol{p})$. Concavity and the condition $f(0) = 0$ also ensure that the maximum value

$$S_f(\boldsymbol{p}^n) = nf(1/n) \tag{4.12}$$

is an *increasing* function of n, since $\boldsymbol{p}$ in (4.11) may in particular also stand for $\boldsymbol{p} = (\frac{1}{n-1} \ldots, \frac{1}{n-1}, 0)$. It is in fact verified that $dS_f(\boldsymbol{p}^n)/dn = f(1/n) - f'(1/n)/n > 0$, since strict concavity implies $f(p) - pf'(p) > f(0) = 0$ for $p \in (0, 1]$.

The entropies (4.1) satisfy therefore three of the four Khinchin axioms [17] that specify the von Neumann entropy, namely:

I) $S_f(\boldsymbol{p})$ is a continuous function of the distribution $\boldsymbol{p}$

II) $S_f(\boldsymbol{p})$ is maximum for the uniform distribution

III) $S_f(\boldsymbol{p})$ remains unchanged if an event with zero probability is added, *i.e.*, $S_f(p_1, \ldots, p_n, 0) = S_f(p_1, \ldots, p_n)$.

The fourth axiom of additivity is obviously not fulfilled in general. It is possible to show [25] that for product distributions, the forms (4.1) are *subadditive* (*superadditive*) if $pf''(p)$ is a *decreasing (increasing)* function of p for all $p \in (0, 1)$:

$$S_f(\boldsymbol{p} \otimes \boldsymbol{q}) \underset{\geq}{\overset{\leq}{}} S_f(\boldsymbol{p}) + S_f(\boldsymbol{q}) \text{ if } (pf''(p))' \underset{\geq}{\overset{\leq}{}} 0\,, \tag{4.13}$$

where $\boldsymbol{p}^{AB} \equiv \boldsymbol{p} \otimes \boldsymbol{q}$ denotes the joint distribution of two independent systems, with probabilities $p_{ij}^{AB} = p_i q_j$, such that $S_f(\boldsymbol{p} \otimes \boldsymbol{q}) = \sum_{i,j} f(p_i q_j)$. Moreover, for a general joint distribution we may write $p_{ij}^{AB} = p_j^{B/i} p_i^A$, where $p_i^A = \sum_j p_{ij}^{AB}$ is the marginal and $p_j^{B/i}$ the conditional probability, and Eq. (4.13) is to be replaced by [25]

$$S_f(\boldsymbol{p}^{AB}) \underset{\geq}{\overset{\leq}{}} S_f(\boldsymbol{p}^A) + \sum_i p_i^A S_f(\boldsymbol{p}^{B/i}) \text{ if } (pf''(p))' \underset{\geq}{\overset{\leq}{}} 0\,, \tag{4.14}$$

where $S_f(\boldsymbol{p}^{B/i}) = \sum_j f(p_j^{B/i})$ is the entropy of the conditional distribution. For instance, in the Tsallis case (4.5), we have $[pf''(p)]' = -q(q-1)p^{q-2} > 0$ (< 0) if $0 < q < 1$ $(q > 1)$, whence the Tsallis entropies are superadditive (subadditive) for $0 < q < 1$ $(q > 1)$.

Additivity, *i.e.*, $S_f(\boldsymbol{p} \otimes \boldsymbol{q}) = S_f(\boldsymbol{p}) + S_f(\boldsymbol{q})$ or in general,

$$S_f(\boldsymbol{p}^{AB}) = S(\boldsymbol{p}^A) + \sum_i p_i^A S_f(\boldsymbol{p}^{B/i})\,,$$

will then hold amongst the forms (4.1) only if $(pf''(p))' = 0\ \forall\ p \in (0, 1)$ [25]. Together with concavity and the boundary conditions $f(0) = f(1) = 0$, this equation *leads immediately to the Shannon form* $f(p) = -k\,p \ln p$, $k > 0$.

5 Majorization and generalized entropies

A fundamental question that immediately arises is if $\boldsymbol{p} \prec \boldsymbol{p}'$ implies $S_f(\boldsymbol{p}) \geq S_f(\boldsymbol{p}')$ for all forms (4.1). If not valid, it would be questionable to assign an entropic meaning to these forms. But this property does hold [4, 11, 12]: We have

$$\boldsymbol{p} \prec \boldsymbol{p}' \Rightarrow S_f(\boldsymbol{p}) \geq S_f(\boldsymbol{p}')\,, \tag{5.1}$$

for *any* f of the present form. Hence, all entropies $S_f(\boldsymbol{p})$ reflect the greater lack of knowledge associated with a probability $\boldsymbol{p}$ that is majorized by $\boldsymbol{p}'$. Moreover, Eq. (5.1) allows an *entropic* interpretation of majorization: If $\boldsymbol{p} \prec \boldsymbol{p}'$, not only the Shannon entropy but also *any* entropy of the form (4.1), is larger for the distribution $\boldsymbol{p}$.

Eq. (5.1) is again a direct consequence of concavity (and the condition $f(0) = 0$). Since by Eq. (2.5), $\boldsymbol{p} \prec \boldsymbol{p}'$ implies that $\boldsymbol{p}$ is a convex combination of permutations of $\boldsymbol{p}'$ (for distributions of the same length), then

$$S_f(\boldsymbol{p}) \geq \sum_{\alpha} q_\alpha S_f(P_\alpha(\boldsymbol{p}')) = S_f(\boldsymbol{p}')\,. \tag{5.2}$$

The condition $f(0) = 0$ ensures the validity of the previous inequality also when $\boldsymbol{p}$ and $\boldsymbol{p}'$ have different lengths. Note also that writing $p_i^s = s_i - s_{i-1}$, with $s_i = \sum_{j=1}^{i} p_j^s$ (and $s_0 = 0$), we have $\partial S_f(\boldsymbol{p})/\partial s_i = f'(p_i^s) - f'(p_{i+1}^s) \leq 0$ for $i = 1, \ldots, n-1$ (since $p_i^s \geq p_{i+1}^s$). $S_f(\boldsymbol{p})$ is thus a decreasing function of the partial sums s_i. If at least one of the inequalities in (2.2) is strict, we then have $S_f(\boldsymbol{p}) > S_f(\boldsymbol{p}')$.

The above proof actually shows that Eq. (5.1) holds for any concave function (*i.e.*, not necessarily of the form (4.1)), at least for distributions of identical length. Functions of $\boldsymbol{p}$ which satisfy the inequality (5.1) are denoted in general as *Schur concave functions* [4], which include standard concave functions but are in general not restricted to them [4].

Nonetheless, a remarkable feature of the set of forms (4.1) is that *they also suffice to warrant majorization:* Although the converse relation in (5.1) does not hold if valid for a single S_f, it does hold if valid *for all* S_f of the previous form:

$$S_f(\boldsymbol{p}) \geq S_f(\boldsymbol{p}')\ \forall\ S_f \ \Rightarrow \boldsymbol{p} \prec \boldsymbol{p}'\,. \tag{5.3}$$

An example of a *smooth* sufficient set of entropic forms was given in ref. [12] and is based on the functions (Fig. 3)

$$f_t^\alpha(p) = g_t(p-\alpha) - (1-p)g_t(-\alpha) - p g_t(1-\alpha)\,, \tag{5.4}$$

$$g_t(x) = -\frac{1}{2t}\ln\cosh(tx)\,, \tag{5.5}$$

where $\alpha \in (0,1)$ and $t > 0$, such that for $t \to \infty$, $g_t(x) \to -|x|/2$ and $f_t^\alpha(p) \to p(1-\alpha)$ if $p \leq \alpha$ and $\alpha(1-p)$ if $p \geq \alpha$. These functions are strictly concave for $t > 0$ and satisfy $f_t^\alpha(0) = f_t^\alpha(1) = 0$. If $\boldsymbol{p} \not\prec \boldsymbol{p}'$ it can be shown [12] that $S_{f_t^\alpha}(\boldsymbol{p}) < S_{f_t^\alpha}(\boldsymbol{p}')$ for some $\alpha \in (0,1)$ and sufficiently large t. Of course, setting $t = \infty$, *i.e.*, $f_\infty^\alpha(p) = -\frac{1}{2}|p-\alpha| + p(\frac{1}{2}-\alpha) + \frac{1}{2}\alpha$, with $\alpha \in (0,1)$, would also suffice [4], but such $f_\infty^\alpha(p)$ is not strictly concave nor smooth.

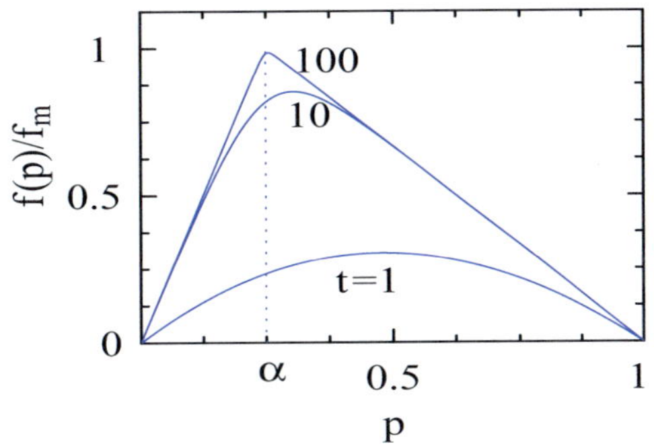

Figure 3: The entropic functions (5.4), for $\alpha = 1/4$ and different values of t. $f_m = \alpha(1-\alpha)$ denotes the maximum value in the $t \to \infty$ limit.

For instance, in the simple example (3.4), we obtain $S_{f_t^\alpha}(\boldsymbol{p}) < S_{f_t^\alpha}(\boldsymbol{p}')$ for $\alpha < 0.3$ and sufficiently large t (for instance, $t > 9.63$ at $\alpha = 0.25$, and $\alpha \in (0.2, 0.3)$ at $t = \infty$). Note that the absence of majorization in the example (3.4) *cannot* be detected with the Tsallis entropy either (we still obtain $S_q(\boldsymbol{p}) > S_q(\boldsymbol{p}')\ \forall\ q > 0$). See also Section 10 for a more complete and physical example.

Eq. (5.3) allows then *a fully entropic (and hence sorting-free) formulation of the majorization relations.* As an example, Eq. (5.1) allows to write [11, 12]

$$S_f(\boldsymbol{p}^{AB}) \geq S_f(\boldsymbol{p}^A)\,,\ S_f(\boldsymbol{p}^{AB}) \geq S_f(\boldsymbol{p}^B)\,, \tag{5.6}$$

for *any* S_f of the form (4.1), indicating that the positivity of the difference $S_f(\boldsymbol{p}^{AB}) - S_f(\boldsymbol{p}^A)$ is a general feature of the entropies (4.1) and not just a peculiarity of the Shannon entropy: The joint distribution is more disordered than its marginals according to *any* entropic measure S_f. Moreover, it is actually simple to prove Eq. (5.6) directly through the concavity of f [11] (for $p \geq 0, q \geq 0$, $f(p+q) \leq f(p) + f(q)$ if f is concave and $f(0) = 0$), implying $S_f(\boldsymbol{p}^A) = \sum_i f(\sum_j p_{ij}^{AB}) \leq \sum_{i,j} f(p_{ij}^{AB}) = S_f(\boldsymbol{p}^{AB})$. This shows that $\boldsymbol{p}^{AB} \prec \boldsymbol{p}^A$, according to Eq. (5.3).

6 Generalized maxent distributions

Let us now consider the maximization of the general forms (4.1) subject to a standard linear constraint

$$E = \langle H \rangle = \sum_i p_i \varepsilon_i\,, \tag{6.1}$$

in addition to normalization ($1 = \langle 1 \rangle = \sum_i p_i$). Here H is a random variable taking the value ε_i on event i. It may, for instance, represent the Hamiltonian in the case of a statistical description of a physical system [33, 34], in which case ε_i is the energy of system state i and E the corresponding average energy. The maximum can be obtained through the unconstrained maximization of

$$\bar{S}_f(\boldsymbol{p}) = S_f(\boldsymbol{p}) - \alpha\langle 1 \rangle - \beta\langle H \rangle \tag{6.2}$$

$$= \sum_i f(p_i) - \lambda_i p_i, \quad \lambda_i = \alpha + \beta\varepsilon_i \,, \tag{6.3}$$

where α and β are Lagrange multipliers. Due to the concavity of f, $f(p) - \lambda p$ will possess a unique maximum $p(\lambda)$ located either at the border or at the value determined by $f'(p) = \lambda$, *i.e.*, [25, 26]

$$p_i = p(\lambda_i) = \begin{cases} (f')^{-1}(\lambda_i)\,, & f'(1) < \lambda_i < f'(0) \\ 0\,, & \lambda_i \geq f'(0) \end{cases}, \tag{6.4}$$

where f'^{-1} denotes the inverse of the decreasing function f'. Eq. (6.4) is then a non-increasing function of λ_i, begin strictly decreasing for $\lambda_i < f'(0)$:

$$p_i' \equiv p'(\lambda_i) = \begin{cases} 1/f''(p_i) < 0\,, & f'(1) < \lambda_i < f'(0) \\ 0\,, & \lambda_i \geq f'(0) \end{cases}. \tag{6.5}$$

Hence, for $\beta > 0$, p_i is a *non-increasing function of the energy* ε_i, strictly decreasing for $\varepsilon_i < (f'(0) - \alpha)/\beta$ and exhibiting a "*cut-off*" $p_i = 0$ for $\varepsilon_i > (f'(0) - \alpha)/\beta$, arising whenever $f'(0)$ is *finite* [25]. When $f'(p) \to \infty$ for $p \to 0$, as in the Shannon case, no cut-off arises and p_i vanishes only in the limit $\lambda_i \to \infty$.

For instance, in the Shannon case [33, 34], $f'^{-1}(\lambda) = e^{-\lambda-1}$ and Eq. (6.4) reduces, of course, to the well-known exponential distribution (Boltzmann-Gibbs distribution)

$$p_i = e^{-(\tilde{\alpha}+\beta\varepsilon_i)} = e^{-\beta\varepsilon_i}/Z\,, \tag{6.6}$$

where $\tilde{\alpha} = \alpha + 1$ and $Z = e^{\tilde{\alpha}} = \sum_i e^{-\beta\varepsilon_i}$ the standard partition function. When ε_i represents an energy, Z becomes the standard partition function of the system, in which case $\beta = 1/kT$, with T the absolute temperature and k Boltzmann's constant [33, 34].

On the other hand, the Tsallis entropy (4.5) leads to the general power-law distribution [18, 19, 20, 25]

$$p_i = [1 - \tfrac{q-1}{q}(\tilde{\alpha} + \beta\varepsilon_i)]_+^{\frac{1}{q-1}}\,, \tag{6.7}$$

where $[u]_+ \equiv \frac{1}{2}(u + |u|) = u$ (0) for $u > 0$ (≤ 0), while the entropy (4.7) to the general exponential distribution [25, 41]

$$p_i = e^{-\gamma^q(\tilde{\alpha}+\beta\varepsilon_i)^q}\,, \tag{6.8}$$

where $\gamma = \Gamma^q(1/q + 1)$. Both distributions (6.7) and (6.8) reduce to (6.6) for $q \to 1$.

Finally, the form (4.8) leads to the *logarithmic* distribution [24, 26, 30]

$$p_i = \ln[1 - (e^q - 1)(\alpha + \beta\varepsilon_i - h_c)]/q\,, \tag{6.9}$$

for $h_c - 1 \leq \alpha + \beta\varepsilon_i < h_c$, with $h_c = f'(0) = \frac{1}{q} - \frac{1}{e^q - 1}$.

Let us remark [25] that since $[pf''(p)]' \propto \frac{d^2}{d\lambda^2}\ln p(\lambda)$ (Eq. (6.5)), it turns out from Eqs. (4.13)–(4.14) that a subadditive (superadditive) S_f corresponds to a *concave (convex)* $\ln p(\lambda)$. The Shannon entropy leads, accordingly, to a *linear* (in λ) $\ln p(\lambda)$.

The generalized distributions (6.4) are verified to share with (6.6) the same basic properties [25, 13], with the parameter $T = \beta^{-1}$ playing the role of an effective temperature for $\beta > 0$: For $T \to +\infty$, p_i always approaches the uniform distribution $p_i = p(\alpha) = 1/n\ \forall\ i$, whereas for $T \to 0^+$, $p_i \to 0$ if ε_i is larger than the minimum energy ε_1 (in this limit $\lambda_i \to \infty$ for $\varepsilon_i > \varepsilon_1$). The main qualitative difference is the strict vanishing $p_i = 0$ for $\lambda_i > f'(0)$ when $f'(0)$ is finite. In the previous examples, such a cut-off arises for $q > 1$ in (6.7).

Let us briefly quote the generalized thermodynamic relationships associated with the generalized maxent distribution (6.4), which resemble those of the standard Shannon based statistics [22, 23, 24, 25, 26] (see also section 10 in chapter 2). Considering $S_f(\boldsymbol{p})$ at the constrained maximum as a function of $E = \langle H\rangle$ (at fixed normalization $\langle 1\rangle = 1$) we obtain

$$\frac{\partial S_f}{\partial E} = \beta, \quad \frac{\partial^2 S_f}{\partial E^2} = -[(-\sum_i p_i')(\langle H^2\rangle_{\tilde{\boldsymbol{p}}} - \langle H\rangle^2_{\tilde{\boldsymbol{p}}})]^{-1} \leq 0\,, \tag{6.10}$$

where $\tilde{\boldsymbol{p}} = (\tilde{p}_1, \ldots, \tilde{p}_n)$, with $\tilde{p}_i = p_i'/\sum_i p_i'$ and $p_i' \equiv p'(\lambda_i)$, is a probability distribution and $\langle H\rangle_{\tilde{\boldsymbol{p}}} = \sum_i \varepsilon_i \tilde{p}_i$ denotes the average with respect to $\tilde{\boldsymbol{p}}$. Thus, while the first order thermodynamic relation is formally exactly the same as that of standard Shannon based statistics, the second relation in (6.10) indicates that S_f remains a *concave* function of E, again as in standard Shannon statistics. The difference is that the curvature is now related to the energy fluctuation with respect to the associated distribution $\tilde{p}$. In the Shannon case $\tilde{p}_i = p_i$ and the latter becomes the standard fluctuation. The condition $\tilde{\boldsymbol{p}} = \boldsymbol{p}$ is in fact the same as that determining additivity $((pf''(p))' = 0$, Eq. (4.13)) [25].

The generalized free energy can now be defined as $F = -T\tilde{S}_f = \langle H\rangle - TS_f$, where $T = \beta^{-1}$ as before and

$$\tilde{S}_f = \bar{S}_f + \alpha\langle 1\rangle\,, \tag{6.11}$$

with α determined from the fixed normalization constraint $\sum_i p(\lambda_i) = 1$. Eq. (6.11) satisfies

$$\frac{\partial\tilde{S}_f}{\partial\beta} = -E\,,\ \frac{\partial^2\tilde{S}_f}{\partial\beta^2} = -\frac{\partial E}{\partial\beta} = (-\sum_i p'_i)(\langle H^2\rangle_{\tilde{\boldsymbol{p}}} - \langle H\rangle^2_{\tilde{\boldsymbol{p}}}) \geq 0\,. \tag{6.12}$$

The first relation is again the same as in Shannon based statistics, whereas the second indicates that $\tilde{S}_f$ is again a convex function of β, *i.e.*, that the average energy is an *increasing* function of $T = \beta^{-1}$. The specific heat $C_v = \frac{\partial E}{\partial T} = -\beta^2\frac{\partial E}{\partial\beta}$, though positive, is now related to the energy fluctuation with respect to the associated distribution $\tilde{\boldsymbol{p}}$.

The present scheme can be easily generalized to the case of several constraints $\langle O_\alpha\rangle = M_\alpha$, $\alpha = 1,\ldots,m$ [25, 26, 32]. We should just replace βH by $\sum_\alpha \lambda_\alpha O_\alpha$ in Eq. (6.2), with λ_α suitable Lagrange multipliers associated with the m constraints. Other types of constraints for the Tsallis and other entropies, non-linear in $\boldsymbol{p}$, have also been considered (see for instance [19, 20, 24] for more details).

7 Majorization and generalized maxent distributions

In the same way as (6.6), the maxent distribution (6.4) becomes flatter as $T = \beta^{-1}$ (assumed positive in what follows) increases. The natural question that now arises is the relation of this increasing flatness with entropy increase and majorization [13]. Considering the maximum entropy $S_f \equiv S_f(\boldsymbol{p}(\lambda))$ as a function of T (at fixed standard normalization), a straightforward calculation leads to (see Eqs. (6.10)–(6.12))

$$\frac{\partial S_f}{\partial T} = \beta^3(-\sum_i p'_i)(\langle H^2\rangle_{\tilde{\boldsymbol{p}}} - \langle H\rangle^2_{\tilde{\boldsymbol{p}}}) \geq 0\,. \tag{7.1}$$

This shows that S_f is always an *increasing* function of the temperature $T = \beta^{-1} > 0$: The generalized entropy of the corresponding maxent distribution increases with temperature, as in the standard Shannon-based statistics [25, 13]. In the latter, $p'_i = -p_i$ and hence $\tilde{\boldsymbol{p}} = \boldsymbol{p}$, so that $\langle H^2\rangle_{\tilde{\boldsymbol{p}}} - \langle H^2\rangle_{\tilde{\boldsymbol{p}}}$ reduces to the dispersion determined by the original maxent probabilities.

But even more important is the following general property [13]: *The distributions (6.4) always become more mixed as T increases:*

$$\boldsymbol{p}(T) \prec \boldsymbol{p}(T') \text{ if } T \geq T' > 0\,, \tag{7.2}$$

where $\boldsymbol{p}(T) = (p(\lambda_1), \ldots, p(\lambda_n))$ denotes the maxent distribution (6.4) as a function of $T = \beta^{-1}$ (of course at fixed normalization $\sum_i p_i = 1$) for *any* S_f of the prescribed form. This implies that *any* entropy $S_f(\boldsymbol{p}(T))$ increases as T increases, and not just that used in the construction of the generalized maxent state (6.4). Eq. (7.1) can now be seen as just a particular consequence of the fundamental property (7.2). In particular, as T increases, the standard exponential distribution (6.6) becomes more mixed and *any* entropy $S_f(\boldsymbol{p}(T))$, including of course the Shannon entropy, increases as T increases.

A proof of (7.2) is as follows [13]: At fixed normalization,

$$\frac{\partial p_i}{\partial T} = -\beta^2 p_i'(\varepsilon_i - \langle H \rangle_{\tilde{\boldsymbol{p}}})\,, \tag{7.3}$$

implying that the partial sums $s_i = \sum_{j=1}^{i} p_j$ satisfy (we now assume the standard increasing energy ordering, such that $\boldsymbol{p}(T)$ becomes directly sorted in decreasing order and $\boldsymbol{p}^s(T) = \boldsymbol{p}(T)$)

$$\frac{\partial s_i}{\partial T} = -\beta^2 \sum_{j=1}^{i} p_j'(\varepsilon_j - \langle H \rangle_{\tilde{\boldsymbol{p}}}) \leq 0\,, \tag{7.4}$$

since $-p_j' \geq 0$ and $\sum_{i=1}^{n} p_i'(\varepsilon_i - \langle H \rangle_{\tilde{\boldsymbol{p}}}) = 0$. Hence, all maxent distributions lead to a proper disorder increase (in the sense of majorization) as the temperature parameter increases [13]. This makes T a proper *mixing parameter* (see below). In other words, the paths in probability space that these generalized thermal states follow as T increases *are all characterized by increasing mixedness.*

8 Mixing parameters in generalized distributions

Given now an arbitrary probability distribution $p(\lambda)$ depending on a parameter λ, we say that λ is a *mixing parameter* if [36]

$$\boldsymbol{p}(\lambda) \prec \boldsymbol{p}(\lambda') \text{ if } \lambda \geq \lambda'\,. \tag{8.1}$$

For instance, we have previously shown that the effective temperature $T = \beta^{-1}$ in the generalized maxent distribution (6.4) is actually a mixing parameter (Eq. (7.2)) *for any* S_f.

Let us now consider discrete distributions of the form

$$p_i(\lambda) = g(\varepsilon_i, \lambda)/Z_g(\lambda), \quad Z_g(\lambda) = \sum_i g(\varepsilon_i, \lambda)\,, \tag{8.2}$$

where $g(\varepsilon, \lambda)$ is assumed to be an arbitrary smooth positive decreasing (or in general non-increasing) function of ε, depending on a parameter λ. A sufficient condition which ensures that λ is a mixing parameter for this $\boldsymbol{p}(\lambda)$ is that $\partial \ln g/\partial\lambda$ be a non-decreasing function of ε [36]:

$$\frac{\partial^2 \ln g}{\partial\varepsilon\partial\lambda} \geq 0\,. \tag{8.3}$$

As an example, given a distribution $\boldsymbol{p}$, the parameter $\lambda = 1/q > 0$ of the *escort* distribution $\boldsymbol{p}^q$ [17, 20], of elements

$$p_i^q/\sum_i p_i^q\,, \tag{8.4}$$

is a mixing parameter for $\boldsymbol{p}^q$. This can be verified directly through concavity [2], but it can easily be demonstrated *via* Eq. (8.3): Writing p_i^q in the form (8.2), with $g(\varepsilon, \lambda) = p^{1/\lambda}(\varepsilon)$ and $p(\varepsilon)$ a non-increasing function of ε, we obtain

$$\frac{\partial^2 \ln g}{\partial\varepsilon\partial\lambda} = -\frac{1}{\lambda^2 p}p'(\varepsilon) \geq 0\,.$$

Of course, any other decreasing function of q is also a mixing parameter.

As application, let us consider the general power-law distribution, which can be conveniently written in the form (8.2) with [37]

$$g(\varepsilon, q, T) = [1 - (1-q)\varepsilon/T]_+^{\frac{1}{1-q}}\,, \tag{8.5}$$

where $\varepsilon \geq 0$, $T > 0$, $q \in \Re$ and $[u]_+ \equiv \frac{1}{2}(u + |u|)/2$, as in Eq. (6.7). Eq. (8.5) is a positive decreasing function of ε (with a cut-off for $\varepsilon > T/(1-q)$ if $q < 1$), which approaches the Boltzmann distribution (6.6) for $q \to 1$. Eq. (8.3) is, remarkably, fulfilled for *both* $\lambda = T > 0$ and $\lambda = q$ ($\frac{\partial^2 \ln g}{\partial\varepsilon\partial q} = \varepsilon\frac{\partial^2 \ln g}{\partial\varepsilon\partial T} = \varepsilon(T - (1-q)\varepsilon)^{-2}$), such that [37]

$$\boldsymbol{p}(q, T) \prec \boldsymbol{p}(q, T') \text{ if } T > T' > 0\,, \tag{8.6}$$

$$\boldsymbol{p}(q, T) \prec \boldsymbol{p}(q', T) \text{ if } q > q'\,. \tag{8.7}$$

This warrants the increase of *any* entropy $S_f(\boldsymbol{p}(q, T))$ for increasing T o q.

Other representations of (8.5) in terms of mixing parameters are also feasible. For instance, $\gamma = |1-q|$ and $\mu = T/|1-q| > 0$, such that

$$g(\gamma, \mu) = [1 - s\varepsilon/\mu]_+^{s/\gamma}\,, \tag{8.8}$$

with $s = \pm 1$, are as well mixing parameters [36, 37]. All different Tsallis distributions derived from the Tsallis entropy (4.5) through its maximization subject to different forms of constraints [18, 19, 20], including of course Eq. (6.7), can be written in the form (8.5) through suitable substitutions [36, 37]. In particular, Eq. (6.7) is recovered for $q \to 2 - q$ and $T \to T(q - \tilde{\alpha}(q-1))$.

9 Majorization and generalized entropies in the quantum case

In quantum mechanics, the state of a system is characterized by a *density operator* ρ (also denoted as density matrix or statistical operator) which replaces the probability distribution characteristic of a statistical description of a classical system. The density operator is a positive and hence hermitian operator, whose eigenvalues form a probability distribution $\boldsymbol{p}$:

$$\rho = \sum_i p_i |i\rangle\langle i|, \quad p_i \geq 0, \ \mathrm{Tr}\,\rho = \sum_i p_i = 1\,. \tag{9.1}$$

Here $|i\rangle$ denotes the normalized orthogonal eigenvectors of ρ, with $|i\rangle\langle i|$ the corresponding orthogonal projector, and p_i its eigenvalues, such that $\langle i|\rho|j\rangle = p_i\delta_{ij}$. The eigenvalue p_i represents the probability of finding the system in the pure quantum state $|i\rangle$. In an arbitrary orthonormal basis, Eq. (9.1) becomes

$$\rho = \sum_{\alpha,\alpha'} \langle\alpha'|\rho|\alpha\rangle |\alpha'\rangle\langle\alpha|, \quad \langle\alpha'|\rho|\alpha\rangle = \sum_i p_i \langle\alpha'|i\rangle\langle i|\alpha\rangle\,, \tag{9.2}$$

where the diagonal elements

$$q_\alpha \equiv \langle\alpha|\rho|\alpha\rangle = \sum_i p_i |\langle\alpha|i\rangle|^2\,, \tag{9.3}$$

form again a probability distribution ($q_\alpha \geq 0$, $\sum_\alpha q_\alpha = 1$) and represent the probabilities of finding the system in the state $|\alpha\rangle$ when measuring an observable diagonal in this basis.

We say that ρ is pure when $\rho^2 = \rho$, *i.e.*, when ρ becomes an orthogonal projector: $\rho^2 = \rho \Leftrightarrow \rho = |j\rangle\langle j|$ for some normalized state $|j\rangle$, corresponding to $p_i = \delta_{ij}$ in Eq. (9.2). We then say that the system is in the pure state $|j\rangle$. In all other cases ($\rho^2 \neq \rho$) the system is said to be in a *mixed* state. The density operators form a convex set: if ρ_α are density operators, their mixture (convex combination) $\rho = \sum_\alpha q_\alpha \rho_\alpha$, with $q_\alpha \geq 0$, $\sum_\alpha q_\alpha = 1$, is also a density operator, *i.e.*, it is also a positive normalized operator, even when the different ρ_α do not commute.

Given two density operators ρ, ρ', we will say that $\rho \prec \rho'$, *i.e.*, that ρ is majorized by ρ' (or that ρ is more mixed than ρ') if the inequalities (2.1) are all satisfied for their eigenvalues sorted in decreasing order:

$$\rho \prec \rho' \text{ iff } \sum_{j=1}^{i} p_i^s \leq \sum_{j=1}^{i} p'^s_i \,, i = 1, \ldots, n-1 \,, \tag{9.4}$$

for $\rho = \sum_i p_i |i\rangle\langle i|$, $\rho' = \sum_i p'_i |i'\rangle\langle i'|$ and $\sum_{i=1}^n p_i = \sum_{i=1}^n p'_i = 1$. If the dimensions of ρ and ρ' differ, the same definition can be applied after completing with zeros the smallest set of eigenvalues.

The meaning of the inequalities (9.4) is of course the same as in the classical case: If $\rho \prec \rho'$, the eigenvalues of ρ are more spread out than those of ρ'. In particular, in the quantum case Eq. (2.4) becomes

$$I_n/n \prec \rho \prec |\Psi\rangle\langle\Psi| \;\; \forall \, \rho \,, \tag{9.5}$$

where I_n/n represents the *maximally mixed state*, *i.e.*, that with eigenvalues $p_i = 1/n$ for $i = 1, \ldots, n$ (I_n denotes the identity operator of dimension n), and $|\Psi\rangle$ denotes an arbitrary pure state. Any mixed state ρ is thus more mixed than any pure state, and always majorizes the maximally mixed state.

The theorem (2.5) remains valid as follows [4]:

$$\rho \prec \rho' \text{ iff } \rho = \sum_\alpha q_\alpha U_\alpha \rho' U_\alpha^\dagger \,, \tag{9.6}$$

where U_α are now unitary operators ($U_\alpha U_\alpha^\dagger = I_n$) and q_α are again probabilities ($q_\alpha \geq 0$, $\sum_\alpha q_\alpha = 1$). In other words, $\rho \prec \rho'$ if and only if ρ is a convex superposition of unitaries of ρ'. Note that a permutation is a particular case of a unitary operator.

An important example of a majorization relation in quantum physics is the following: The diagonal of ρ in a given orthonormal basis $\{|\alpha\rangle\}$, which is also a density operator (Eq. (9.3)), is always more mixed than ρ:

$$\rho_d = \sum_\alpha \langle\alpha|\rho|\alpha\rangle |\alpha\rangle\langle\alpha| \Rightarrow \rho_d \prec \rho \,. \tag{9.7}$$

This property is sometimes denoted as Schur's theorem [4] and indicates that the least spread of the diagonal elements is obtained when the chosen basis is that of its eigenvectors.

Let us now examine the connection of majorization with entropy in the quantum case. The quantum extension of the generalized entropy (4.1) is defined as [26]

$$S_f(\rho) = \mathrm{Tr}\, f(\rho) = \sum_{i=1}^{n} f(p_i)\,, \tag{9.8}$$

being fully determined by its eigenvalues. It is therefore left unaltered by any unitary transformation U ($UU^\dagger = I_n$): $S_f(\rho) = S_f(\rho')$ if $\rho' = U\rho U^\dagger$. In particular, the von Neumann entropy of a density operator [2] is the Shannon entropy of its eigenvalues:

$$S(\rho) = -\mathrm{Tr}\,\rho \ln \rho\,. \tag{9.9}$$

Eq. (9.8) satisfies, for any strictly concave f fulfilling $f(0) = f(1) = 0$, $S_f(\rho) \geq 0$, with $S_f(\rho) = 0$ if and only if ρ is pure ($\rho^2 = \rho$). Similarly, for fixed dimension n, the maximum of $S_f(\rho)$ for any S_f is attained at the maximally mixed state I_n/n, since its eigenvalues are $p_i = 1/n\ \forall\ i$. Due to the concavity of f, the concavity of S_f also holds in the quantum case [26]: For any convex combination, we have

$$S_f(\sum_\alpha q_\alpha \rho_\alpha) \geq \sum_\alpha q_\alpha S_f(\rho_\alpha)\,, \tag{9.10}$$

if the ρ_α are density operators and $q_\alpha \geq 0$, $\sum_\alpha q_\alpha = 1$.

Concavity and Eq. (9.6) allow then to extend Eq. (5.1) to the quantum realm: We have

$$\rho \prec \rho' \Rightarrow S_f(\rho) \geq S_f(\rho')\,, \tag{9.11}$$

$\forall\ S_f$ of the previous form. Moreover, by the same previous arguments, we have

$$S_f(\rho) \geq S_f(\rho')\ \forall\ S_f\ \Rightarrow \rho \prec \rho'\,. \tag{9.12}$$

Eqs. (9.7) and (9.11) imply, for instance,

$$S_f(\rho_d) \geq S_f(\rho) \tag{9.13}$$

$\forall\ S_f$, a property which can be also directly shown through the concavity of f and S_f. This just reflects the fact that the lack of information cannot decrease, and will in general increase, when conserving only the information about the diagonal elements of ρ in a given basis. Obviously $\rho_d = \rho$ only if the chosen basis is a basis of eigenvectors of ρ. We remark again that while Eq. (9.13) implies $S(\rho_d) \geq S(\rho)$ for the von Neumann entropy, this last inequality does not imply Eq. (9.13).

10 Quantum entanglement and majorization: The generalized entropic criterion for separability

The application of majorization ideas to quantum systems has become very important in the field of quantum information [6], where it has led to deep insights into the relation between quantum entanglement and local disorder [6]. We will here describe one of these applications, namely *the disorder criterion for separability* [9] and the ensuing *generalized entropic criterion for separability* [11, 12].

In quantum mechanics, the state of a bipartite quantum system $A+B$, composed of two subsystems A and B, is described by a joint density operator ρ_{AB}, which is an operator defined in the tensor product space $H_A \otimes H_B$ of the Hilbert spaces of the subsystems, of dimension $d_{AB} = d_A d_B$. The *reduced* density operators are defined as

$$\rho_A = \mathrm{Tr}_B\, \rho_{AB}\,, \quad \rho_B = \mathrm{Tr}_A\, \rho_{AB}\,, \tag{10.1}$$

and correspond to the marginal distributions in classical systems. Here Tr_B, Tr_A denote *partial* trace [6]. These reduced density matrices completely determine the expectation values of any *local* observable, *i.e.*, an observable of the form $O_A = o_A \otimes I_B$ or $O_B = I_A \otimes o_B$ of subsystem A or B respectively:

$$\langle O_A \rangle = \mathrm{Tr}\, \rho_{AB}\, O_A = \mathrm{Tr}_A\, \rho_A\, o_A\,. \tag{10.2}$$

A *product state* is a state of the form

$$\rho_{AB} = \rho_A \otimes \rho_B\,, \tag{10.3}$$

which corresponds to independent marginal distributions: if p_i^A, p_j^B denote the eigenvalues of ρ_A and ρ_B, then $p_{ij}^{AB} = p_i^A p_j^B$ are the eigenvalues of ρ_{AB}. Of course, Eq. (10.3) is just a very particular case of a joint density operator, representing the state of two independent systems.

A *separable* state [42] is one in which the joint density is a convex combination (*i.e.*, a statistical mixture) of product densities:

$$\rho_{AB} \text{ separable } \Leftrightarrow \rho = \sum_\alpha q_\alpha \rho_A^\alpha \otimes \rho_B^\alpha\,, \; q_\alpha \geq 0, \; \sum_\alpha q_\alpha = 1\,. \tag{10.4}$$

Otherwise, *i.e.*, if ρ cannot be written in the form (10.4) with all $q_\alpha \geq 0$, the system is said to be *entangled*. The underlying meaning is that an entangled state cannot be generated by local operations and classical communications (LOCC) [42], in contrast with the separable states (10.4): The state (10.4) can in principle be prepared by generating, *via* local operations and classical communication between the two systems, product states $\rho_A^\alpha \otimes \rho_B^\alpha$ with probabilities q_α.

For a pure state $\rho_{AB} = |\Psi_{AB}\rangle\langle\Psi_{AB}|$, separability reduces to the condition $|\Psi_{AB}\rangle = |\Psi_A\rangle \otimes |\Psi_B\rangle$, *i.e.*, $\rho_{AB} = \rho_A \otimes \rho_B$ with $\rho_A = |\Psi_A\rangle\langle\Psi_A|$, $\rho_B = |\Psi_B\rangle\langle\Psi_B|$. An entangled pure state is then one which cannot be written as a product of local states. For mixed states, however, product states are just a particular case of separable state.

One of the most striking features of quantum systems is precisely entanglement. In an entangled pure state, even though the whole system is in a definite state $|\Psi_{AB}\rangle$, the subsystems are not in pure states but rather in *mixed states* $\rho_A = \mathrm{Tr}_B|\Psi_{AB}\rangle\langle\Psi_{AB}|$, $\rho_B = \mathrm{Tr}_A|\Psi_{AB}\rangle\langle\Psi_{AB}|$, which have *non-zero* entropy despite the fact that $S(\rho_{AB}) = 0$. Such a situation is clearly *impossible* in classical systems, *i.e.*, in any composite system described by a joint probability distribution, since $S(\boldsymbol{p}^{AB})$ is always larger than $S(\boldsymbol{p}^A)$ or $S(\boldsymbol{p}^B)$ (Eq. (3.5)). If $S(\boldsymbol{p}^{AB}) = 0$ then $S(\boldsymbol{p}^A) = S(\boldsymbol{p}^B) = 0$. The same holds for a general entropy S_f: If $S_f(\boldsymbol{p}^{AB}) = 0$ then $S_f(\boldsymbol{p}^A) = S_f(\boldsymbol{p}^B) = 0$.

When ρ_{AB} is pure, ρ_A and ρ_B are isospectral [6, 2] and therefore possess exactly *the same entropy* S_f. The standard entanglement entropy of a bipartite pure state $|\Psi_{AB}\rangle$ is precisely defined as the von Neumann entropy of any of its subsystems:

$$E(|\Psi_{AB}\rangle\langle\Psi_{AB}|) = S(\rho_A) = S(\rho_B)\,. \tag{10.5}$$

We may similarly define the generalized entanglement entropy of a pure state as [11, 15, 16]

$$E_f(|\Psi_{AB}\rangle\langle\Psi_{AB}|) = S_f(\rho_A) = S_f(\rho_B)\,. \tag{10.6}$$

Therefore, the classical majorization property between the joint probability distribution and its marginals (Eq. (2.6)) *no longer holds in quantum systems:* In general, for an arbitrary mixed state,

$$\rho_{AB} \not\prec \rho_A, \quad \rho_{AB} \not\prec \rho_B$$

and it may well happen, conversely, that $\rho_A \prec \rho_{AB}$, $\rho_B \prec \rho_{AB}$, as occurs in pure entangled states, where $S_f(\rho_{AB}) = 0$ but $S_f(\rho_A) = S_f(\rho_B) > 0$. Hence, the quantum extension of Eq. (5.6) no longer holds in general, and $S_f(\rho_{AB})$ may be larger or smaller than the marginal entropies $S_f(\rho_A)$ and $S_f(\rho_B)$.

Nevertheless, the *disorder criterion for separability* [9] states that when ρ_{AB} is *separable*, the joint density operator ρ_{AB} is *more mixed* than the reduced densities ρ_A and ρ_B, exactly as occurs in *classical* systems (Eq. 2.6):

$$\rho_{AB} \text{ separable } \Rightarrow \rho_{AB} \prec \rho_A,\; \rho_{AB} \prec \rho_B\,. \tag{10.7}$$

This indicates a classical nature of the correlations in separable states.

Eq. (10.7) leads to the *generalized entropic criterion for separability* [11, 12]:

$$\rho_{AB} \text{ separable } \Rightarrow S_f(\rho_{AB}) \geq S_f(\rho_A),\; S_f(\rho_{AB}) \geq S_f(\rho_B)\,, \tag{10.8}$$

for *any* S_f of the present form (4.1). This fundamental property generalizes previous criteria based on the von Neumann entropy [38] and on the Tsallis entropy [39, 40], which are just particular cases of Eq. (10.8). It is stronger than the latter, being equivalent to (10.7) and in fact the *strongest* disorder based criterion (see below). Eq. (10.8) implies

$$S_f(\rho_A) > S_f(\rho_{AB}) \Rightarrow \rho_{AB} \text{ entangled}$$

such that violation of the classical inequality $S_f(A,B) \geq S_f(A)$ for some S_f already ensures ρ_{AB} is entangled.

The criterion (10.8) is not sufficient for separability (except for the case of pure states), *i.e.*, weak entangled mixed states exist where ρ_{AB} is still more mixed than its reduced densities ρ_A and ρ_B, *i.e.*, where entanglement does not manifest itself in the violation of the classical majorization relation (10.7) or equivalently, the classical general entropic relation (10.8). Nevertheless, the determination of separability is in general a hard problem and the disorder criterion provides a simple and tractable characterization of separability, which reflects the classical nature of the underlying global and local (*i.e.*, reduced) probability distributions. Moreover, it can be shown that it is the *strongest* criterion based on the knowledge of just the spectrum of ρ_{AB} and of one of its reduced densities [6].

Illustration. Let us illustrate these criteria in a specific example [12, 29]. We will consider a two-qudit system (*i.e.*, one where each component has Hilbert space dimension D, as in the case of two spins s where $D = 2s + 1$), in a state which is a mixture of two antisymmetric states and a random background:

$$\rho_{AB} = x|01^-\rangle\langle 01^-| + y|02^-\rangle\langle 02^-| + \frac{1-x-y}{D^2} I_{D^2}\,, \tag{10.9}$$

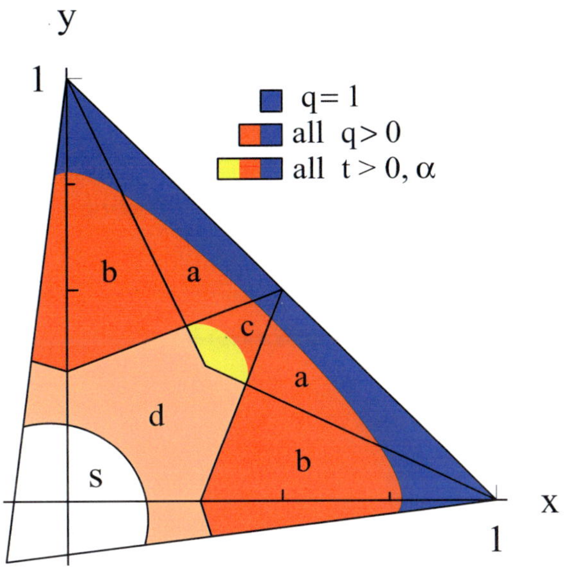

Figure 4: The allowed values of the parameters x and y in the state (10.9) for $D = 3$. The state is entangled in all colored regions and separable (*i.e.*, of the form (10.4)) in white sector S. The blue region is that where entanglement is detected by violation of the von Neumann based separability criterion (10.12), the red and blue regions those where it is detected by violation of the Tsallis entropy based criterion (10.13), and the red, blue and yellow regions those where it is detected by violation of the generalized entropic criterion (10.8), equivalent to the disorder criterion (10.7). Sectors a, b and c, delimited by solid black lines, indicate respectively those where both, just the first and just the second classical majorization inequalities (10.14) are violated. Sector c contains the yellow region not covered by the criterion (10.13).

where $|ij^-\rangle = (|ij\rangle - |ji\rangle)/\sqrt{2}$ and $|ij\rangle \equiv |i\rangle \otimes |j\rangle$, with $\{|i\rangle, i = 0, \ldots, D-1\}$ denoting an orthonormal basis for each qudit. We assume $D \geq 3$. Positivity of ρ_{AB} implies $x + y \leq 1$, $x - (D^2-1)y \leq 1$, $y - (D^2-1)x \leq 1$, which leads for $D = 3$ to the triangle shown in Fig. 4. The pure state limit corresponds to the points $(x, y) = (1, 0)$ or $(0, 1)$. Negative values of x, y indicate depletion of antisymmetric states with respect to the fully mixed state.

It can be shown [12] that this system is separable for

$$r \leq \frac{1}{\sqrt{2}\cos(\theta - \pi/4) + D^2/2}, \tag{10.10}$$

where $(x, y) = r(\cos\theta, \sin\theta)$, being otherwise entangled. This leads to the white and colored sectors of Fig. 4.

The reduced densities for this state are identical and given by

$$\rho_A = \rho_B = \frac{x+y}{2}|0\rangle\langle 0| + \frac{x}{2}|1\rangle\langle 1| + \frac{y}{2}|2\rangle\langle 2| + \frac{1-x-y}{D} I_D\,. \tag{10.11}$$

We then show in Fig. 4 the sectors within the entangled region where the inequalities of i) the von Neumann based separability criterion,

$$\rho_{AB} \text{ separable } \Rightarrow S(\rho_{AB}) \geq S(\rho_A)\,, \tag{10.12}$$

ii) the Tsallis set based separability criterion

$$\rho_{AB} \text{ separable } \Rightarrow S_q(\rho_{AB}) \geq S_q(\rho_A)\ \forall\, q > 0\,, \tag{10.13}$$

and iii) the generalized entropic criterion (10.8), are broken (at least for some q in ii) or some f in iii)). The latter coincides with the full disorder criterion and can be implemented using the generalized entropic forms (5.4). We also show the regions (a, b, c) where at least one of the first two majorization inequalities in (9.4),

$$s_1^{AB} \leq s_1^A\,,\ \ s_2^{AB} \leq s_2^A\,, \tag{10.14}$$

where s_i^{AB}, s_i^A denote the partial sums associated with ρ^{AB} and ρ^A, are broken (the remaining ones cannot be broken by the state (10.9)).

It is then seen that the von Neumann based criterion is rather weak, *i.e.*, its violation takes place only in the region close to the outer limit, while the Tsallis based criterion is able to detect entanglement in a much larger region, which exhausts sectors a and b, *i.e.*, the regions where the first inequality in (10.14) is violated. This is a general result [11]: Whenever the first majorization inequality in (2.1) is not fulfilled, there is some $q > 0$ such that $S_q(\mathbf{p}) < S_q(\mathbf{p}')$. The Tsallis based criterion fails in some part of sector c where only the second inequality in (10.14) is broken. This sector can of course be covered using the generalized entropies (5.4).

It is also seen that there is a rather large region (sector d) where the state is entangled but the entanglement does not break any classical entropic or majorization inequality. In this region the entanglement can therefore be considered "weak", in the sense that it cannot be detected with the knowledge of the eigenvalues of ρ_{AB} and one of its reduced states. The positive partial transposition criterion of Peres and Horodecki [43, 44], though also not sufficient in the general case, allows here to detect entanglement in sector d. Other cases were analyzed in refs. [12, 28, 29].

Other fundamental applications of majorization theory to the theory of quantum information can be found in refs. [6–14]. More recently, Eq. (9.7) has been used to derive a generalized entropic measure of quantum correlations for mixed states [15, 16, 45, 46].

11 Conclusions

We have reviewed, in a tutorial level, the fundamental features of majorization, and determined its relation with generalized entropic forms. As we have seen, majorization provides a rigorous concept of disorder that cannot be captured by any single choice of entropy. The use of generalized entropic forms allows, however, to express the majorization relations through an equivalent generalized entropic inequality, which is to be fulfilled by *all* entropies of the form (4.1) and not just by a single choice. A smooth sufficient set of entropic functions (Eqs. (5.4–5.5)) has also been provided. Several entropic forms and the ensuing generalized maxent distributions have been as well described.

We have then shown that majorization relations arise naturally in very general and distinct scenarios, implying that the actual disorder increase in these situations is more fundamental than that indicated by a single entropic inequality. For instance, all "thermal" distributions derived from the generalized entropies (4.1), including of course the Boltzmann-Gibbs distribution (6.6), do become more mixed, in the precise sense determined by majorization, as the corresponding temperature increases [13], while power-law distributions also exhibit strict majorization properties with respect to the index q.

We have finally described the essentials of the extension of the majorization and entropy concepts to quantum systems. As opposed to "classical" systems, quantum systems are characterized by density matrices or operators rather than probability distributions, and majorization and entropy become both essential for characterizing some of the most basic and remarkable features of these systems, like quantum entanglement. We have seen in particular that all separable states in bipartite quantum systems are more mixed globally than locally, *i.e.*, the whole system is more mixed than its constituents, as occurs in all classical systems [9, 11, 12]. However, this property can be violated in entangled states. Majorization then leads to a generalized entropic criterion for separability, *i.e.*, a sufficient entropic criterion for entanglement, which generalizes previous von Neumann and Tsallis entropy based criteria and provides the strongest disorder based criterion.

Distinct natural processes seem then to prefer a strict disorder increase in a wide variety of situations, a fact which stems ultimately from the averaging (2.5) or (9.6) that underlies any majorization relation. Several applications of majorization and generalized entropic inequalities in different fields are therefore currently under investigation.

Conflict of interest: None.

Acknowledgments

The authors acknowledge support from CONICET (NC) and CIC (RR) of Argentina.

References

[1] G.H. Hardy, J.E. Littlewood, G. Polya, *Inequalities,* Cambridge University Press, Cambridge, 1988.

[2] A. Wehrl, General properties of entropy, Rev. Mod. Phys. **50**, 221 1978.

[3] A.W. Marshall, I. Olkin, *Inequalities: Theory of Majorization and its Applications,* Academic Press, New York, 1979.

[4] R. Bhatia, *Matrix Analysis,* Springer, Berlin, 1997.

[5] C.E. Shannon, A mathematical theory of communication, Bell System Tech. **27**, 379, 623 (1948).

[6] M.A. Nielsen, I.L. Chuang, *Quantum Computation and Quantum Information*, Cambridge University Press, Cambridge, UK, 2000.

[7] M.A. Nielsen, Probability distributions consistent with a mixed state, Phys. Rev. A **62**, 052308 (2000).

[8] M.A. Nielsen, Conditions for a class of entanglement transformations, Phys. Rev. Lett. 83, 436 (1999).

[9] M.A. Nielsen, J. Kempe, Separable states are more disordered globally than locally, Phys. Rev. Lett. **86**, 5184 (2001).

[10] J.I. Latorre, M.A. Martín Delgado, Majorization arrow in quantum-algorithm design, Phys. Rev. **A 66**, 022305 (2002).

[11] R. Rossignoli, N. Canosa, Generalized entropic criterion for separability, Phys. Rev. **A 66**, 042306 (2002).

[12] R. Rossignoli, N. Canosa, Violation of majorization relations in entangled states and their detection by means of generalized entropic forms, Phys. Rev. **A 67**, 042302 (2003).

[13] R. Rossignoli, N. Canosa, Limit temperature for entanglement in generalized statistics, Phys. Lett. **A 323**, 22 (2004).

[14] R. Orús, Entanglement and majorization in 1+1-dimensional quantum systems, Phys. Rev. **A 71**, 0052327 (2005); ibid A 73, 019904(E) (2006).

[15] R. Rossignoli, N. Canosa, L. Ciliberti, Generalized entropic measures of quantum correlations, Phys. Rev. **A 82**, 052342 (2010).

[16] R. Rossignoli, N. Canosa, L. Ciliberti, Quantum correlations and least disturbing local measurements, Phys. Rev. **A 84**, 052329 (2011).

[17] C. Beck, F. Schlögl, *Thermodynamics of Chaotic Systems: An Introduction,* Cambridge University Press, Cambridge, UK (1993).

[18] C. Tsallis, Possible generalizations of Boltzmann Gibbs statistics, J. Stat. Phys. **52**, 479 (1988).

[19] E.M.F. Curado, C. Tsallis, Generalized statistical mechanics: Connection with thermodynamics, J. Phys. **A 24**, L69 (1991); **24**, 3187(E) (1991), **25**, 1019 (1992).

[20] C. Tsallis, *Introduction to Non-Extensive Statistical Mechanics,* Springer, New York, 2009.

[21] A.R. Plastino A. Plastino, From Gibbs microcanonical ensemble to Tsallis generalized canonical distribution, Phys. Lett. **A 193**, 140 (1994).

[22] R.S. Mendes, Some general relations in arbitrary thermostatistics, Phys. **A 242**, 299 (1997).

[23] A. Plastino, A.R. Plastino, Universality of thermodynamics' Legendre transform structure, Phys. Lett. **A 226**, 257 (1997).

[24] E.M.F. Curado, General aspects of the thermodynamics formalism, Braz. J. Phys. **29**, 36 (1999).

[25] R. Rossignoli, N. Canosa, Non-additive entropies and quantum statistics, Phys. Lett. **A 264**, 148 (1999).

[26] N. Canosa, R. Rossignoli, Generalized non-additive entropies and quantum entanglement, Phys. Rev. Lett. **88**, 170401 (2002).

[27] G. Kaniadakis, Statistical mechanics in the context of special relativity, Phys. Rev. **E 66**, 056125 (2002).

[28] R. Rossignoli, N. Canosa, Generalized entropies and quantum entanglement, Phys. **A 329**, 371 (2003).

[29] R. Rossignoli, N. Canosa, Generalized disorder measures and the detection of quantum entanglement, Phys. **A 344**, 637 (2004).

[30] E.M.F. Curado, F. Nobre, On the stability of analytic entropic forms, Phys. **A 335**, 94 (2004).

[31] G.A. Tsekouras, C. Tsallis, Generalized entropy arising from a distribution of q indices, Phys. Rev. **E 71**, 046144 (2005).

[32] N. Canosa, R. Rossignoli, General non-additive entropic forms and the inference of quantum density operators, Phys. **A 348**, 121 (2005).

[33] E.T. Jaynes, Information theory and statistical mechanics, Phys. Rev. **106**, 620 (1957).

[34] E.T. Jaynes, Information theory and statistical mechanics II, Phys. Rev. **108**, 171 (1957).

[35] C.H. Bennett, G. Brassard, C. Crepeau, R. Jozsa, A. Peres, W.K. Wootters, Teleporting an unknown quantum state via dual classical and Einstein-Podolsky-Rosen channels, Phys. Rev. Lett. **70**, 1895 (1993).

[36] N. Canosa, R. Rossignoli, M. Portesi, Majorization properties of generalized thermal distributions, Phys. **A 368**, 435 (2006).

[37] N. Canosa, R. Rossignoli, M. Portesi, Majorization relations and disorder in generalized statistics, Phys. **A 371**, 126 (2006).

[38] R. Horodecki, M. Horodecki, Information-theoretic aspects of inseparability of mixed states, Phys. Rev. **A 54**, 1838 (1996).

[39] C. Tsallis, S. Lloyd, M. Baranger, Peres criterion for separability through nonextensive entropy, Phys. Rev. **A 63**, 042104 (2001).

[40] S. Abe, A.K. Rajagopal, Quantum entanglement inferred by the principle of maximum nonadditive entropy Phys. Rev. **A 60**, 3461 (1999).

[41] C. Anteneodo, A.R. Plastino, Maximum entropy approach to stretched exponential probability distributions, J. Phys. **A 32**, 1089 (1999).

[42] R.F. Werner, Quantum states with Einstein-Podolsky-Rosen correlations admitting a hidden-variable model, Phys. Rev. **A 40**, 4277 (1989).

[43] A. Peres, Separability criterion for density matrices, Phys. Rev. Lett. **77**, 1413 (1996).

[44] M. Horodecki, P. Horodecki, R. Horodecki, Separability of mixed states: necessary and sufficient conditions Phys. Lett. **A 223**, 1 (1996).

[45] N. Canosa, L. Ciliberti, R. Rossignoli, Quantum discord and related measures of quantum correlations in finite XY chains, Int. J. Mod. Phys. **B 27**, 1345033 (2012).

[46] L. Ciliberti, N. Canosa, R. Rossignoli, Discord and Information Deficit in the XX chain, Phys. Rev. **A 88**, 012119 (2013).

CHAPTER 6

Distances Measures for Probability Distributions

Pedro W. Lamberti [a,*] and Ana P. Majtey [b]

[a] Facultad de Matemática, Astronomía y Física, Universidad Nacional de Córdoba and CONICET, Ciudad Universitaria, 5000 Córdoba, Argentina

[b] Instituto Carlos I de Física Teórica y Computacional and Departamento de Física Atómica Molecular y Nuclear, Universidad de Granada, Granada, E-18071, Spain

Abstract

Many problems of statistical and quantum mechanics can be established in terms of a distance between probability distributions. The present work is devoted to review some of the most relevant applications of the notion of distance in the context of statistical mechanics, quantum mechanics and information theory. Although we make a general overview of the most frequently used distances between probability distributions, we will center our presentation in a distance known as the Jensen-Shannon divergence both in their classical and quantum versions. For the classical one we present its main properties and we discuss its relevance as a segmentation tool for symbolic sequences. In the quantum case we show that the quantum Jensen-Shannon divergence is an adequate measure of entanglement.

Keywords: Probabilty distributions; Divergences; Geometry; Quantum mixed states; Entanglement.

1 The notion of distance

The subject matters of this chapter are two fundamental mathematical concepts: probability and distance. On one side probability plays a central role in some of the most important theories of contemporaneous physics: statistical mechanics, quantum mechanics, and information theory. Probability theory is the mathematical analysis of random events, that is, events that do not have *deterministic regularity* but that have some *statistical regularity* [1]. On the other side, as was pointed out by Riemann in his pioneer works, one of the basic geometrical notion is the concept of (local) distance between points of a given abstract space (usually a differentiable manifolds). According to his conceptions, the distance

* Corresponding author. E-mail: lamberti@famaf.unc.edu.ar

Andres M. Kowalski, Raul D. Rossignoli and Evaldo M. F. Curado (Eds)

function (local metric) determines all the geometrical properties of the space. This fact has been essential for the relationship between geometry and physics which has blossomed in a remarkable manner along the past century. A notorious example of this is the General Theory of Relativity [2].

In Riemannian geometry the notion of distance is built from a inner product defined between tangent vectors to the differentiable manifold. Under adequate suppositions, the local Riemannian metric gives to the differential manifold the structure of a metric space, that is, a space provided with a metric [3]. To be precise, let us recall that a metric over an arbitrary set X (a topological space, a Hilbert space, a differential manifold, *etc.*), is a non-negative function D from the cartesian product $X \times X$ into the real numbers set that verifies:

$$D(x,y) = 0 \ iff \ x = y \tag{1}$$

$$D(x,y) = D(y,x) \ \forall \ x,y \in X \ \ (symmetric) \tag{2}$$

$$D(x,z) \leq D(x,y) + D(y,z), \forall x,y,z \in X (triangle \ inequality) \tag{3}$$

If the triangle inequality is not valid, we will name the function D as a distance. When distances between (discrete) probability of distributions are investigated, the set X is given by a subset of $\mathbb{R}^{N+1}$:

$$X = M_+^1(N+1) \equiv \{(p_1, \ldots, p_{N+1}), p_i \in \mathbb{R}, p_i \geq 0, \sum_i p_i = 1\}.$$

Here we are considering a random variable R, which can take one of the values $r_1, ..., r_{N+1}$ and p_j is the probability of occurrence of value r_j: $p_j = Prob(R = r_j)$. The set $M_+^1(N+1)$ can be thought as a differentiable manifold with local coordinates θ^i such that:

$$\theta^1 = p_1, \ \ \theta^2 = p_2, \ \ \ldots \theta^N = p_N \tag{4}$$

This manifold can be identified with the simplex defined by [6]:

$$\sum_{i=1}^{N+1} \theta^i = 1, \ \ \theta^i > 0.$$

The notion of distance between two probability distributions was initially developed by Mahalonobis [4]. Along the years several distances between probabilities distributions have been introduced. In the next we will review some of them as well as some examples where they are used. For example, in a formal context, the concept of convergence is of fundamental importance in probability theory. Several of its main results are established as limit theorems. Among the most important of these results, is Bernoullis law of large numbers and the central limit theorem. In the probability theory context, several distances between two probability distributions are found in the literature. Even though the most frequently used are the Minkowskian metrics

$$d_\nu(P, P') \equiv \left(\sum_j |p_j - p'_j|^\nu \right)^{\frac{1}{\nu}} \quad \nu \geq 1 \tag{5}$$

several important results are expressed in terms of other metrics, for instance the Lévy-Prokhorov metric and the Banach-Lipschitz metric [1]. In the expression (5) p_j and p'_j are the event j's probabilities of occurrence for each of the probability distributions P and P', respectively. The case $\nu = 2$ corresponds to the Euclidean one.

The use of the Euclidean distance in the context of probability theory has been criticized by Wootters in an illuminating communication because, in using it, one is ignoring the fact that we are dealing with a space of probability-distributions and thus disregarding the stochastic nature of the distribution $P = p_j$ [5]. The concept of statistical distance originates in a quantum mechanical context. One uses it primarily to distinguish among different preparations of a given quantum state, and, more generally, to ascertain to what an extent two such states differ from one another. The concomitant considerations being of an intrinsic statistical nature, the concept can be applied to any probabilistic space. The main idea underlying this notion of distance is that of adequately taking into account statistical fluctuations inherent to any finite sample. As a result of the associated statistical errors, the observed frequencies of occurrence of the various possible outcomes typically differ somewhat from the actual probabilities, with the result that, in a given fixed number of trials, two preparations are indistinguishable if the difference between the actual probabilities is smaller than the size of a typical fluctuation. The Wootters's distance has the expression:

$$d_W(P, P') = \arccos \left(\sum_j \sqrt{p_j p'_j} \right) \tag{6}$$

In statistics the natural Riemannian metric is the Fisher metric. In the manifold with coordinates θ^i given by (4), let P and P' be two very close points in S and let θ and $\theta + d\theta$

be, respectively their coordinates. The Fisher metric between P and P' is given by

$$ds_F^2 = \sum_{i,j=1}^{N} g_{ij} d\theta^i d\theta^j$$

with

$$g_{ij} = \frac{1}{\theta^i}\delta_{ij} + \frac{1}{\theta^{N+1}} \quad i,j = 1,...,N \tag{7}$$

Incidentally we mention that the Wootters's distance is the geodesic distance corresponding to the Fisher metric [6].

A frequently used quantity to compare two probability distributions, arose in information theory, is the Kullback-Leibler divergence [7]:

$$K(P, P') \equiv \sum_j p_j \log_2 \left(\frac{p_j}{p'_j} \right), \tag{8}$$

which does not verify properties (2) and (3) of the metric definition. More important even, this "distance" is not always well defined. Kullback -Leibler divergence emerges as the exponent in the probability of error in a hypothesis test between distribution P and P' [8].

In spite of these previous examples of distances between probability distributions are very relevant, they do not exhaust the list of all the distances introduced in statistical and probabilistic studies (for a more complete list, see for example, reference [9]).

At this point a question naturally emerges: why do we need distances between probability distributions? The importance of suitable measures of distance between probability distributions arises because of the role they play in problems of statistics and physics. In the first case, problems of inference and discrimination, require of a notion of distance. In the second, in several problems of statistical and quantum mechanics we need to compare "objects" (for example symbolic sequences, signals or classical and quantum states). Applications of distances range from the development of error detecting and error correcting codes [10], hypothesis testing in statistics [11], and philogenetic studies [12, 13], on the one hand, and to compare the amount of information carried by quantum states [14], on the other. In the realm of statistical physics many relevant problems are stated in terms of distances between probability distributions. Here we will comment two examples of them.

1. Lesche introduced quite an interesting property of Boltzmann-Gibbs entropy, $H(P) = -\sum_j p_j \log p_j$. He named this property stability [15, 16]- One states that a physical

functional of a probability distribution, $\Gamma(P)$, is stable if and only if, for any given $\varepsilon > 0$, a $\delta_\varepsilon > 0$ exists such that $d_1(P, P') < \delta_\varepsilon \Rightarrow \frac{\Gamma(p_i)-\Gamma(p_i')}{\Gamma_{max}} < \varepsilon$. Within such a context, Lesche showed that (i) the Boltzmann-Gibbs entropy is stable but (ii) the Renyi entropy $S_\alpha^R = \log \sum_j \frac{p_j^\alpha}{\alpha-1}$ does not exhibit this feature for any $\alpha \neq 1$. Recently, the possibility of investigating the concomitant issue for (a) metrics other than $d_1(P, P')$ and (b) a non-extensive entropies, has also been investigated [17]

2. Another instance is provided by the López-Ruiz, Mancini and Calvet (LMC) measure of statistical complexity, which is based on the notion of "disequilibrium" [18]. The LMC definition of statistical complexity is given by $C^{LMC} = QH$ where Q stands for the so-called "disequilibrium": $Q(p_n) = Q_0 D(p_n, p_n^e)$ with D an appropriate measure of distance between the distribution p_n and the "equilibrium" one $p_n^e \equiv (\frac{1}{N+1}, \ldots, \frac{1}{N+1})$. Q_0 is a normalization constant, and H is the Shannon entropy. In the original work of López-Ruiz, Mancini and Calvet the square of the Euclidean distance $d_2^2(p_n, p_n^e) = \sum_{n=1}^N \{p_n - \frac{1}{N}\}^2$ was used but the use of an alternative disequilibrium measure has been proposed by other authors [19, 20].

This listing can be enlarged by including examples taken from classical and quantum information theory (see instance [21, 22, 23]).

During last years our research interest was focused in investigating an alternative to the previously mentioned distances: the Jensen Shannon divergence (JSD). Recently this quantity has been exhaustively applied in several problems of interest in theoretical physics and related topics.

Next sections are devoted to review the main properties of the classical and quantum versions of this distance.

2 Jensen-Shannon divergence

2.1 Classical JSD

The JSD was introduced in 1987 by C. Rao and generalized by J. Lin in 1991 as an alternative to the Kullback-Leibler divergence [24, 25]. To overcome the already mentioned limitations of this divergence, they proposed the quantity

$$\begin{aligned} JS(P,P') &\equiv \frac{1}{2}\left[K\left(P,\frac{P+P'}{2}\right)+K\left(P',\frac{P+P'}{2}\right)\right] \\ &= -\frac{1}{2}\left[\sum_j p_j \log\left(\frac{2p_j}{p_j+p'_j}\right)+\sum_j p'_j \log\left(\frac{2p'_j}{p_j+p'_j}\right)\right], \end{aligned} \tag{9}$$

as a measure of the distance between the probability distributions P and P'. In terms of the Shannon entropy $H[P] = -\sum_j p_j \log p_j$, expression (9) can be rewritten in the form:

$$JS(P,P') = H\left(\frac{P+P'}{2}\right) - \frac{1}{2}H(P) - \frac{1}{2}H(P') \tag{10}$$

By using logarithm in basis two, this quantity has units of bits.

The JSD is positive definite, symmetric, bounded and, unlike the Kullback-Leibler divergence, always well defined (in the sense that it can be evaluated even when $p_j \neq 0$ and $p'_j = 0$ for some j). More important yet, its square root verifies the triangle inequality (3) (but JS does not!!). Therefore, it can be concluded that $\sqrt{JS(P,P')}$ is a true metric for the probability distribution space [26].

The JSD can be interpreted in the frame of information theory [28] and in a Bayesian probabilistic sense [25]. This last fact allows to give to the JSD a very interesting physical interpretation: it has been proposed as a measure of the time asymmetry in microscopic systems [23] Incidentally, it is worth mentioning that for P close to P' the JSD coincides (up to numerical factors) with the Wootters and the Fubini-Study distances and, for the continuous case, is related to the Fisher information measure [27].

Lin has proposed a generalization of (9) for an arbitrary number of probability distributions [25]. Let $P^{(k)}$, $k = 1..M$, a set of M probability distributions. The JSD among these probabilities distributions is defined as:

$$JS[P^{(1)},..,P^{(M)}] = H[\sum_k \pi^{(k)} P^{(k)}] - \sum_k^M \pi^{(k)} H[P^{(k)}], \tag{11}$$

where the numbers $\pi^{(k)}$, $k = 1..M$, $\sum_k \pi^{(k)} = 1$, are weights properly chosen that allow to assign different relevance to each probability distribution.

The JSD has been extensively applied in different contexts, for example, as a tool for segmenting symbolic sequences [28, 29], for analyzing literary texts and musical score [30], as the disequilibrium for defining a non-triviality measure for chaotic maps [31], to study

EEG records [32], as a complexity measure for genomic sequences [33], to compare electron distributions in atoms [34], to define an alternative distinguishability criterium between pure quantum states [27], as a measure of discernibility between wave functions [35] and to compare atomic electronic distributions [36, 37].

As it was mentioned above, the JSD has shown to be an excellent tool for segmenting symbolic sequences, that is, to separate a given symbolic sequence in stationary subsequences. Let $\mathcal{A} = \{a_1, a_2, \ldots, a_M\}$ be an alphabet of M symbols and let $S = a_{m_1} a_{m_2} \cdots a_{m_L}$ be a sequence of characters from $\mathcal{A}$ which we think as partitioned in two sequences S_1 and S_2, each one built up from different probability distributions of occurrence of symbols a. Let us denote by $\alpha_i(\beta_i)$ the probability of occurrence of symbol a_i in S_1 (S_2). The lengths of S_1 and S_2 are L_1 and L_2 respectively and let $\pi_1 = L_1/L$ and $\pi_2 = L_2/L$ play the role of weights. In this way we can associate to the sequence S a value of the JSD:

$$JS^{(\pi_1,\pi_2)}(S,\mathcal{A}) = \sum_{j=1}^{M}(\pi_1\alpha_j + \pi_2\beta_j)\log(\pi_1\alpha_j + \pi_2\beta_j) + \pi_1\sum_{j=1}^{M}\alpha_j\log\alpha_j + \pi_2\sum_{j=1}^{M}\beta_j\log\beta_j$$

For a given (non-stationary) sequence which is divided in two disjoint subsequences, the quantities α_i and β_i are not directly observables. The observables are the frequencies of occurrence of symbols a_i in subsequence S_k:

$$f_i^{(k)} = \frac{N_i^{(k)}}{L_k} \tag{12}$$

where $N_i^{(k)}$ is the number of symbols a_i in the subsequence S_k. Thus in evaluating the divergence JS for a real sequence we first must calculate the frequencies $f_i^{(k)}$. In practice the segmentation procedure consists in separating a non-stationary sequence in two or more stationary sequences. In Ref. [28] a recursive segmentation algorithm has been developed. The basic idea is the following: let us suppose that we define a moving cursor along the complete sequence. For each position of the cursor (let us say, J), it results in two subsequences, one to the left and other to the right of the cursor. For each sub-sequence we can evaluate the occurrence frequencies given by (12). At the points where the maximum values of the DJS are attained, can be identified as the fusion points of the subsequences. For more details on this subject, see [28, 29].

2.2 Quantum JSD

Distances between quantum states play a central role in quantum information theory (see, for example, chapter 9 of reference [14]). Trace distance, fidelity and the Bures's metric are frequently used as measures of distance between quantum states. Let us recall that a quantum state is mathematically represented by a density operator acting over the Hilbert space of the system, that is, by an hermitian operator, positive definite and with trace equal to one.

An extensively referenced distance between quantum states, is the relative entropy, which is the natural generalization of the Kullback-Leibler divergence, eq. (8), to the framework of quantum theory [38, 39]:

$$K(\varrho\|\sigma) = Tr(\varrho \log \varrho - \varrho \log \sigma) \tag{13}$$

with ϱ and σ density operators, and where Tr stands for the trace operator. This measure has several inconveniences: it is not symmetric, not bounded and not always well defined (for example, $K(\varrho\|\sigma)$ is not defined when σ represents a pure state). To be precise it should be said that in order that $K(\varrho\|\sigma) < \infty$ it is necessary that $ker(\sigma) \subset ker(\varrho)$, where ker denotes the kernel of the operator (the kernel of an operator is the subspace spanned by eigenvectors corresponding to null eigenvalues).

Following the steps that lead to the definition of the JSD from the Kullback-Leibler divergence (first line of eq. (9)), we can introduce the quantum JSD in the form [40]:

$$JS(\varrho\|\sigma) \equiv \frac{1}{2}\left[K\left(\varrho\|\frac{\varrho+\sigma}{2}\right) + K\left(\sigma\|\frac{\varrho+\sigma}{2}\right)\right] \tag{14}$$

which, in terms of the von-Neumann entropy, $H_N(\varrho) = -Tr(\varrho \log \varrho)$, can be rewritten in the form

$$JS(\varrho\|\sigma) = H_N\left(\frac{\varrho+\sigma}{2}\right) - \frac{1}{2}H_N(\varrho) - \frac{1}{2}H_N(\sigma) \tag{15}$$

This measure is positive definite, symmetric, bounded, and always well defined; in particular the condition over the kernels of the operators ϱ and σ can be removed. It has all the adequate dynamical properties for a distance between quantum states [40].

If ϱ and σ are density operators with complete sets of eigenvectors $\{|r_i\rangle\}$ and $\{|s_i\rangle\}$, such that $\varrho = \sum_i r_i|r_i\rangle\langle r_i|$ and $\sigma = \sum_i s_i|s_i\rangle\langle s_i|$ then the QJSD between ϱ and σ is expressed as

$$JS(\varrho\|\sigma) = \frac{1}{2}\left[\sum_{k,i} |\langle t_k|r_i\rangle|^2 r_i \log\left(\frac{2r_i}{\tau_k}\right) + \sum_{k,i} |\langle t_k|s_i\rangle|^2 s_i \log\left(\frac{2s_i}{\tau_k}\right)\right]$$

with $\tau_k = \sum_i r_i |\langle t_k | r_i \rangle|^2 + \sum_i s_i |\langle t_k | s_i \rangle|^2$ and $|t_k\rangle$ a complete set of normalized eigenvectors of $\varrho + \sigma$.

Recently the metric character of $\sqrt{JS(\varrho \| \sigma)}$ has been proved [41, 42].

When the quantum JSD is generalized as a distance among several quantum states, according to the recipe (11), the resulting quantity has a very noticeable interpretation: it gives the upper bound of the mutual information that can be gathered between two quantum computing operators (Holevo bound) [40].

To conclude this chapter, we present an application of the QJSD to the definition of a measure of the degree of entanglement.

3 QSJD as a measure of entanglement

Here we investigate the ability of the QJSD to serve as a measure of the degree of entanglement. In this section we study its properties as an entanglement measure and we apply it to quantify the degree of entanglement of different families of two-qubit mixed states.

One of the most distinctive features of quantum mechanics is that of entanglement, which in recent years has been the focus of an intense and increasing research activity (see for instance [14, 43, 44, 45, 46] and references therein). From a fundamental point of view entanglement has a key role in our understanding of quantum mechanics [47, 48] and on the other hand it has important technological applications such as quantum computation [14] and quantum metrology [49]. A state of a composite quantum system is called entangled if it can not be expressed as a convex sum of factorizable pure states. Otherwise, the state is called separable.

The canonical entanglement measure for bipartite pure states is the entropy of entanglement, however the amount of entanglement associated with a given mixed state is rather more complex. Many measures have been defined to quantify the amount of entanglement of mixed states, all of them trying to capture different features of this phenomenon.

Let us now enumerate the properties that any adequate measure of entanglement should satisfy [43]:

- Discrimination: $\mathcal{E}(\varrho) = 0 \ iff \ \varrho$ is separable.
- Monotonicity: the measure does not increase under local general measurements and classical communication, for every completely positive map Φ, $\mathcal{E}(\Phi\varrho) \leq \mathcal{E}(\varrho)$

- Convexity: $\mathcal{E}(x\varrho + (1-x)\sigma) \leq x\mathcal{E}(\varrho) + (1-x)\mathcal{E}(\sigma)$, with $x \in [0,1]$.

One widely used measure of entanglement is the entanglement of formation, however only for two-qubit systems exist a closed mathematical expression that allowed us to compute it.

$$E_F(\varrho) = h\left(\frac{1+\sqrt{1-C^2(\varrho)}}{2}\right) \tag{16}$$

with $h(x) = x\log_2 x - (1-x)\log_2(1-x)$. Entanglement of formation is a strictly monotonic function of the squared concurrence

$$C(\varrho) = max\{0, \lambda_1 - \lambda_2 - \lambda_3 - \lambda_4\}, \tag{17}$$

where λ_i are in decreasing order the square root of the eigenvalues of the matrix $\varrho(\sigma_y \otimes \sigma_y)\varrho^*(\sigma_y \otimes \sigma_y)$, being ϱ^* the element-wise complex conjugate of ϱ, and σ_y the Pauli matrix. The computation of the concurrence for mixed states with dimension higher than two is extremely hard.

There exist entanglement measures based on several distance measures. The underlying idea is that the entanglement of a given state ϱ can be properly determined by the minimum distance between ϱ and the set of separable states. The most popular measure used to induce a *geometrical* measure of entanglement is the quantum relative entropy [50, 51]. As was already said the quantum relative entropy is asymmetric, unbounded and it is even not a proper metric. Different distance measures have also been used such as the trace distance [52] the Bures metric ($\mathcal{E}_B(\varrho)$) [50], and the Hilbert-Schmidt metric [53]. These distances, while fulfilling most of the properties required to any adequate distance measure, lack of the meaning that an entropic based measure (as the relative entropy) can offer to the quantum information theory field.

We define an entanglement measure $\mathcal{E}(\varrho)$ as the minimum QJSD from the state ϱ to the set $\mathcal{D}$ of the disentangled states.

$$\mathcal{E}_{JS}(\varrho) = \min_{\sigma\in\mathcal{D}} JS(\varrho, \sigma). \tag{18}$$

The QJSD fulfills the conditions for an adequate entanglement measure and as it was mentioned it has interesting application in quantum information theory [40].

Here we normalize the entanglement measure by a trivial re-scaling to compare different quantities of interest, *i.e.*, $\mathcal{E}(\varrho) = \mathcal{E}(\varrho)/\mathcal{E}(|\psi^-\rangle\langle\psi^-|)$.

In order to investigate the behavior of the QJSD as an entanglement measure we consider now three well-known families of two-qubits mixed states, namely,

- Werner ones [54],
- maximally entangled mixed ones [55] (MEM), and
- parametric down-conversion [56] (PDC) states.

All of these are diagonal in the Bell basis. We perform an optimization procedure in order to find the separable state that lies in the closest proximity to the state of interest ϱ. Such task is performed by following a simulated annealing minimization procedure, starting from the maximally mixed state ϱ_{MM} and taking advantage of the fact that $\mathcal{D}$ is a convex set. Appropriately perturbing ϱ_{MM} we look for the minimal QJSD (and Bures metric) between ϱ and some state in $\mathcal{D}$.

In Fig. 1 we depict, *versus* the linear entropy S_L (the degree of mixedness), i) the concurrence and the two normalized quantities: ii) our $\mathcal{E}_{JS}$ and iii) Bures metric induced $\mathcal{E}_B$ measure. This is done for the three families enumerated above. Note that $\mathcal{E}_{JS}$ curve lies always below the other two for the three families, except for the extreme (trivial) situations. The behavior here described mimics that of the relative entropy of entanglement vs. S_L. In that instance, the relative entropy-induced entanglement is strictly smaller than that of formation, as analytically demonstrated in [51].

The main point is the fact that the measure induced by the QJSD is easier to compute than that induced by the Bures metric. This is so because of the square root in the definition of Bures' metric, which forces cumbersome matrix diagonalization and basis' changes in order to compute $\mathcal{E}_B$. Instead, so as to calculate the QJSD one i) only needs to solve the eigenvalue-equation for three matrices (eigenvectors are not necessary) and ii) the optimization procedure is much more efficient than in the Bures situation.

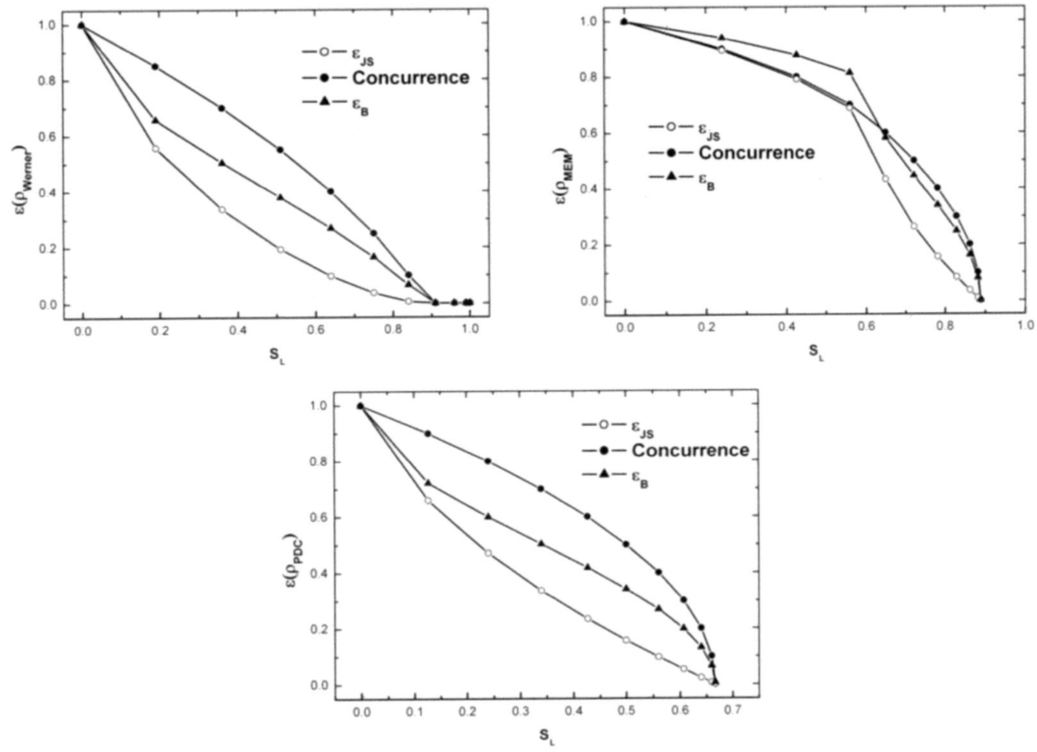

Figure 1: Comparison between the concurrence, the JS- measure, and the measure $\mathcal{E}_B$ induced by the Bures metric. The task is accomplished for (a) Werner, (b) MEM, and (c) PDC states.

Conflict of interest: None.

Acknowledgments

The authors are grateful to the Secretaria de Ciencia y Técnica de la Universidad Nacional de Córdoba, Argentina, for financial assistance. APM acknowledges support from Project FQM-2445, Spain.

References

[1] A.N. Shiryaev, *Probability*, Second Edition, Springer-Verlag, New York, 1996.

[2] R.M. Wald, *General Relativity*, The University of Chicago Press, Chicago, 1984.

[3] M. Berger, *A Panoramic View of Riemannian Geometry*, Springer-Verlag, Heidelberg, 2003.

[4] P.C. Mahalanobis, On the generalized distance in statistics,*Proc. Nat. Inst. Sci.*, India **2**, 49-55, 1936.

[5] W. Wootters, Statistical Distance and Hilbert Space, *Phys. Rev D* **23**, 357-362, 1981.

[6] S. Amari *Differential-Geometrical Methods in Statisitcs*, Lectures notes in Statistics, Springer-Verlag, Heidelberg, 1990.

[7] S. Kullback and R.A. Leibler, On Information and Sufficiency, *Ann. Math. Stat.* **22**, 79-86, 1951.

[8] T. Cover and J.A. Thomas, *Elements of Information Theory*, Second Edition, Wiley-Interscience, New Jersey, 2006.

[9] E. Deza and M. M. Deza, *Dictionary of Distances.* Fifth Edition, Elsevier, Holland (2006).

[10] R.V. Hamming, Error detecting and error correcting codes, *Bell Sys. Tech. Journal* **29**, 147-160, 1950.

[11] P.M. Robinson, Consistent nonparametric entropy-based testing, *Rev. Econ. Stud* **88**, 437-453, 1991.

[12] M. Li, X. Chen, X. Li, B. Ma and P. M. N. Vitányi, The similarity metric, *IEEE Trans. Inf. Theory* **50**, 3250-3264, 2004.

[13] M. Li, J.H. Badger, X. Chen, S. Kwong, P. Kearney and H. Zhang, An information-based sequence distance and its application to whole mitochondrial genome, *Bioinformatics* **17**, 149-154, 2001.

[14] M. Nielsen and I. Chuang, *Quantum Computation and Quantum Information*, Cambridge University Press, Cambridge, 2000.

[15] B. Lesche, Inestability of Rényi entropy, *J. Stat. Phys.* **27**, 419-423, 1982.

[16] B. Lesche, Rényi entropies and observables, *Phys. Rev. E* **70**, 017102, 2004.

[17] C. Tsallis and E. Briganti, Continuum Mech. Thermodyn. **16**, 223-235, 2004.

[18] R. Lopez-Ruiz, H. Mancini and X. Calbet, A statistical measure of complexity, *Phys. Lett. A* **209**, 321-326 (1995).

[19] P.W. Lamberti, M.T. Martin, A. Plastino, O.A. Rosso, Intensive entropic non-triviality measure, *Physica A* **334**, 119, 2004.

[20] O. A. Rosso, H. A. Larrondo, M. T. Martin, A. Plastino, and M. A. Fuentes, Distinguishing noise from chaos, *Phys. Rev. Lett.* **99**, 154102, 2007.

[21] A. Daffertshofer, A. R. Plastino and A. Plastino, Classical no-cloning theorem, *Phys. Rev. Lett.* **88**, 210601 (2002).

[22] V. Vedral, M.B. Plenio, M.A. Rippin and P.L. Knight, Quantifying entanglement, *Phys. Rev. Lett* **78**, 2275-2278, 1997.

[23] E. H. Feng and G. E. Crooks, Length of time's arrow, *Phys. Rev. Lett.* **101**, 090602, 2008.

[24] C.R. Rao, *Differential Geometry in Statistical Inference, IMS-Lectures Notes* **10** 217-225, 1987.

[25] J. Lin, Divergence Measures based on the Shannon Entropy, *IEEE Trans. Inf. Theory* **37**, 145-151, 1991.

[26] J. Schindelin and D. Endres, A new metric for probability distributions, *IEEE Trans. Inform. Theory* **49**, 1858-1860 (2003).

[27] A. Majtey, P.W. Lamberti, M.T. Martin and A. Plastino, Wootters's distance revisited: a new distinguishability criterium *Eur. Phys. J. D* **32**, 413-418 (2005).

[28] I. Grosse, P. Bernaola-Galvan, P. Carpena, R. Roman Roldan, J. Oliver & H.E. Stanley, Analysis of symbolic sequences using the Jensen-Shannon divergence, *Phys. Rev. E* **65**, 041905, 2002.

[29] P.W. Lamberti and A.Majtey, Non-logarithmic Jensen-Shannon divergence *Physica A*, **329**, 81-86, 2003.

[30] D. Zanette, Segmentation and context of literary and musical sequences, *Complex Systems*, **17**, 279-293, 2007.

[31] P.W. Lamberti, M.T. Martin, A. Plastino and O. Rosso, *Physica A* **334**, 119-126, 2004.

[32] M.E. Pereyra, P.W. Lamberti and O. Rosso, Wavelet Jensen-Shannon divergence as a tool for studying the dynamics of frequency band components in EEG epileptic seizure, *Physica A* **379**, 122-132, 2007.

[33] R. Román-Roldán, P. Bernaola-Galván and J. Oliver, Sequence compositional complexity of DNA through an entropic segmentation method *Phys. Rev. Lett.* **80**, 1344-1347, 1998.

[34] K. Ch. Chatzisavvas, Ch. C. Moustakidis and C.P. Panos, *J. Chem. Phys.* **123**, 174111 (2005).

[35] M. Casas, P.W. Lamberti, A. Plastino, A.R. Plastino and G. Roston, *Il Nuovo Cimento* **120B**, 521-532, 2005.

[36] J. Antoln, J.C. Angulo, S. Lpez-Rosa, *J. Chem. Phys* **130**, 074550, 2009.

[37] J.C. Angulo, J. Antoln, S. Lpez-Rosa, R. O. Esquivel, *Physica A* **389**, 899, 2010.

[38] G. Lindblad, Entropy, information and quantum measurements, *Comm. Math. Phys.* **33**, 305-322, 1973.

[39] V. Vedral, The role of relative entropy in quantum information theory, *Rev. Mod. Phys.* **74**, 197-234, 2002.

[40] A. Majtey, P.W. Lamberti and D. Prato, Jensen-Shannon divergence as a measure of distinguishability between mixed quantum states, *Phys. Rev. A* **72** 052310, 2005.

[41] P.W. Lamberti, A. Majtey, A. Borras, M. Casas and A. Plastino, The metric character of quantum Jensen-Shannon divergence, *Phys. Rev. A* **77**, 052311, 2008.

[42] J. Briët and P. Harremoës, Properties of classical and quantum Jensen-Shannon divergence *Phys. Rev. A* **79**, 052311, 2009.

[43] I. Bengtsson and K. Życzkowski, *Geometry of Quantum States: An Introduction to Quantum Entanglement*, Cambridge University Press, Cambridge, 2006.

[44] H.-K. Lo, S. Popescu, T. Spiller (Eds.), *Introduction to Quantum Computation and Information*, World Scientific, River Edge, 1998.

[45] D. Bouwmeester, A. Ekert, A. Zeilinger (Eds.), *The Physics of Quantum Information*, Springer, Berlin, 2000.

[46] M. Schlosshauer, Decoherence, the measurement problem, and interpretations of quantum mechanics, *Rev. Mod. Phys.* **76**, 1267, 2005.

[47] J. Gemmer, M. Michel, and G. Mahler, *Quantum Thermodynamics* Springer-Verlag, Berlin, 2004.

[48] S. Popescu, A.J. Short, and A. Winter, Entanglement and the Foundations of Statistical Mechanics, *Nature Physics* **2**, 754, 2006.

[49] V. Giovannetti, S. Lloyd, and L. Maccone, Quantum Metrology, *Phys. Rev. Lett.* **96**, 010401, 2006.

[50] V. Vedral, M.B. Plenio, K. Jacobs and P.L. Knight, Statistical Inference, Destinguishability of Quantum States, and Quantum Entanglement, *Phys. Rev. A* **56**, 4452 (1997).

[51] V. Vedral and M.B. Plenio, Entanglement measures and purification procedures, *Phys. Rev. A* **57** 1619, 1998.

[52] J. Eisert, K. Audenaert, M. B. Plenio, Remarks on entanglement measures and non-local state distinguishability, *J. Phys. A* **36**, 5605, 2003.

[53] C. Witte, M. Trucks, Entanglement measure induced By the Hilbert-Schmidt norm, *Phys. Lett. A* **257**, 14, 1999.

[54] R.F. Werner, Quantum states with Einstein-Rosen-Podolsky correlations admitting a hidden- variable model, *Phys. Rev. A* **40**, 4277, 1989.

[55] W.J. Munro, D.F.V. James, A.G. White, P.G. Kwiat, Phys. Rev. A **64**, 030302, 2001. M. Barbieri, F. De Martini, G. Di Nepi, and P. Mataloni, Phys. Rev. Lett. **92**, 177901, 2004.

[56] A. Cabello, A. Feito, A. Lamas-Linares, Bell's inequalities with realistic noise for polarization-entangled photons, *Phys. Rev. A* **72**, 052112, 2005.

Concepts and Recent Advances in Generalized Information Measures and Statistics, 2013, 147-168

CHAPTER 7

A Statistical Measure of Complexity

Ricardo López-Ruiz [a,*], Héctor Mancini [b] and Xavier Calbet [c]

[a] DIIS and BIFI, Facultad de Ciencias, Universidad de Zaragoza, E-50009 Zaragoza, Spain

[b] Department of Physics and Applied Mathematics, Institute of Physics, Universidad de Navarra, 31080 Pamplona, Spain

[c] EUMETSAT, Eumetsat Allee 1, D-64295 Darmstadt, Germany

Abstract

In this chapter, a statistical measure of complexity is introduced and some of its properties are discussed. Also, some straightforward applications are shown.

Keywords: Measures of Complexity; Information Theory; Statistical Indicators; Nonequilibrium Systems.

1 Shannon Information

Entropy plays a crucial theoretical role in physics of macroscopic equilibrium systems. The probability distribution of accessible states of a constrained system in equilibrium can be found by the inference principle of maximum entropy [1]. The macroscopic magnitudes and the laws that relate them can be calculated with this probability distribution by standard statistical mechanics techniques.

The same scheme could be thought for extended systems far from equilibrium, but in this case we do have neither a method to find the probability distribution nor the knowledge of the relevant magnitudes bringing the information that can predict the system's behavior. It is not the case, for instance, with the metric properties of low dimensional chaotic systems by means of the Lyapunov exponents, invariant measures and fractal dimensions [2].

Shannon information or entropy H [3] can still be used as a magnitude in a general situation with N accessible states:

$$H = -K \sum_{i=1}^{N} p_i \log p_i \qquad (1)$$

* Corresponding author. E-mail: rilopez@unizar.es

Andres M. Kowalski, Raul D. Rossignoli and Evaldo M. F. Curado (Eds)

with K a positive real constant and p_i the normalized associated probabilities, $\sum_{i=1}^{N} p_i = 1$. An isolated system in equilibrium presents equiprobability, $p_i = 1/N$ for all i, among its accessible states and this is the situation of maximal entropy,

$$H_{max} = K \log N. \tag{2}$$

If the system is out of equilibrium, the entropy H can be expanded around this maximum H_{max}:

$$H(p_1, p_2, \ldots, p_N) = K \log N - \frac{NK}{2} \sum_{i=1}^{N} \left(p_i - \frac{1}{N}\right)^2 + \ldots = H_{max} - \frac{NK}{2} D + \ldots \tag{3}$$

where the quantity $D = \sum_i (p_i - 1/N)^2$, that we call *disequilibrium*, is a kind of distance from the actual system configuration to the equilibrium. If the expression (3) is multiplied by H we obtain:

$$H^2 = H \cdot H_{max} - \frac{NK}{2} H \cdot D + K^2 f(N, p_i), \tag{4}$$

where $f(N, p_i)$ is the entropy multiplied by the rest of the Taylor expansion terms, which present the form $\frac{1}{N} \sum_i (Np_i - 1)^m$ with $m > 2$. If we rename $C = H \cdot D$,

$$C = cte \cdot H \cdot (H_{max} - H) + K\bar{f}(N, p_i), \tag{5}$$

with $cte^{-1} = NK/2$ and $\bar{f} = 2f/N$. The idea of distance for the disequilibrium is now clearer if we see that D is just the real distance $D \sim (H_{max} - H)$ for systems in the vicinity of the equiprobability. In an ideal gas we have $H \sim H_{max}$ and $D \sim 0$, then $C \sim 0$. Contrarily, in a crystal $H \sim 0$ and $D \sim 1$, but also $C \sim 0$. These two systems are considered as classical examples of simple models and are extrema in a scale of disorder (H) or disequilibrium (D) but those should present null complexity in a hypothetic measure of *complexity*. This last asymptotic behavior is verified by the variable C (Fig. 1) and C has been proposed as a such magnitude [4]. We formalize this simple idea recalling the recent definition of *LMC complexity* in the next section.

Let us see another important property [5] arising from relation (5). If we take the time derivative of C in a neighborhood of equilibrium by approaching $C \sim H(H_{max} - H)$, then we have

$$\frac{dC}{dt} \sim -H_{max} \frac{dH}{dt}. \tag{6}$$

The irreversibility property of H implies that $\frac{dH}{dt} \geq 0$, the equality occurring only for the equipartition, therefore

$$\frac{dC}{dt} \leq 0. \tag{7}$$

Hence, in the vicinity of H_{max}, LMC complexity is always decreasing on the evolution path towards equilibrium, independently of the kind of transition and of the system under study. This does not forbid that complexity can increase when the system is very far from equilibrium. In fact this is the case in a general situation as it can be seen, for instance, in the gas system presented in Ref. [6].

2 A Statistical Complexity Measure

On the most basic grounds, an object, a procedure, or system is said to be "complex" when it does not match patterns regarded as simple. This sounds rather like an oxymoron but common knowledge tells us what is simple and complex: simplified systems or idealizations are always a starting point to solve scientific problems. The notion of "complexity" in physics [7, 8] starts by considering the perfect crystal and the isolated ideal gas as examples of simple models and therefore as systems with zero "complexity". Let us briefly recall their main characteristics with "order", "information" and "equilibrium".

A perfect crystal is completely ordered and the atoms are arranged following stringent rules of symmetry. The probability distribution for the states accessible to the perfect crystal is centered around a prevailing state of perfect symmetry. A small piece of "information" is enough to describe the perfect crystal: the distances and the symmetries that define the elementary cell. The "information" stored in this system can be considered minimal. On the other hand, the isolated ideal gas is completely disordered. The system can be found in any of its accessible states with the same probability. All of them contribute in equal measure to the "information" stored in the ideal gas. It has therefore a maximum "information". These two simple systems are extrema in the scale of "order" and "information". It follows that the definition of "complexity" must not be made in terms of just "order" or "information".

It might seem reasonable to propose a measure of "complexity" by adopting some kind of distance from the equiprobable distribution of the accessible states of the system [4]. Defined in this way, "disequilibrium" would give an idea of the probabilistic hierarchy of the system. "Disequilibrium" would be different from zero if there are privileged, or more probable, states among those accessible. But this would not work. Going back to the two examples we began with, it is readily seen that a perfect crystal is far from an equidistribution among the accessible states because one of them is totally prevailing, and so "disequilibrium" would be maximum. For the ideal gas, "disequilibrium" would be zero by construction. Therefore such a distance or "disequilibrium" (a measure of a probabilistic hierarchy) cannot be directly associated with "complexity".

In Fig. 1 we sketch an intuitive qualitative behavior for "information" H and "disequilibrium" D for systems ranging from the perfect crystal to the ideal gas. As indicated in the former section, this graph suggests that the product of these two quantities could be used as a measure of "complexity": $C = H \cdot D$. The function C has indeed the features and asymptotic properties that one would expect intuitively: it vanishes for the perfect crystal and for the isolated ideal gas, and it is different from zero for the rest of the systems of particles. We will follow these guidelines to establish a quantitative measure of "complexity".

Before attempting any further progress, however, we must recall that "complexity" cannot be measured univocally, because it depends on the nature of the description (which always involves a reductionist process) and on the scale of observation. Let us take an example to illustrate this point. A computer chip can look very different at different scales. It is an entangled array of electronic elements at microscopic scale but only an ordered set of pins attached to a black box at a macroscopic scale.

We shall now discuss a measure of "complexity" based on the statistical description of systems. Let us assume that the system has N accessible states $\{x_1, x_2, ..., x_N\}$ when observed at a given scale. We will call this an N-system. Our understanding of the behavior of this system determines the corresponding probabilities $\{p_1, p_2, ..., p_N\}$ (with the condition $\sum_{i=1}^{N} p_i = 1$) of each state ($p_i > 0$ for all i). Then the knowledge of the underlying physical laws at this scale is incorporated into a probability distribution for the accessible states. It is possible to find a quantity measuring the amount of "information". As presented in the former section, under to the most elementary conditions of consistency, Shannon [3] determined the unique function $H(p_1, p_2, ..., p_N)$ given by expression (1), that accounts for the "information" stored in a system, where K is a positive constant. The quantity H is called *information*. The redefinition of information H as some type of monotone function of the Shannon entropy can be also useful in many contexts. In the case of a crystal, a state x_c would be the most probable $p_c \sim 1$, and all others x_i would be very improbable, $p_i \sim 0$ $i \neq c$. Then $H_c \sim 0$. On the other side, equiprobability characterizes an isolated ideal gas, $p_i \sim 1/N$ so $H_g \sim K \log N$, *i.e.*, the maximum of information for a N-system. (Notice that if one assumes equiprobability and $K = \kappa \equiv$ *Boltzmann constant*, H is identified with

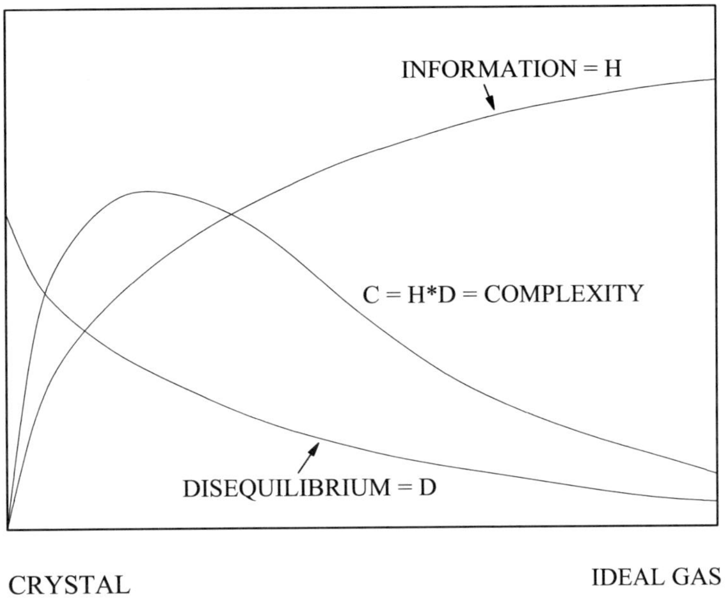

Figure 1: Sketch of the intuitive notion of the magnitudes of "information" (H) and "disequilibrium" (D) for the physical systems and the behavior intuitively required for the magnitude "complexity". The quantity $C = H \cdot D$ is proposed to measure such a magnitude.

the thermodynamic entropy, $S = \kappa \log N$). Any other N-system will have an amount of information between those two extrema.

Let us propose a definition of *disequilibrium* D in a N-system [9]. The intuitive notion suggests that some kind of distance from an equiprobable distribution should be adopted. Two requirements are imposed on the magnitude of D: $D > 0$ in order to have a positive measure of "complexity" and $D = 0$ on the limit of equiprobability. The straightforward solution is to add the quadratic distances of each state to the equiprobability as follows:

$$D = \sum_{i=1}^{N} \left(p_i - \frac{1}{N} \right)^2 . \tag{8}$$

According to this definition, a crystal has maximum disequilibrium (for the dominant state, $p_c \sim 1$, and $D_c \to 1$ for $N \to \infty$) while the disequilibrium for an ideal gas vanishes ($D_g \sim 0$) by construction. For any other system D will have a value between these two extrema.

We now introduce the definition of *complexity* C of a N-system [4, 10]. This is simply the interplay between the information stored in the system and its disequilibrium:

$$C = H \cdot D = -\left(K \sum_{i=1}^{N} p_i \log p_i\right) \cdot \left(\sum_{i=1}^{N} \left(p_i - \frac{1}{N}\right)^2\right) . \tag{9}$$

This definition fits the intuitive arguments. For a crystal, disequilibrium is large but the information stored is vanishingly small, so $C \sim 0$. On the other hand, H is large for an ideal gas, but D is small, so $C \sim 0$ as well. Any other system will have an intermediate behavior and therefore $C > 0$.

As was intuitively suggested, the definition of complexity (9) also depends on the *scale*. At each scale of observation a new set of accessible states appears with its corresponding probability distribution so that complexity changes. Physical laws at each level of observation allow us to infer the probability distribution of the new set of accessible states, and therefore different values for H, D and C will be obtained. The straightforward passage to the case of a continuum number of states, x, can be easily inferred. Thus we must treat with probability distributions with a continuum support, $p(x)$, and normalization condition $\int_{-\infty}^{+\infty} p(x)dx = 1$. Disequilibrium has the limit $D = \int_{-\infty}^{+\infty} p^2(x)dx$ and the complexity could be defined by:

$$C = H \cdot D = -\left(K \int_{-\infty}^{+\infty} p(x) \log p(x) dx\right) \cdot \left(\int_{-\infty}^{+\infty} p^2(x) dx\right) . \tag{10}$$

Other possibilities for the continuous extension of C are also possible. For instance, a successful attempt of extending the LMC complexity for continuous systems has been performed in Ref. [11]. When the number of states available for a system is a continuum then the natural representation is a continuous distribution. In this case, the entropy can become negative. The positivity of C for every distribution is recovered by taking the exponential of H [12]. If we define $\hat{C} = \hat{H} \cdot D = e^H \cdot D$ as an extension of C to the continuous case interesting properties characterizing the indicator $\hat{C}$ appear. Namely, its invariance under translations, rescaling transformations and replication convert $\hat{C}$ in a good candidate to be considered as an indicator bringing essential information about the statistical properties of a continuous system.

Direct simulations of the definition give the values of C for general N-systems. The set of all the possible distributions $\{p_1, p_2, ..., p_N\}$ where an N-system could be found is sampled. For the sake of simplicity H is normalized to the interval $[0, 1]$. Thus $H =$

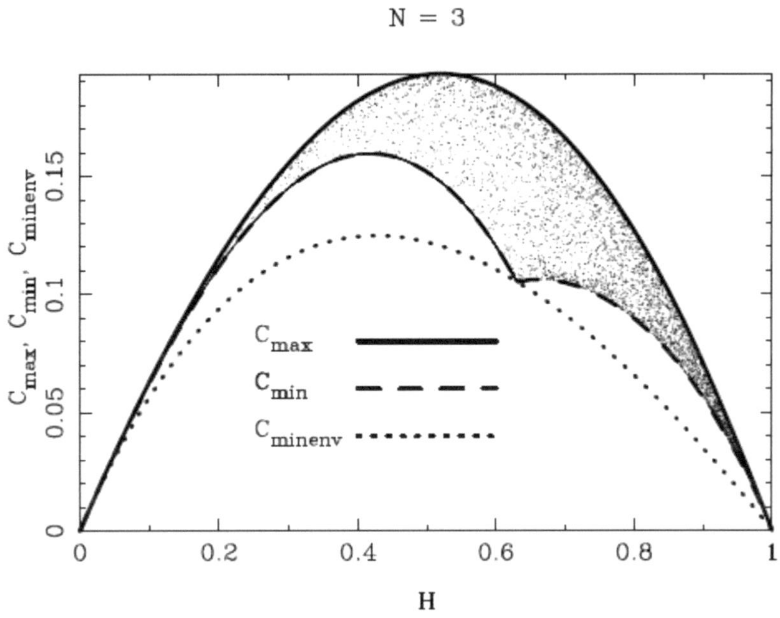

Figure 2: In general, dependence of complexity (C) on normalized information (H) is not univocal: many distributions $\{p_i\}$ can present the same value of H but different C. This is shown in the case $N = 3$.

$\sum_{i=1}^{N} p_i \log p_i / \log N$. For each distribution $\{p_i\}$ the normalized information $H(\{p_i\})$, and the disequilibrium $D(\{p_i\})$ (Eq. 8) are calculated. In each case the normalized complexity $C = H \cdot D$ is obtained and the pair (H, C) stored. These two magnitudes are plotted on a diagram $(H, C(H))$ in order to verify the qualitative behavior predicted in Fig. 1. For $N = 2$ an analytical expression for the curve $C(H)$ is obtained. If the probability of one state is $p_1 = x$, that of the second one is simply $p_2 = 1 - x$. Complexity vanishes for the two simplest 2-systems: the crystal ($H = 0$; $p_1 = 1$, $p_2 = 0$) and the ideal gas ($H = 1$; $p_1 = 1/2$, $p_2 = 1/2$). Let us notice that this curve is the simplest one that fulfills all the conditions discussed in the introduction. The largest complexity is reached for $H \sim 1/2$ and its value is: $C(x \sim 0.11) \sim 0.151$. For $N > 2$ the relationship between H and C is not univocal anymore. Many different distributions $\{p_i\}$ store the same information H but have different complexity C. Fig. 2 displays such a behavior for $N = 3$. If we take the maximum complexity $C_{max}(H)$ associated with each H a curve similar to the one for a 2-system is recovered. Every 3-system will have a complexity below this line and upper the line of $C_{min}(H)$ and also upper the minimum envelope complexity $C_{\rm minenv}$. These lines will be analytically found in a next section. In Fig. 3 curves $C_{max}(H)$ for the cases $N = 3, \ldots, 10$ are also shown. Let us observe the shift of the complexity-curve peak to smaller values of entropy for rising N. This fact agrees with the intuition telling us that the biggest complexity (number of possibilities of 'complexification') be reached for lesser entropies for the systems with bigger number of states.

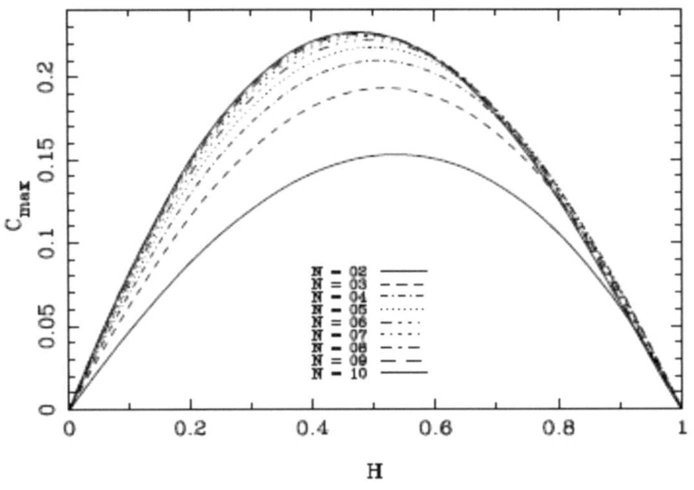

Figure 3: Complexity ($C = H \cdot D$) as a function of the normalized information (H) for a system with two accessible states ($N = 2$). Also curves of maximum complexity (C_{max}) are shown for the cases: $N = 3, \ldots, 10$.

Let us return to the point at which we started this discussion. Any notion of complexity in physics [7, 8] should only be made on the basis of a well defined or operational magnitude [4, 10]. But two additional requirements are needed in order to obtain a good definition of complexity in physics: (1) the new magnitude must be measurable in many different physical systems and (2) a comparative relationship and a physical interpretation between any two measurements should be possible.

Many different definitions of complexity have been proposed to date, mainly in the realm of physical and computational sciences. Among these, several can be cited: algorithmic complexity (Kolmogorov-Chaitin) [13, 14], the Lempel-Ziv complexity [15], the logical depth of Bennett [16], the effective measure complexity of Grassberger [17], the complexity of a system based in its diversity [18], the thermodynamical depth [19], the ϵ-machine complexity [20] , the physical complexity of genomes [21], complexities of formal grammars, *etc.* The definition of complexity (9) proposed in this section offers a new point of view, based on a statistical description of systems at a given *scale*. In this scheme, the knowledge of the physical laws governing the dynamic evolution in that scale is used to find its accessible states and its probability distribution. This process would immediately indicate the value of complexity. In essence this is nothing but an interplay between the information stored by the system and the *distance from equipartition* (measure of a probabilistic hierarchy between the observed parts) of the probability distribution of its accessible states. Besides giving the main features of a "intuitive" notion of complexity, we will show in this chapter that we can go one step further and that it is possible to compute this quantity in relevant physical situations [6, 22, 23]. The most important point is that the new definition successfully enables us to discern situations regarded as complex.

3 LMC Complexity: Extremal Distributions

Now we proceed to calculate the distributions which maximize and minimize the LMC complexity and its asymptotic behavior [6].

Let us assume that the system can be in one of its N possible accessible states, i. The probability of the system being in state i will be given by the discrete distribution function, $f_i \geq 0$, with the normalization condition $I \equiv \sum_{i=1}^{N} f_i = 1$. The system is defined such that, if isolated, it will reach equilibrium, with all the states having equal probability, $f_e = \frac{1}{N}$. Since we are supposing that H is normalized, $0 \leq H \leq 1$, and $0 \leq D \leq (N-1)/N$, then complexity, C, is also normalized, $0 \leq C \leq 1$.

When an isolated system evolves with time, the complexity cannot have any possible value in a C *versus* H map as it can be seen in Fig. 2, but it must stay within certain bounds, $C_{\max}$ and $C_{\min}$. These are the maximum and minimum values of C for a given H. Since $C = D \cdot H$, finding the extrema of C for constant H is equivalent to finding the extrema of D.

There are two restrictions on D: the normalization, I, and the fixed value of the entropy, H. To find these extrema undetermined Lagrange multipliers are used. Differentiating expressions of D, I and H, we obtain

$$\frac{\partial D}{\partial f_j} = 2(f_j - f_e)\,, \qquad (11)$$

Table 1: Probability values, f_j, that give a maximum of disequilibrium, $D_{\max}$, for a given H.

Number of states with f_j	f_j	Range of f_j
1	$f_{\max}$	$\frac{1}{N} \ldots 1$
$N-1$	$\frac{1-f_{\max}}{N-1}$	$0 \ldots \frac{1}{N}$

$$\frac{\partial I}{\partial f_j} = 1, \tag{12}$$

$$\frac{\partial H}{\partial f_j} = -\frac{1}{\ln N}(\ln f_j + 1)\,. \tag{13}$$

Defining λ_1 and λ_2 as the Lagrange multipliers, we get:

$$2(f_j - f_e) + \lambda_1 + \lambda_2(\ln f_j + 1)/\ln N = 0\,. \tag{14}$$

Two new parameters, α and β, which are a linear combinations of the Lagrange multipliers are defined:

$$f_j + \alpha \ln f_j + \beta = 0\,, \tag{15}$$

where the solutions of this equation, f_j, are the values that minimize or maximize the disequilibrium.

In the maximum complexity case there are two solutions, f_j, to Eq. (15) which are shown in Table 1. One of these solutions, $f_{\max}$, is given by

$$H = -\frac{1}{\ln N}\left[f_{\max}\ln f_{\max} + (1-f_{\max})\ln\left(\frac{1-f_{\max}}{N-1}\right)\right]\,, \tag{16}$$

and the other solution by $(1-f_{\max})/(N-1)$. The maximum disequilibrium, $D_{\max}$, for a fixed H is

$$D_{\max} = (f_{\max} - f_e)^2 + (N-1)\left(\frac{1-f_{\max}}{N-1} - f_e\right)^2\,, \tag{17}$$

and thus, the maximum complexity, which depends only on H, is

$$C_{\max}(H) = D_{\max} \cdot H\,. \tag{18}$$

The behavior of the maximum value of complexity *versus* $\ln N$ was computed in Ref. [24].

Table 2: Probability values, f_j, that give a minimum of disequilibrium, D_{min}, for a given H.

Number of states with f_j	f_j	Range of f_j
n	0	0
1	f_{min}	$0 \ldots \frac{1}{N-n}$
$N-n-1$	$\frac{1-f_{\text{min}}}{N-n-1}$	$\frac{1}{N-n} \cdots \frac{1}{N-n-1}$

n can have the values $0, 1, \ldots N-2$.

Equivalently, the values, f_j, that give a minimum complexity are shown in Table 2. One of the solutions, f_{min}, is given by

$$H = -\frac{1}{\ln N}\left[f_{\text{min}} \ln f_{\text{min}} + (1 - f_{\text{min}}) \ln\left(\frac{1 - f_{\text{min}}}{N - n - 1}\right)\right], \tag{19}$$

where n is the number of states with $f_j = 0$ and takes a value in the range $n = 0, 1, \ldots, N-2$. The resulting minimum disequilibrium, D_{min}, for a given H is,

$$D_{\text{min}} = (f_{\text{min}} - f_{\text{e}})^2 + (N - n - 1)\left(\frac{1 - f_{\text{min}}}{N - n - 1} - f_{\text{e}}\right)^2 + n f_{\text{e}}^2 . \tag{20}$$

Note that in this case $f_j = 0$ is an additional hidden solution that stems from the positive restriction in the f_i values. To obtain these solutions explicitly we can define x_i such that $f_i \equiv {x_i}^2$. These x_i values do not have the restriction of positivity imposed to f_i and can take a positive or negative value. If we repeat the Lagrange multiplier method with these new variables a new solution arises: $x_j = 0$, or equivalently, $f_j = 0$. The resulting minimum complexity, which again only depends on H, is

$$C_{\text{min}}(H) = D_{\text{min}} \cdot H . \tag{21}$$

As an example, the maximum and minimum of complexity, C_{max} and C_{min}, are plotted as a function of the entropy, H, in Fig. 4 for $N = 4$. Also, in this figure, it is shown the minimum envelope complexity, $C_{\text{minenv}} = D_{\text{minenv}} \cdot H$, where D_{minenv} is defined below. In Fig. 5 the maximum and minimum disequilibrium, D_{max} and D_{min}, *versus* H are also shown.

As shown in Fig. 5 the minimum disequilibrium function is piecewise defined, having several points where its derivative is discontinuous. Each of these function pieces corresponds to a different value of n (Table 2).In some circumstances it might be helpful to work

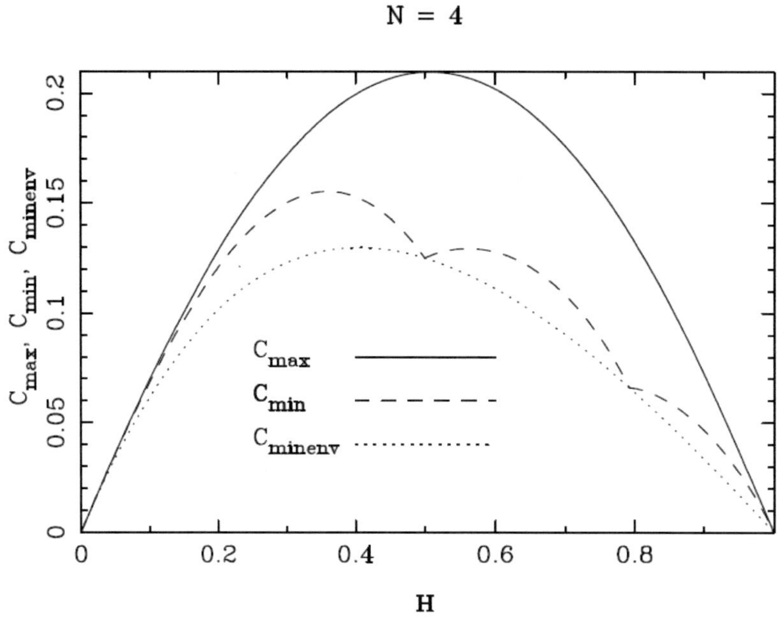

Figure 4: Maximum, minimum, and minimum envelope complexity, $C_{\max}$, $C_{\min}$, and C_{minenv} respectively, as a function of the entropy, H, for a system with $N = 4$ accessible states.

with the "envelope" of the minimum disequilibrium function. The function, D_{minenv}, that traverses all the discontinuous derivative points in the $D_{\min}$ *versus* H plot is

$$D_{\text{minenv}} = e^{-H \ln N} - \frac{1}{N}, \tag{22}$$

and is also shown in Fig. 5.

When N tends toward infinity the probability, $f_{\max}$, of the dominant state has a linear dependence with the entropy,

$$\lim_{N\to\infty} f_{\max} = 1 - H, \tag{23}$$

and thus the maximum disequilibrium scales as $\lim_{N\to\infty} D_{\max} = (1 - H)^2$. The maximum complexity tends to

$$\lim_{N\to\infty} C_{\max} = H \cdot (1 - H)^2. \tag{24}$$

The limit of the minimum disequilibrium and complexity vanishes, $\lim_{N\to\infty} D_{\text{minenv}} = 0$, and thus

$$\lim_{N\to\infty} C_{\min} = 0. \tag{25}$$

In general, in the limit $N \to \infty$, the complexity is not a trivial function of the entropy, in the sense that for a given H there exists a range of complexities between 0 and $C_{\max}$, given

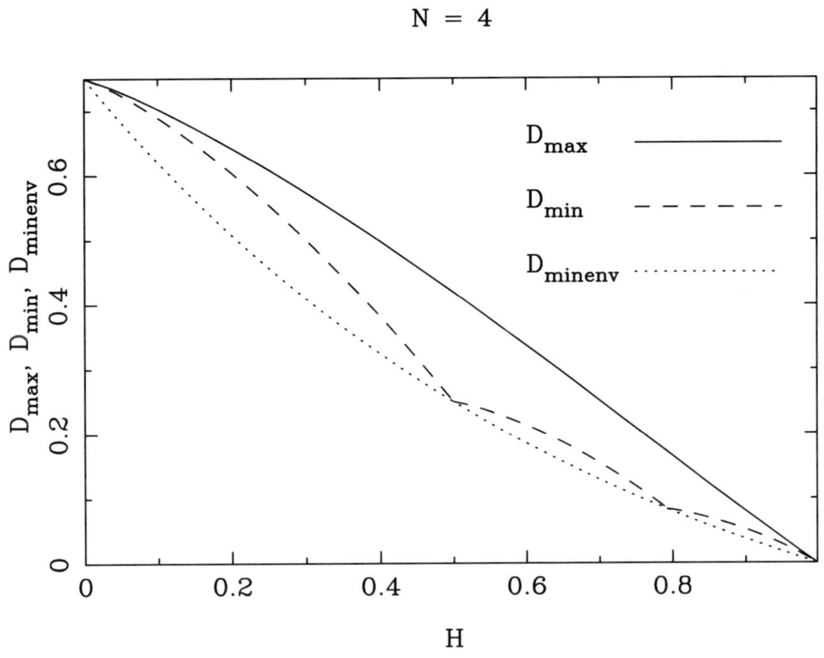

Figure 5: Maximum, minimum, and minimum envelope disequilibrium, $D_{\rm max}$, $D_{\rm min}$, and $D_{\rm minenv}$ respectively, as a function of the entropy, H, for a system with $N = 4$ accessible states.

by Eqs. (25) and (24), respectively.

In particular, in this asymptotic limit, the maximum of $C_{\rm max}$ is found when $H = 1/3$, or equivalently $f_{\rm max} = 2/3$, which gives a maximum of the maximum complexity of $C_{\rm max} = 4/27$. This value was numerically calculated in Ref. [24].

4 Rényi Entropies and LMC Complexity

Generalized entropies were introduced by Rényi [25] in the form of

$$I_q = \frac{1}{1-q} \log \left(\sum_{i=1}^{N} p_i^q \right), \tag{26}$$

where q is an index running over all the integer values. By differentiating I_q with respect to q a negative quantity is obtained independently of q, then I_q monotonously decreases when q increases.

The Rényi entropies are an extension of the Shannon information H. In fact, H is obtained in the limit $q \to 1$:

$$H = I_1 = \lim_{q \to 1} I_q = -\sum_{i=1}^{N} p_i \log p_i, \tag{27}$$

where the constant K of Eq. (1) is considered to be the unity. The disequilibrium D is also related with $I_2 = -\log\left(\sum_{i=1}^{N} p_i^2\right)$. We have that

$$D = \sum_{i=1}^{N} p_i^2 - \frac{1}{N} = e^{-I_2} - \frac{1}{N}, \tag{28}$$

then the LMC complexity is

$$C = H \cdot D = I_1 \cdot \left(e^{-I_2} - \frac{1}{N}\right). \tag{29}$$

The behavior of C in the neighborhood of H_{max} takes the form

$$C \sim \frac{1}{N}(\log^2 N - I_1 I_2), \tag{30}$$

The obvious generalization of the Rényi entropies for a normalized continuous distribution $p(x)$ is

$$I_q = \frac{1}{1-q} \log \int [p(x)]^q dx. \tag{31}$$

Hence,

$$H = I_1 = -\int p(x) \log p(x) dx, \tag{32}$$

$$D = e^{-I_2} = \int [p(x)]^2 dx. \tag{33}$$

The dependence of $\hat{C} = e^H \cdot D$ with I_1 and I_2 yields

$$\log \hat{C} = (I_1 - I_2). \tag{34}$$

This indicates that a family of different indicators could derive from the differences established among Rényi entropies with different q-indices [5]. Let us remark at this point the coincidence of the indicator $\log \hat{C}$ with the quantity S_{str} introduced by Varga and Pipek as a meaningful parameter to characterize the shape of a distribution. They apply this formalism to the Husimi representation, *i.e.*, to the projection of wave functions onto the coherent state basis [26]. A further generalization of the LMC complexity measure as function of the Rényi entropies has been introduced in Ref. [27].

The invariance of $\hat{C}$ under rescaling transformations implies that this magnitude is conserved in many different processes. For instance, the initial Gaussian-like distribution will continue to be Gaussian in a classical diffusion process. Then $\hat{C}$ is constant in time: $\frac{d\hat{C}}{dt} = 0$, and we have:

$$\frac{dI_1}{dt} = \frac{dI_2}{dt}. \tag{35}$$

The equal losing rate of I_1 and I_2, *i.e.*, the synchronization of both quantities, is the cost to be paid in order to maintain the shape of the distribution associated to the system and, hence, all its statistical properties will remain unchanged during its time evolution.

5 Some Applications

If by complexity it is to be understood that property present in all systems attached under the epigraph of 'complex systems', this property should be reasonably quantified by the measures proposed in the different branches of knowledge. In our case, the main advantage of LMC complexity is its generality and the fact that it is operationally simple and do not require a big amount of calculations [28]. This advantage has been worked out in different examples, such as the study of the time evolution of C for a simplified model of an isolated gas, the "tetrahedral gas" [6] or also in the case of a more realistic gas of particles [29, 30], the slight modification of C as an effective method by which the complexity in hydrological systems can be identified [31], the attempt of generalize C in a family of simple complexity measures [32, 33, 34], some statistical features of the behavior of C for DNA sequences [35] or earthquake magnitude time series [36], some wavelet-based informational tools used to analyze the brain electrical activity in epilectic episodes in the plane of coordinates (H, C) [37], a method to discern complexity in two-dimensional patterns [38] or some calculations done on quantum systems [39, 40, 41, 42, 43]. As an example, we show in the next subsections some straightforward calculation of the LMC complexity [44].

5.1 Canonical ensemble

Each physical situation is closely related to a specific distribution of microscopic states. Thus, an isolated system presents equipartition, by hypothesis: the microstates compatible with a macroscopic situation are equiprobable [45]. The system is said to be in equilibrium. For a system surrounded by a heat reservoir the probability of the microstates associated to the thermal equilibrium follow the Boltzmann distribution. Let us try to analyze the behavior of C in an ideal gas in thermal equilibrium. In this case the probability p_i of each accesible state is given by the Boltzmann distribution:

$$p_i = \frac{e^{-\beta E_i}}{Q_N}, \tag{36}$$

$$Q_N = \int e^{-\beta E(p,q)} \frac{d^{3N}p d^{3N}q}{N! h^{3N}} = e^{-\beta A(V,T)}, \tag{37}$$

where Q_N is the partition function of the canonical ensemble, $\beta = 1/\kappa T$ with κ the Boltzmann constant and T the temperature, V the volume, N the number of particles, $E(p,q)$ the hamiltonian of the system, h is the Planck constant and $A(V,T)$ the Helmholtz potential.

Calculation of H and D gives us:

$$H(V,T) = (1 + T\frac{\partial}{\partial T})(\kappa \log Q_N) = S(V,T), \tag{38}$$

$$D(V,T) = e^{2\beta\,[A(V,T) - A(V,T/2)]}. \tag{39}$$

Note that Shannon information H coincides with the thermodynamic entropy S when K is identified with κ. If a system verifies the relation $U = C_v T$ (U the internal energy, C_v the specific heat) the complexity takes the form:

$$C(V,T) \sim cte(V) \cdot S(V,T) e^{-S(V,T)/\kappa} \tag{40}$$

that matches the intuitive function proposed in Fig. 1.

5.2 Gaussian and exponential distributions

Gaussian distribution: Suppose a continuum of states represented by the x variable whose probability density $p(x)$ is given by the normal distribution of variance σ:

$$p(x) = \frac{1}{\sigma\sqrt{2\pi}} \exp\left(-\frac{x^2}{2\sigma^2}\right). \tag{41}$$

After calculating H and D, the expression for C is the following:

$$C_g = H \cdot D = \frac{K}{2\sigma\sqrt{\pi}}\left(\frac{1}{2} + \log(\sigma\sqrt{2\pi})\right). \tag{42}$$

If we impose the additional condition $H \geq 0$, then $\sigma \geq \sigma_{min} = (2\pi e)^{-1/2}$. The highest complexity is reached for a determined width: $\bar{\sigma} = \sqrt{(e/2\pi)}$.

Exponential distribution: Consider an exponential distribution of variance γ:

$$p(x) = \begin{cases} \frac{1}{\gamma} e^{-x/\gamma} & x > 0, \\ 0 & x < 0. \end{cases} \tag{43}$$

The same calculation gives us:

$$C_e = \frac{K}{2\gamma}(1 + \log \gamma), \tag{44}$$

with the condition $H \geq 0$ imposing $\gamma \geq \gamma_{min} = e^{-1}$. The highest complexity corresponds in this case to $\bar{\gamma} = 1$.

Remark that for the same width than a Gaussian distribution ($\sigma = \gamma$), the exponential distribution presents a higher complexity ($C_e/C_g \sim 1.4$).

5.3 Complexity in a two-level laser model

One step further, combining the results obtained in the former sections, is now done. We calculate LMC complexity for an unrealistic and simplified model of laser [46].

Let us suppose a laser of two levels of energy: $E_1 = 0$ and $E_2 = \epsilon$, with N_1 atoms in the first level and N_2 atoms in the second level, and the condition $N_1 + N_2 = N$ (the total number of atoms). Our aim is to sketch the statistics of this model and to introduce the results of photon counting [47] that produces an asymmetric behavior of C as function of the population inversion $\eta = N_2/N$. In the range $\eta \in (0, 1/2)$ spontaneous and stimulated emission can take place, but only in the range $\eta \in (1/2, 1)$ the condition to have lasing action is reached, because the population must be, at least, inverted, $\eta > 1/2$.

The entropy S of this system vanishes when N_1 or N_2 is zero. Moreover, S must be homogenous of first order in the extensive variable N [48]. For the sake of simplicity we approach S by the first term in the Taylor expansion:

$$S \sim \kappa \frac{N_1 N_2}{N} = \kappa N \eta (1 - \eta). \tag{45}$$

The internal energy is $U = N_2 \epsilon = \epsilon N \eta$ and the statistical temperature is:

$$T = \left(\frac{\partial S}{\partial U} \right)_N^{-1} = \frac{\epsilon}{\kappa} \frac{1}{(1 - 2\eta)}. \tag{46}$$

Note that for $\eta > 1/2$ the temperature is negative as corresponds to the stimulated emission regime dominating the actual laser action.

We are now interested in introducing qualitatively the results of laser photon counting in the calculation of LMC complexity. It was reported in [47] that the photo-electron distribution of laser field appears to be poissonian. In the continuous limit the Poisson distribution is approached by the normal distribution [49]. The width σ of this energy distribution in the canonical ensemble is proportional to the statistical temperature of the system. Thus, for a switched on laser in the regime $\eta \in [1/2, 1]$, the width of the gaussian energy distribution can be fitted by choosing $\sigma \sim -T \sim 1/(2\eta - 1)$ (recall that $T < 0$ in this case). The range of variation of σ is $[\sigma_\infty, \sigma_{min}] = [\infty, (2\pi e)^{-1/2}]$. Then we obtain:

$$\sigma \sim \frac{(2\pi e)^{-1/2}}{2\eta - 1}. \tag{47}$$

By replacing this expression in Eq. (42), and rescaling by a factor proportional to entropy, $S \sim \kappa N$, (in order to give to it the correct order of magnitude), LMC complexity for a population inversion in the range $\eta \in [1/2, 1]$ is reobtained:

$$C_{laser} \simeq \kappa N \cdot (1 - 2\eta) \log(2\eta - 1). \tag{48}$$

We can consider at this level of discussion $C_{laser} = 0$ for $\eta < 1/2$. Regarding the behavior of this function, it is worth noticing the value $\eta_2 \simeq 0.68$ where the laser presents the highest complexity. By following theses ideas, if the width, σ, of the experimental photo-electron distribution of laser field is measured, the population inversion parameter, η, would be given by Eq. (47). In a next step, the LMC complexity of the laser system would be obtained by Eq. (48).

It is necessary to remark that a model helps us to approach the reality and provides invaluable guidance in the goal of a finer understanding of a physical phenomenon. From this point of view the present calculation evidently only tries to enlighten the problem of calculating the LMC complexity of a physical system *via* an unrealistic but simplified model.

6 Conclusions

A definition of complexity (LMC complexity) based on a probabilistic description of physical systems has been explained. This definition contains basically an interplay between the *information* contained in the system and the *distance to equipartition* of the probability distribution representing the system. Besides giving the main features of an intuitive notion of complexity, we show that it allows to successfully discern situations considered as complex in systems of a very general interest. Also, its relationship with the Shannon information and the generalized Rényi entropies has been shown to be explicit. Moreover it has been possible to establish the decrease of this magnitude when a general system evolves from a near-equilibrium situation to the equipartition.

From a practical point of view, we are convinced that this statistical complexity measure provides a useful way of thinking [50] and it can help in the future to gain more insight on the physical grounds of models with potential biological interest.

Conflict of interest: None.

Acknowledgments

None.

References

[1] E.T. Jaynes, E.T.: Information theory and statistical mechanics. Phys. Rev. **106**, 620-630 (1957).

[2] R. Badii, R., Politi, A.: Complexity. Hierarchical Structures and Scaling in Physics. Cambridge University Press, Cambridge (1997).

[3] Shannon, C.E., Weaver, W.: The Mathematical Theory of Communication. University of Illinois Press, Urbana, Illinois (1949).

[4] López-Ruiz, R., Mancini, H.L., Calbet, X.: A statistical measure of complexity. Phys. Lett. A **209**, 321-326 (1995).

[5] López-Ruiz, R.: Shannon information, LMC complexity and Rényi entropies: a straightforward approach. Biophys. Chem. **115**, 215 (2005)..

[6] Calbet, X., López-Ruiz, R.: Tendency toward maximum complexity in a non-equilibrium isolated system. Phys. Rev. E **63**, 066116 (9pp) (2001).

[7] Anderson, P.W.: Is complexity physics? Is it science? What is it?. Physics Today, 9-11, July (1991).

[8] Parisi, G.: Statistical physics and biology. Physics World, **6**, 42-47, Setember (1993).

[9] Nicolis, G., Prigogine, I.: Self-organization in Nonequilibrium Systems. Wiley, New York (1977).

[10] López-Ruiz, R.: On Instabilities and Complexity. Ph. D. Thesis, Universidad de Navarra, Pamplona (1994).

[11] Catalán, R.G., Garay, J., López-Ruiz, R.: Features of the extension of a statistical measure of complexity for continuous systems. Phys. Rev. E **66**, 011102(6) (2002).

[12] Dembo, A., Cover, T.M., Thomas, J.A.: Information theoretic inequalities. IEEE Trans. Information Theory **37**, 1501-1518 (1991).

[13] Kolmogorov, A.N.: Three approaches to the definition of quantity of information. Probl. Inform. Theory **1**, 3-11 (1965).

[14] Chaitin, G.J.: On the length of programs for computing finite binary sequences. J. Assoc. Comput. Mach. **13**, 547-569 (1966); Information, Randomness & Incompleteness. World Scientific, Singapore (1990).

[15] Lempel A., Ziv, J.: On the complexity of finite sequences. IEEE Trans. Inform Theory **22**, 75-81 (1976).

[16] Bennett, C.H.: Information, dissipation, and the definition of organization. Emerging Syntheses in Science, David Pines ed., Santa Fe Institute, Santa Fe, NM, 297-313 (1985).

[17] Grassberger, P.: Toward a quantitative theory of self-generated complexity. Int. J. Theor. Phys. **25**, 907-938 (1986).

[18] Huberman, B.A., Hogg, T.: Complexity and adaptation. Physica D **22**, 376-384 (1986).

[19] LLoyd, S., Pagels, H.: Complexity as thermodynamic depth. Ann. Phys. (N.Y.) **188**, 186-213 (1988).

[20] Crutchfield, J.P., Young, K.: Inferring statistical complexity. Phys. Rev. Lett. **63**, 105-108 (1989).

[21] Adami, C., Cerf, N.T.: Physical complexity of symbolic sequences. Physica D **137**, 62-69 (2000).

[22] Sánchez, J.R., López-Ruiz, R.: A method to discern complexity in two-dimensional patterns generated by coupled map lattices. Physica A **355**, 633-640 (2005).

[23] Escalona-Morán, M., Cosenza, M.G., López-Ruiz, R., García, P.: Statistical complexity and nontrivial collective behavior in electroencephalographic signals. Int. J. Bif. Chaos **20**, special issue on *Chaos and Dynamics in Biological Networks*, Ed. Chávez & Cazelles (2010).

[24] Anteneodo, C., Plastino, A.R.: Some features of the statistical LMC complexity. Phys. Lett. A **223**, 348-354 (1996).

[25] Rényi, A.: Probability Theory. North-Holland, Amsterdam (1970).

[26] Varga, I., Pipek, J.: Rényi entropies characterizing the shape and the extension of the phase space representation of quantum wave functions in disordered systems. Phys. Rev. E **68**, 026202(8) (2003).

[27] López-Ruiz, R., Nagy, Á, Romera, E., Sañudo, J.: A generalized statistical complexity measure: Applications to quantum systems. J. Math. Phys. **50**, 123528(10) (2009).

[28] Perakh, M.: Defining complexity. On Talk Reason, www.talkreason.org/articles/ complexity.pdf, August (2004).

[29] Calbet, X., López-Ruiz, R.: Extremum complexity distribution of a monodimensional ideal gas out of equilibrium. Physica A **382**, 523-530 (2007).

[30] Calbet, X., López-Ruiz, R.: Extremum complexity in the monodimensional ideal gas: the piecewise uniform density distribution approximation. Physica A **388**, 4364-4378 (2009).

[31] Feng, G., Song, S., Li, P., A statistical measure of complexity in hydrological systems. J. Hydr. Eng. Chin. (Hydr. Eng. Soc.) **11**, article no. 14 (1998).

[32] Shiner, J.S., Davison, M., Landsberg, P.T.: Simple measure for complexity. Phys. Rev. E **59**, 1459-1464 (1999).

[33] Martin, M.T., Plastino, A., Rosso, O.A.: Statistical complexity and disequilibrium. Phys. Lett. A **311** (2-3), 126-132 (2003).

[34] Lamberti, W., Martín, M.T., Plastino, A., Rosso, O.A.: Intensive entropic non-triviality measure. Physica A **334**, 119-131 (2004).

[35] Yu, Z., Chen, G.: Rescaled range and transition matrix analysis of DNA sequences. Comm. Theor. Phys. (Beijing China) **33** 673-678 (2000).

[36] Lovallo, M., Lapenna, V., Telesca, L.: Transition matrix analysis of earthquake magnitude sequences. Chaos, Solitons and Fractals **24**, 33-43 (2005).

[37] Rosso, O.A., Martin, M.T., Plastino, A.: Brain electrical activity analysis using wavelet-based informational tools (II): Tsallis non-extensivity and complexity measures. Physica A **320**, 497-511 (2003).

[38] Sánchez, J.R., López-Ruiz, R.: Detecting synchronization in spatially extended discrete systems by complexity measurements. Discrete Dyn. Nat. Soc. **9**, 337-342 (2005).

[39] Chatzisavvas, K.Ch., Moustakidis, Ch.C., Panos, C.P.: Information entropy, information distances, and complexity in atoms. J. Chem. Phys. **123**, 174111 (10 pp) (2005).

[40] Sañudo, J., López-Ruiz, R.: Statistical complexity and Fisher-Shannon information in the H-atom. Phys. Lett. A **372**, 5283-5286 (2008).

[41] Montgomery Jr., H.E., Sen, K.D.: Statistical complexity and Fisher-Shannon information measure of H_2^+. Phys. Lett. A **372**, 2271-2273 (2008).

[42] Kowalski, A.M., Plastino, A., Casas, M.: Generalized complexity and classical-quantum transition. Entropy **11**, 111-123 (2009).

[43] López-Ruiz, R., Sañudo, J.: Evidence of magic numbers in nuclei by statistical indicators. Open Syst. Inf. Dyn. **17**, issue 3, Setember (2010).

[44] López-Ruiz, R.: Complexity in some physical systems. Int. J. of Bifurcation and Chaos **11**, 2669-2673 (2001).

[45] Huang, K.: Statistical Mechanics. John Wiley & Sons, New York (1987).

[46] Svelto, O.: Principles of Lasers. Plenum Press, New York (1949).

[47] Arecchi, F.T.:, Measurement of the statistical distribution of Gaussian and laser sources. Phys. Rev. Lett. **15**, 912-916 (1965).

[48] Callen, H.B.: Thermodynamics and an Introduction to Thermostatistics. J. Whiley & Sons, New York (1985).

[49] Harris, J.W., Stocker, H.: Handbook of Mathematics and Computational Science Springer-Verlag, New York (1998).

[50] "I think the next century will be the century of complexity", Stephen Hawking in *San José Mercury News*, Morning Final Edition, January 23 (2000).

CHAPTER 8

Generalized Statistical Complexity: A New Tool for Dynamical Systems

Osvaldo A. Rosso [a,b,f*], **María Teresa Martín** [c,f], **Hilda A. Larrondo** [d,f], **Andres M. Kowalski** [c,e], **Angelo Plastino** [c,f]

[a] Departamento de Física, Instituto de Ciências Exatas,
Universidade Federal de Minas Gerais
Av. Antônio Carlos, 6627 - Campus Pampulha, 31270-901 Belo Horizonte - MG, Brazil

[b] Chaos & Biology Group, Instituto de Cálculo,
Facultad de Ciencias Exactas y Naturales, Universidad de Buenos Aires
Pabellón II, Ciudad Universitaria, 1428 Ciudad Autónoma de Buenos Aires, Argentina

[c] Instituto de Física, Facultad de Ciencias Exactas, Universidad Nacional de La Plata
C.C. 727, 1900 La Plata, Argentina

[d] Facultad de Ingeniería, Universidad Nacional de Mar del Plata
Juan B. Justo 4302. 7600 Mar del Plata, Argentina

[e] Comisión de Investigaciones Científicas (CICPBA), Argentina

[f] Consejo Nacional de Investigaciones Científicas y Técnicas
(CONICET), Argentina

Abstract

A generalized Statistical Complexity Measure (SCM) is a functional of the probability distribution P associated with the time series generated by a given dynamical system. A SCM is the composition of two ingredients: *i)* an entropy and *ii)* a distance in the probability-space. We address in this review important topics underlying the SCM structure, *viz.*, *a)* the selection of the information measure $\mathcal{I}$; *b)* the choice of the probability metric space and associated distance $\mathcal{D}$, which in this context is called a "disequilibrium" (denoted with the letter $\mathcal{Q}$). $\mathcal{Q}$, indeed the crucial SCM ingredient, is cast in terms of an associated distance $\mathcal{D}$. *c)* The adequate way of picking up the probability distribution P associated with a dynamical system or time series under study, which is indeed a fundamental problem. A good analysis of this topics is essential to get a SCM that quantifies not only randomness but also the presence of correlational structures. In this chapter we specially stress how sensible improvements in the final results

* Corresponding author. E-mail: oarosso@gmail.com

Andres M. Kowalski, Raul D. Rossignoli and Evaldo M. F. Curado (Eds)

can be obtained if the underlying probability distribution is "extracted" *via* appropriate considerations regarding causal effects in the system's dynamics. As an illustration, we show just how these issues affect the description of the celebrated logistic map.

Keywords: Generalized Statistical Complexity; Disequilibrium; Information Theory; Selection of the probability distribution.

1 Statistical Complexity Measures

1.1 Meaning of the concept

Complexity is a subjective concept and there is no universal agreement on a proper definition and quantitative characterization. In the last two decades several complexity measures together with methodologies for their evaluation have been proposed, that capture some aspects of complex systems. We can mention, among several ones, *(a)* those based on data compression algorithms [1] which are related with periodicities in the time series generated by the system; *(b)* those related with optimal predictability like Crutchfield and Young's complexity [2], that measures the amount of information about the past required to predict the future; *(c)* a measure of the self-organization "capacity" of a system [3]; *(d)* the dimension of the concomitant attractor is a measure of complexity for chaotic systems, since it yields the number of active variables [3, 4]; *(e)* the algorithmic-, information-, Kolmogorov-, or Chaitin-complexity [5, 6, 7] based on the size of the smallest computer program that can produce an observed pattern; *(f)* measures based on recurrence plots [8]; *(g)* complexity based on symbolic analysis [9]; *(h)* measures based on Information Theory [10, 11, 12, 13, 14, 15]. An important contribution for the complexity definition we follow in this chapter is due to P. Grassberger [16], who considered the generation of patterns by the dynamical system. The most complex situation in their approach is neither the one with highest Shannon information (random structure) nor the one with lowest (ordered structures). Thus, in Grassberger's view complexity is represented by a mixture of order and disorder, regularity and randomness (for a graphical example, see Fig. 1 of [10]).

An important question is the relation between a complicated dynamics and a complex system. A chaotic map is the typical example of a very simple system with a complicated dynamics. Furthermore if the system itself is already involved enough and/or it is constituted by many different parts, it clearly *may* support a rather intricate dynamics but

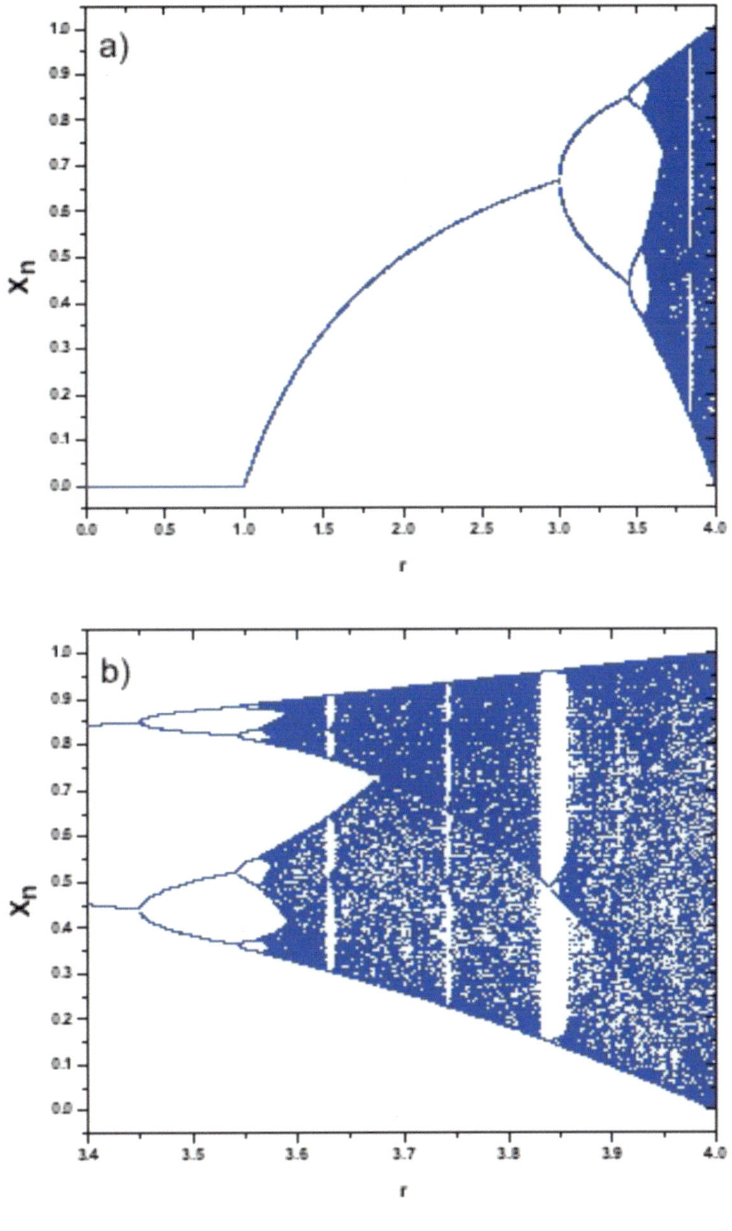

Figure 1: Bifurcation diagram for the logistic map as function of parameter r with step $\Delta r = 0.001$. *a)* $0 < r \leq 4$, *b)* detail for parameter range $3.4 \leq r \leq 4$.

perhaps without the emergence of typical characteristic patterns [17]. Therefore, a complex system does not necessarily generate a complex output. In the previous context, we understand "statistical complexity" as a concept related to patterned structures hidden in the dynamics, emerging from a system which itself can be much simpler than the dynamics it generates [17]. "Statistical complexity" becomes then a kind of indicator for hidden "order". Then one should start by excluding processes that are certainly not complex, such as those exhibiting periodic motion. Additionally, a purely random, like white-noise random process, cannot be considered complex either, notwithstanding its irregular and unpredictable character, because it does not contain any non-trivial, patterned structure.

A special generalized family of statistical complexity measures $\mathcal{C}$ will be the focus of attention here, whose members pretend to be measures of off-equilibrium "regularity", an "order" that is not the one associated with very regular structures (for instance, with crystals), for which the entropy is rather small. Biological life, like EEG signals [18, 19, 20, 21, 22, 23], or the Classical Limit of Quantum Mechanics (CLQM) [24, 25, 26, 27] are typical examples of the kind of regularities or "organization" one has in mind here, associated with relative large entropic values. In comparison with other definitions of complexity, the present family of statistical complexity is rather easy to compute due to it is evaluated in terms of common statistical mechanics' concepts.

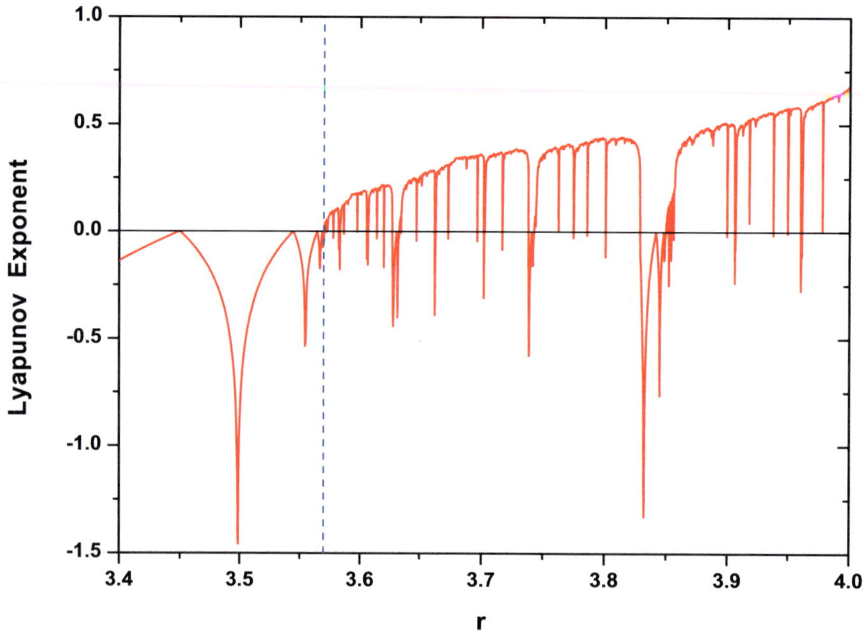

Figure 2: Lyapunov exponent Λ for the logistic map as function of parameter $3.4 \leq r \leq 4$ ($\Delta r = 0.0003$). Vertical line denoted the value r_∞.

In fact, entropy and statistical complexity measures of tonic–clonic epileptic EEG time series support the conjecture that an epileptic focus, in this kind of seizure, triggers a self-organized brain state with both order and maximal complexity [20, 21, 22, 23]. In the study of CLQM using the statistical complexity, dynamical features of quantum-classical transition zone can properly be appreciated and characterized [24, 25, 26, 27]. Other fields where the present statistical complexity family has been used are: *a)* quantification of the performance of Pseudorandom Numbers Generators (PRNG) [28, 29, 30, 31, 32] (see also Chapter 12); *b)* analysis of El Niño-Southern Oscillation (ENSO) [33] (see also Chapter 13); *c)* characterization of erythrocytes (red blood cells) deformation under shear stress [34, 35] (see also Chapter 14); *d)* study of stochastic and coherence resonances [36, 37]; *e)* analysis of dynamical systems with delay [38, 39, 40]; *f)* econophysics time series analysis [41, 42, 43]; *g)* laser dynamics [44, 45, 46]; *h)* characterizarion of fractional Brawnian motion (fBm) and fractional Gaussian noise (fGn) time series [47, 48, 49, 50, 51, 52, 53]; *i)* evolution of complex networks [54]; *j)* distinguishing chaos and noise [55]; amongh others.

The definitions of a statistical complexity measure can be partitioned into three categories [13]. They can either *(a)* grow with increasing disorder (decrease as order increases), *(b)* be quite small for large amounts of the degree of either order or disorder, with a maximum at some intermediate stage, or *(c)* grow with increasing order (decrease as disorder increases). The present contribution deals with a $\mathcal{C}$-family of the second type.

Ascertaining the degree of unpredictability and randomness of a system is not automatically tantamount to adequately grasping the correlational structures that may be present, *i.e.*, to be in a position to capture the relationship between the components of the physical system [2, 12]. These structures strongly influence, of course, the character of the probability distribution function (PDF) that is able to describe the physics one is interested in. Randomness, on the one hand, and structural correlations on the other one, are not totally independent aspects of *this* physics [11, 13, 14, 15]. Certainly, the opposite extremes of *(i)* perfect order and *(ii)* maximal randomness possess no structure to speak of. In between these two special instances a wide range of possible degrees of physical structure exists that should be reflected in the features of the underlying probability distribution P. One would like that they be adequately captured by some functional $\mathcal{F}_{stat}[P]$ in the same fashion that Shannon's information [56] "captures" randomness. A suitable candidate to this effect has come to be called the *statistical complexity measure* [2, 11, 12, 13, 14, 15]. As mentioned above, $\mathcal{F}_{stat}$ should vanish in the two special extreme instances mentioned above. In this work, we advance a rather general functional form for the statistical complexity measure and the adequate way of picking up the probability distribution P associated to a dynamical system or time series under study, which is indeed a fundamental problem. The celebrated logistic map is employed as an illustration-device.

1.2 The functional form for the Statistical Complexity Measure

An information measure $\mathcal{I}$ can primarily be viewed as a quantity that characterizes a given probability distribution. $\mathcal{I}[P]$ is regarded as the measure of the uncertainty associated with the physical processes described by the probability distribution $P = \{p_j, j = 1, \cdots, N\}$, with N the number of possible states of the system under study. From now on we assume that the only restriction on the PDF representing the state of our system is $\sum_{j=1}^{N} p_j = 1$ (micro-canonical representation). If $\mathcal{I}[P] = \mathcal{I}_{min} = 0$ we are in a position to predict with certainty which of the N possible outcomes will actually take place. Our knowledge of the underlying process described by the probability distribution is in this instance maximal. On the other hand, it is our ignorance which is maximized if $\mathcal{I}[P] = \mathcal{I}[P_e] \equiv \mathcal{I}_{max}$; $P_e = \{p_i = 1/N; \forall i\}$, being the uniform distribution.

These two extreme circumstances of *(i)* maximum foreknowledge ("perfect order") and *(ii)* maximum ignorance (or maximum "randomness") can, in a sense, be regarded as "trivial" ones. We define for a given probability distribution P and its associate information measure $\mathcal{I}[P]$, an amount of "disorder" $\mathcal{H}$ in the fashion

$$\mathcal{H}[P] \; = \; \mathcal{I}[P]/\mathcal{I}_{max} \; . \tag{1}$$

We have thus $0 \leq \mathcal{H} \leq 1$. Following the Shannon-Kinchin paradigm we define $\mathcal{I}$ in terms of entropies.

A definition of *Statistical Complexity Measure* (SCM) can not be made in terms of just "disorder" or "information". It might seen reasonable to propose a measure of "statistical complexity" by adopting some kind of distance $\mathcal{D}$ to a reference probability distribution, in particular to the uniform distribution P_e [11, 14, 15] (see also Chapter 6 and 7). This motivates introducing, as a special distance-form, the so-called "disequilibrium-distance" as

$$\mathcal{Q}[P] = \mathcal{Q}_0 \cdot \mathcal{D}[P, P_e] \; , \tag{2}$$

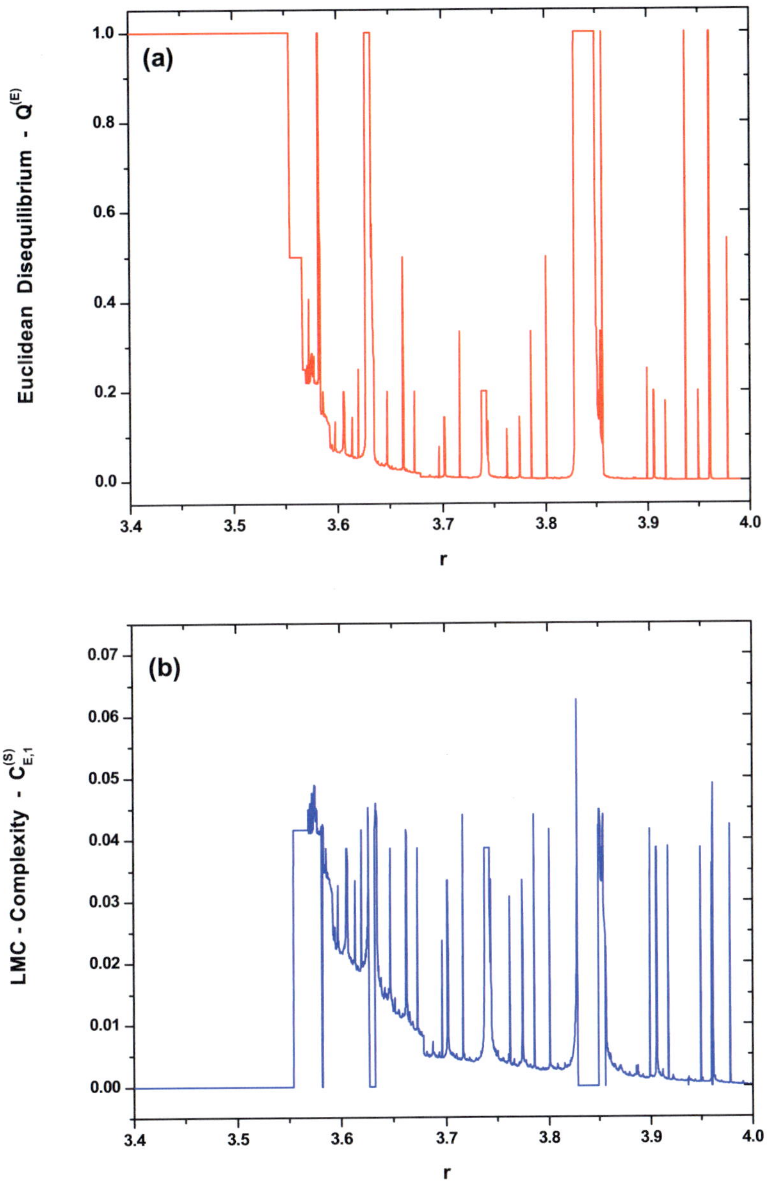

Figure 3: Euclidean disequilibrium $\mathcal{Q}^{(E)}\Big|_{binary}$ *(a)* and, López-Ruiz, Mancini, Calvet (LMC) complexity $\mathcal{C}_{LMC} \equiv \mathcal{C}^{(S)}_{E,1}\Big|_{binary}$ *(b)* for the logistic map (PDF-binary evaluation) as function of parameter r ($3.4 \leq r \leq 4$, $\Delta r = 0.0003$).

where $\mathcal{Q}_0$ is a normalization constant ($0 \leq \mathcal{Q} \leq 1$) with its value equal to the inverse of the maximum possible value of the distance $\mathcal{D}[P, P_e]$. This maximum distance is obtained when one of the components of P, say p_m, is equal to one and the remaining components are equal

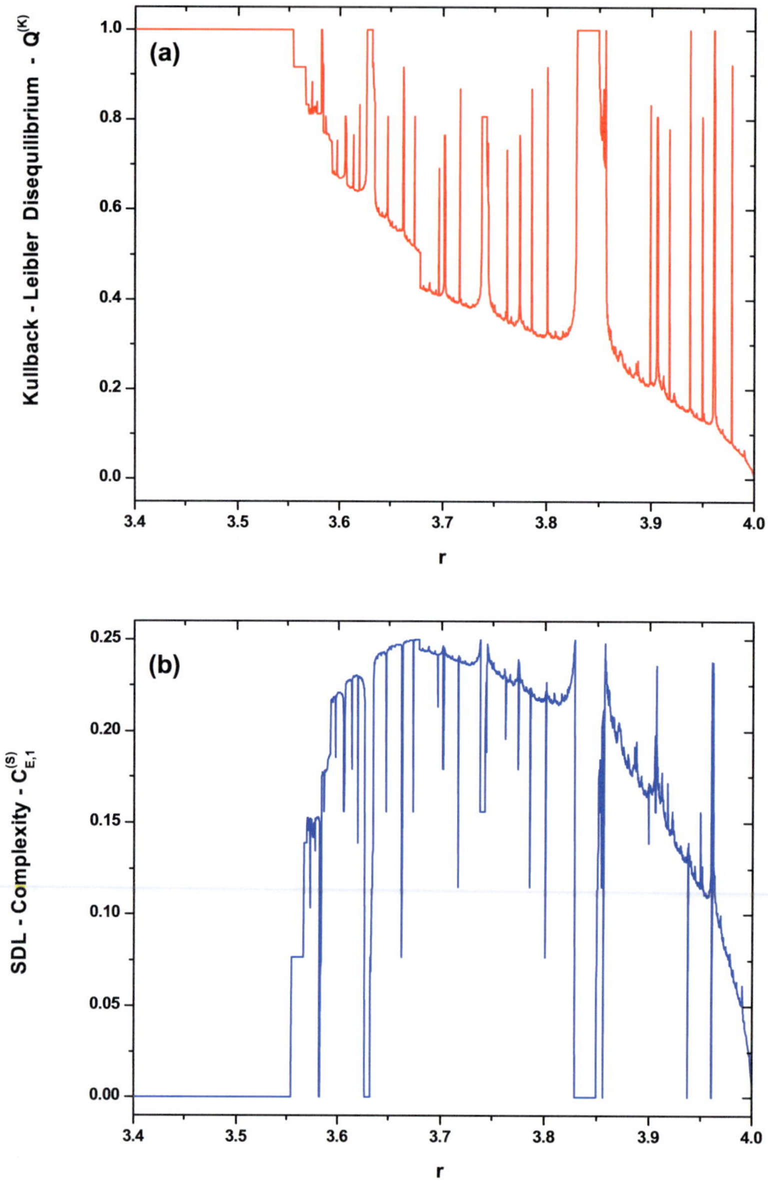

Figure 4: Kullback-Leibler disequilibrium $\mathcal{Q}^{(K)}\Big|_{binary}$ *(a)* and, Shiner, Davison, Landsberg (SDL) complexity $\mathcal{C}_{SDL} \equiv \mathcal{C}^{(S)}_{K,1}\Big|_{binary}$ *(b)* for the logistic map (PDF-binary evaluation) as function of parameter r ($3.4 \leq r \leq 4$, $\Delta r = 0.0003$).

to zero. The disequilibrium-distance $\mathcal{Q}$ would reflect on the systems' "architecture", being different from zero if there exist "privileged", or "more likely" states among the accessible ones.

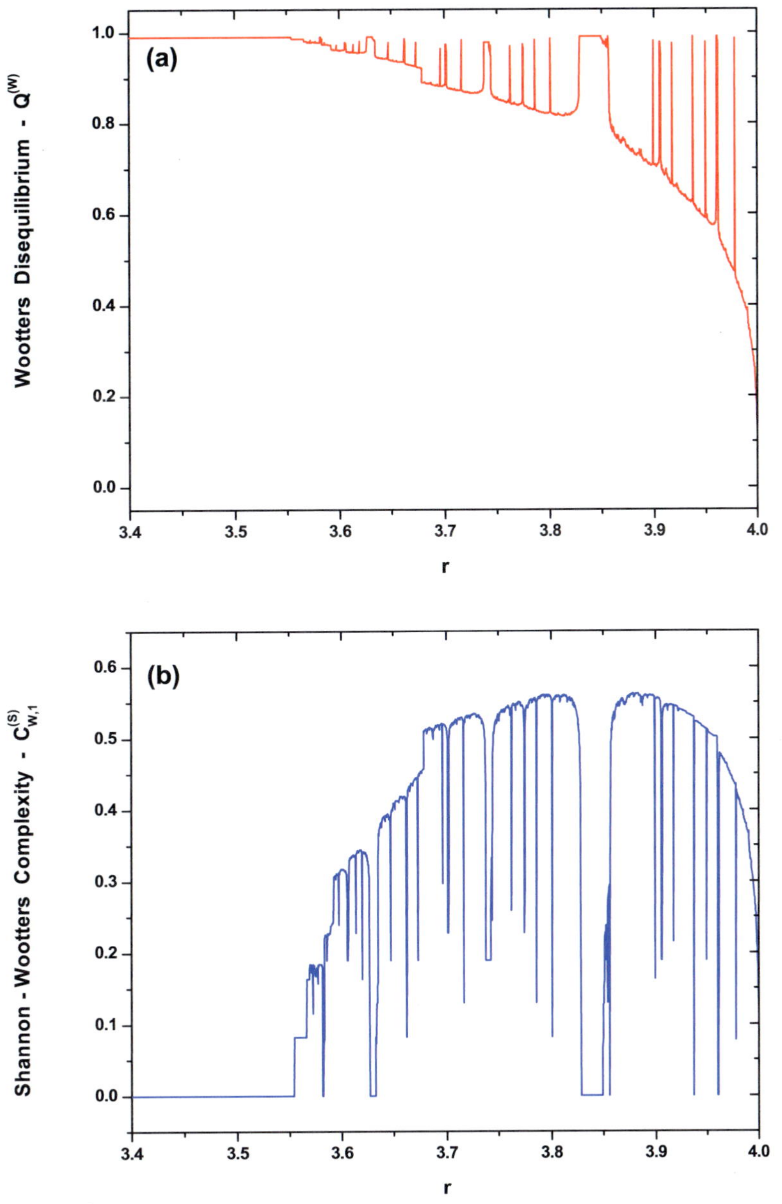

Figure 5: Wootters disequilibrium $\mathcal{Q}^{(W)}\Big|_{binary}$ *(a)* and Shannon-Wootters complexity $\mathcal{C}_W \equiv \mathcal{C}^{(S)}_{W,1}\Big|_{binary}$ *(b)* for the logistic map (PDF-binary evaluation) as function of parameter r ($3.4 \leq r \leq 4$, $\Delta r = 0.0003$).

Consequently, we adopt the following functional product form for the SCM introduced originally by López-Ruiz, Mancini and Calbet (LMC) in 1995 [11] (see Chapter 7):

$$\mathcal{C}[P] \;=\; \mathcal{H}[P] \cdot \mathcal{Q}[P] \; . \tag{3}$$

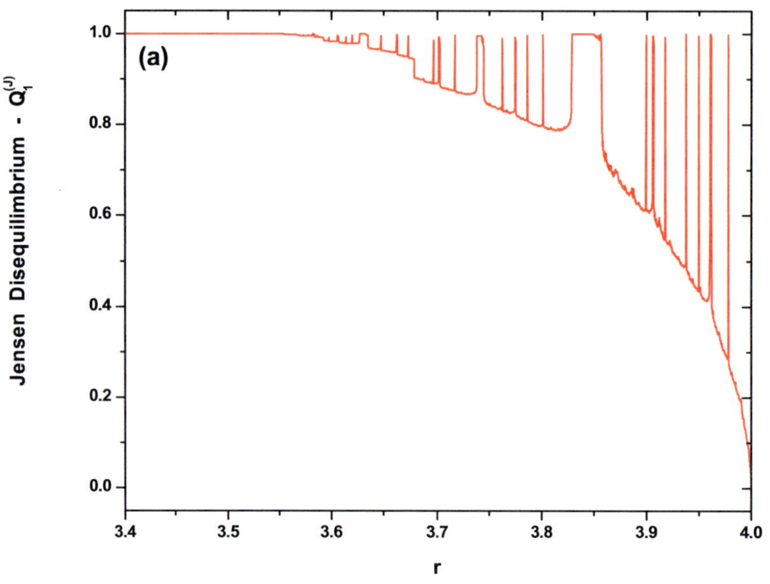

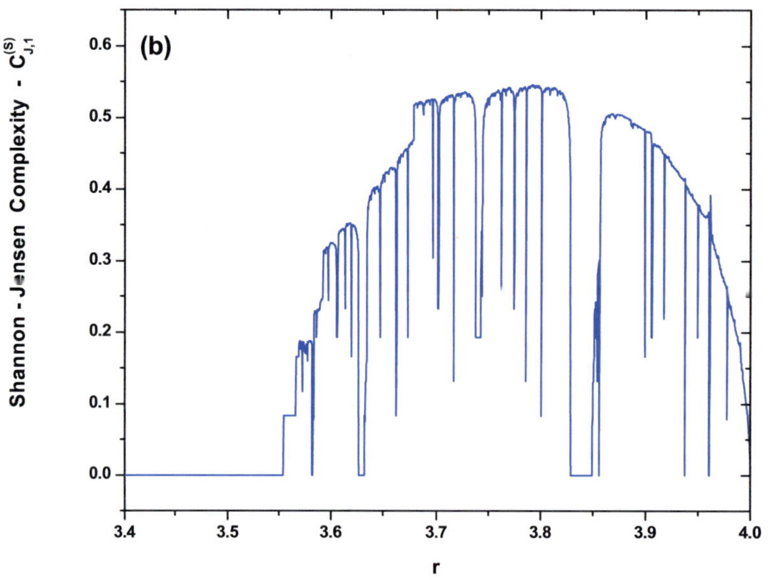

Figure 6: Jensen-Shannon disequilibrium $\mathcal{Q}^{(J)}\Big|_{binary}$ *(a)* and Shannon-Jensen complexity $\mathcal{C}_{JS} \equiv \mathcal{C}_{J,1}^{(S)}\Big|_{binary}$ *(b)* for the logistic map (PDF-binary evaluation) as function of parameter r ($3.4 \leq r \leq 4$, $\Delta r = 0.0003$).

This quantity, on the basis of our statistical description of the system, reflects, at a given "scale" the delicate interplay extant between the amount of information stored in the system and its disequilibrium (the measure of probability hierarchies between the observed parts of its accessible states) [11].

Here we will define $\mathcal{I}$ in terms of entropies (Shannon, Tsallis, escort-Tsallis or Rényi) [57, 58]. As for the choice of the distance $\mathcal{D}$ entering in $\mathcal{Q}$ we face several possibilities, as for instance, *(a)* Euclidean [11] or *(b)* Wootters's statistical one [59, 60]. Moreover, since in analyzing complex signals by recourse to Information Theory tools, entropies, distances, and statistical divergences play a crucial role for prediction, estimation, detection and transmission processes [57, 61, 62], for the selection of $\mathcal{D}$ we consider also *(c)* relative entropies [57, 58, 63] and *(d)* Jensen divergences (Shannon, Tsallis, escort-Tsallis and Rényi) [57, 64].

1.3 Information measures and generalized disorder

According to Information Theory, entropy is a relevant measure of order and disorder for many systems, including dynamical ones [56, 61]. Indeed, Kolmogorov and Sinai converted Shannon's Information Theory into a powerful tool for the study of dynamical systems [65, 66]. The statistical characterization of deterministic sources of apparent randomness performed by many authors during the intervening years has shed much light into the intricacies of dynamical behavior by describing dynamical systems unpredictability using such tools as fractal dimensions, Lyapunov exponents and metric entropy [67].

Thus, for $P \equiv \{p_i, i = 1, \cdots, N\}$ a discrete distribution, we define the information measure $\mathcal{I}$ in the fashion:

a) Shannon Entropy:

$\mathrm{S}_1^{(S)}[P]$ represents Shannon's logarithmic entropy [56], given by

$$\mathrm{S}_1^{(S)}[P] = -\sum_{j=1}^{N} p_j \ \ln(p_j) \ . \tag{4}$$

b) Tsallis Entropy:

In 1988 Tsallis proposed a generalization of the celebrated Shannon – Boltzmann – Gibbs entropic measure [68] (see Chapter 3). The new entropy functional introduced by Tsallis along with its associated generalized thermostatistics is nowadays being hailed as the possible basis of a new theoretical framework appropriate to deal with non-extensive settings. A regularly updated bibliography on Tsallis' formalism and applications can be found at web page http://www.tsallis.cat.cbpf.br/TEMUCO.pdf.

This entropy has the form

$$\mathrm{S}_q^{(T)}[P] = \frac{1}{q-1} \sum_{j=1}^{N} \left[p_j - (p_j)^q\right] , \tag{5}$$

where the entropic index q is any real number. This entropic functional reduces to the standard Shannon – Boltzmann – Gibbs entropy for $q \to 1$ (Eq. 4). $\mathrm{S}_q^{(T)}$ is non-extensive such that

$$\mathrm{S}_q^{(T)}[A+B] = \mathrm{S}_q^{(T)}[A] + \mathrm{S}_q^{(T)}[B] + (1-q) \cdot \mathrm{S}_q^{(T)}[A] \cdot \mathrm{S}_q^{(T)}[B] , \tag{6}$$

where A and B are two independent systems, in the sense that $P(A+B) = P(A) \cdot P(B)$. It is clear that q is an index of non-extensivity.

c) Escort-Tsallis Entropy:

Many relevant mathematical properties of the standard thermostatistics are preserved by Tsallis' formalism or admit natural generalizations. Tsallis pioneering work has stimulated the exploration of the properties of other generalized or alternative information measures. Martín, Plastino and Plastino [69] shown that a new information measure of the Tsallis kind, seems to be more sensitive to a variation of relevant parameters of complex systems than the original measure given by Eq. (5). The new measure is indeed Tsallis' original one, expressed in terms of escort-distributions of order q. Let us briefly review the useful concept of escort probabilities (see Ref. [67] and references therein). To such an end one introduces the following transformation between an original, normalized probability distribution P and a new, related probability distribution P^*, namely, $P \to P^*$ with

$$p_i^* = (p_i)^q \Big/ \left[\sum_j (p_j)^q\right] , \tag{7}$$

q being any real parameter (here we identify it with Tsallis' index). We reiterate: p_i is the original probability distribution one is concerned with. For $q = 1$ we have $P \equiv P^*$ and, obviously, P^* is normalized to unity. General global quantities formed with escort distributions of different order q, such as the different types of information or mean values, will often give more revealing information than those formed with the original distribution only. Changing q is a quite useful device for scanning the structure of the original distribution [67]. Starting now with Tsallis' information measure constructed with some probability distribution P one may think of computing the associated Tsallis measure that results from replacing P by P^*. Then the associated escort-Tsallis information measure, as a functional of the original measure acquires the appearance [18, 69]

$$\mathrm{S}_q^{(G)}[P] = \frac{1}{(q-1)} \left\{ 1 - \left[\sum_{j=1}^{N} (\ p_j\ )^{1/q} \right]^{-q} \right\} . \tag{8}$$

d) Rényi Entropy:

Finally, the Rényi's entropy [70] for a discrete probability distribution P is given by

$$\mathrm{S}_q^{(R)}[P] = \frac{1}{(1-q)} \ln \left\{ \sum_{j=1}^{N} (p_j)^q \right\} . \tag{9}$$

In the $q \to 1$ limit it coincides with Shannon's measure as given by Eq. (4).

In what follows we denote with $\mathrm{S}_q^{(\kappa)}$ any of the above entropies according to the κ assignment: S, T, G or R (Shannon, Tsallis, escort-Tsallis or Rényi) so that $\mathcal{I} \equiv \mathrm{S}_q^{(\kappa)}$. Of course, in Shannon's instance one has $q = 1$. The generalized disorder $\mathcal{H}_q^{(\kappa)}$ based on these measures reads

$$\mathcal{H}_q^{(\kappa)}[P] = \mathrm{S}_q^{(\kappa)}[P] \ / \ \mathrm{S}_{max}^{(\kappa)} . \tag{10}$$

In the $q \to 1$ limit all the above expressions coincide with that for the Shannon-measure. $\mathrm{S}_{max}^{(\kappa)} = \mathrm{S}_q^{(\kappa)}[P_e]$, is the maximum possible value for the information measure one is dealing with, entailing

$$\mathrm{S}_{max}^{(S)} = \mathrm{S}_{max}^{(R)} = \ln N , \tag{11}$$

and

$$\mathrm{S}_{max}^{(T)} = \mathrm{S}_{max}^{(G)} = \frac{1 - N^{1-q}}{q-1} , \tag{12}$$

for $q \in (0, 1) \cup (1, \infty)$.

1.4 Distances and generalized disequilibrium

As for the metrics and its induced distance $\mathcal{D}$ entering in the disequilibrium $\mathcal{Q}$'s definition one faces a panoply of choices [27]. For $P_i \equiv \{p_1^{(i)}, \cdots, p_N^{(i)}\}$, with $i = 1, 2$ discrete proba-

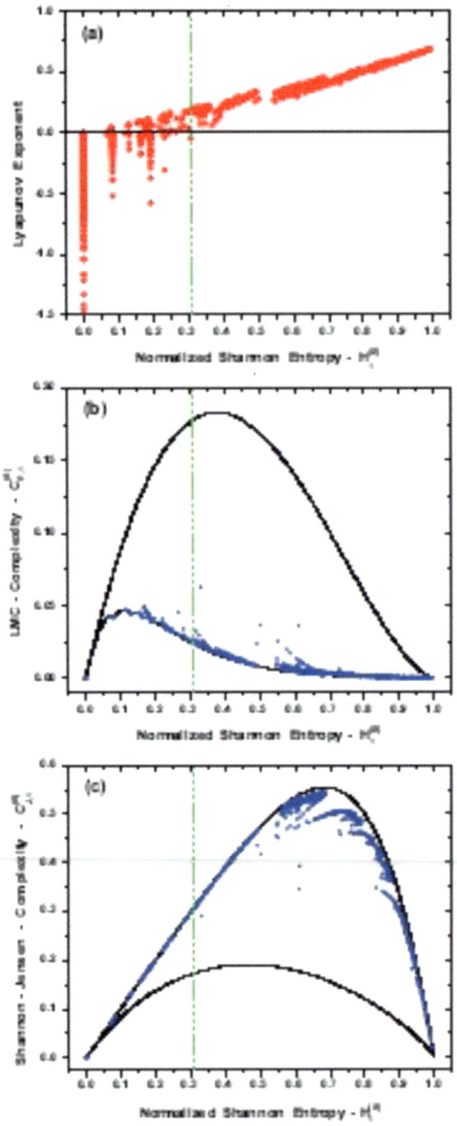

Figure 7: Quantities of interest are plotted as a function of the normalized entropy $\mathcal{H}_1^{(S)}\Big|_{binary}$: *(a)* Lyapunov exponent Λ. Note that the periodic windows are clearly distinguished for $\mathcal{H}_1^{(S)} < H^* \approx 0.3$. *(b)* LMC statistical complexity measure $\mathcal{C}_{LMC} \equiv \mathcal{C}_{E,1}^{(S)}\Big|_{binary}$ and, *(c)* Shannon-Jensen statistical complexity measure $\mathcal{C}_{JS} \equiv \mathcal{C}_{J,1}^{(S)}\Big|_{binary}$. In *(b)* and *(c)* we also display the maximum and minimum possible values of the statistical complexity (continuous lines). The vertical dashed line signals the value H^* for the end of the periodicity zone.

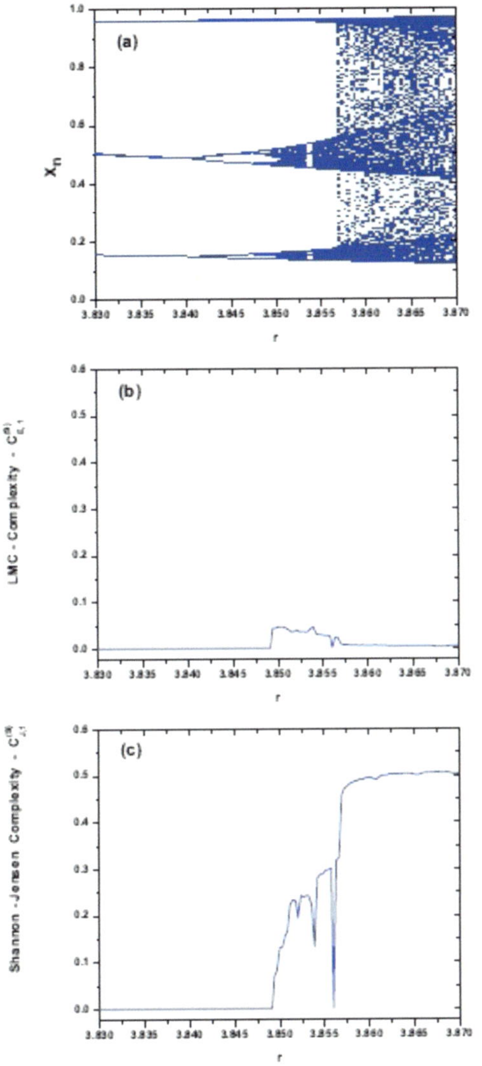

Figure 8: Quantities of interest are plotted against the parameter r in the restricted range that falls within the large 3-periodicity window. *(a)* orbit diagram, *(b)* LMC statistical complexity measure $\mathcal{C}_{LMC} \equiv \mathcal{C}^{(S)}_{E,1}\Big|_{binary}$ and, *(c)* Shannon-Jensen statistical complexity measure $\mathcal{C}_{JS} \equiv \mathcal{C}^{(S)}_{J,1}\Big|_{binary}$.

bility distributions, we limit our considerations here to Euclidean and Wootters distances, Kullback-Leibler relative entropies and Jensen divergences:

a) Euclidean norm $\mathcal{D}_E$ *in* $\mathbb{R}^N$ [11]:

The Euclidean norm is the "natural" choice (the most simple one) for the distance $\mathcal{D}$.

We have

$$\mathcal{D}_E[P_1, P_2] = \|P_1 - P_2\|_E^2 = \sum_{j=1}^{N} \left\{ p_j^{(1)} - p_j^{(2)} \right\}^2 . \quad (13)$$

This will gives the disequilibrium-form for the complexity measure proposed by López-Ruiz, Mancini and Calbet (LMC-complexity measure [11]).

Such straightforward definition of distance has been criticized by Wootters in an illuminating communication [60] because, in using the Euclidean norm, one is ignoring the fact that we are dealing with a space of probability distributions and thus disregarding the stochastic nature of the distribution P.

b) Wootters's distance $\mathcal{D}_W$ [14, 60]:

The concept of "statistical distance" originates in a quantum mechanical context. One uses it primarily to distinguish among different preparations of a given quantum state, and, more generally, to ascertain to what an extent two such states differ from one another. The concomitant considerations being of an intrinsic statistical nature, the concept can be applied to "any" probabilistic space [60]. The main idea underlying this notion of distance is that of adequately taking into account statistical fluctuations inherent to any finite sample. As a result of the associated statistical errors, the observed frequencies of occurrence of the various possible outcomes typically differ somewhat from the actual probabilities, with the result that, in a given fixed number of trials, two preparations are indistinguishable if the difference between the actual probabilities is smaller than the size of a typical fluctuation [60]. The Wootters's distance is given by

$$\mathcal{D}_W[P_1, P_2] = \cos^{-1}\left\{ \sum_{j=1}^{N} \left(p_j^{(1)} \right)^{1/2} \cdot \left(p_j^{(2)} \right)^{1/2} \right\} . \quad (14)$$

Following Basseville [58] two divergence-classes can be built up starting with functionals of the entropy. The first class includes divergences defined as relative entropies, while the second one deals with divergences defined as entropic differences.

c) Kullbak-Leibler relative entropy $\mathcal{D}_K$ [23, 63]:

The relative entropy of P_1 with respect to P_2 associated with Shannon's measure is the relative Kullbak-Leibler Shannon entropy, that in the discrete case reads

$$\mathcal{D}_{K_1^S}[P_1, P_2] = K_1^{(S)}[P_1|P_2] = \sum_{j=1}^{N} p_j^{(1)} \log\left(\frac{p_j^{(1)}}{p_j^{(2)}}\right) . \quad (15)$$

Consider now the probability distribution P and the uniform distribution P_e. The distance between these two distributions, $\mathcal{D}_K$, in Kullback-Leibler Shannon terms, will be

$$\mathcal{D}_{K_1^S}[P, P_e] = K_1^{(S)}[P|P_e] = \mathrm{S}_1^{(S)}[P_e] - \mathrm{S}_1^{(S)}[P] . \quad (16)$$

The relative entropy associated with Tsallis and escort-Tsallis entropy are called the relative Kullback-Tsallis and the Kullback-escort-Tsallis entropies, (or q-Kullback measures). The q-Kullback entropies of P_1 with respect to P_2 (both discrete distributions) are

$$\mathcal{K}_q^{(T)}[P_1|P_2] = \frac{1}{(q-1)} \sum_{j=1}^{N} \left(p_j^{(1)}\right)^q \cdot \left\{\left(p_j^{(2)}\right)^{1-q} - \left(p_j^{(1)}\right)^{1-q}\right\} , \quad (17)$$

$$\mathcal{K}_q^{(G)}[P_1|P_2] = \frac{1}{(q-1)} \sum_{j=1}^{N} \frac{p_j^{(1)}}{(A[P_1])^q} \cdot \left\{\left[\frac{\left(p_j^{(2)}\right)^{1/q}}{A[P_2]}\right]^{1-q} - \left[\frac{\left(p_j^{(1)}\right)^{1/q}}{A[P_1]}\right]^{1-q}\right\} , \quad (18)$$

for $q \neq 1$ and $A[P] = \sum_{j=1}^{N} (p_j)^{1/q}$.

We shall call $\mathcal{D}_{\mathcal{K}_q^T}$ and $\mathcal{D}_{\mathcal{K}_q^G}$ the accompanying distance to the uniform distribution P_e

$$\mathcal{D}_{\mathcal{K}_q^T} = \mathcal{K}_q^{(T)}[P|P_e] = N^{q-1}\,(\mathrm{S}_q^{(T)}[P_e] - \mathrm{S}_q^{(T)}[P]) , \quad (19)$$

$$\mathcal{D}_{\mathcal{K}_q^G} = \mathcal{K}_q^{(G)}[P|P_e] = N^{q-1}\,(\mathrm{S}_q^{(G)}[P_e] - \mathrm{S}_q^{(G)}[P]) . \quad (20)$$

The relative entropy associated with Rényi's measure (also a q-Kullback measure) is the relative Kullback-Rényi entropy. The q-Kullback (Rényi) entropy P_1 vis-a-vis P_2 (discrete distributions) is

$$\mathcal{K}_q^{(R)}[P_1|P_2] = \frac{1}{(q-1)} \ln\left\{\sum_{j=1}^{N} \left(p_j^{(1)}\right)^q \left(p_j^{(2)}\right)^{1-q}\right\} , \quad (21)$$

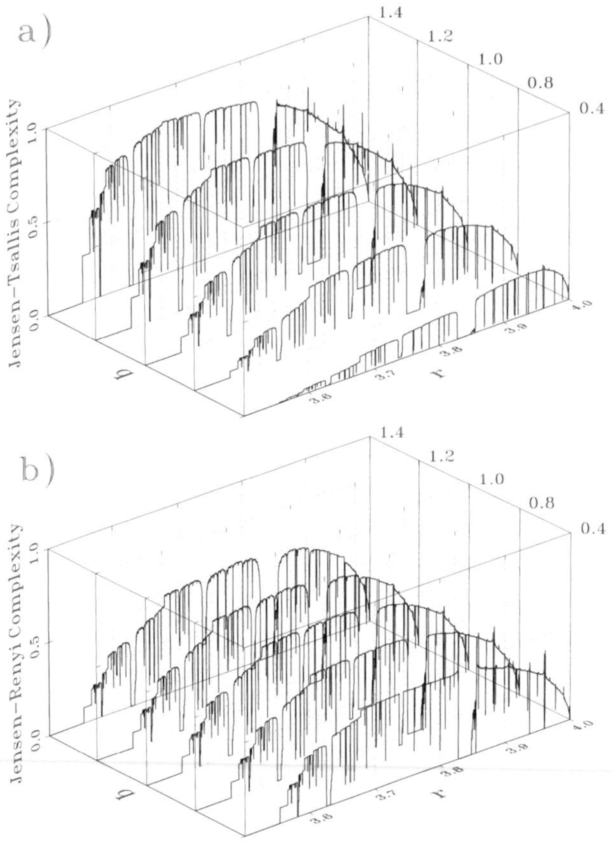

Figure 9: *a)* Tsallis-Jensen statistical complexity measure $\mathcal{C}^{(T)}_{\mathcal{J},q}\big|_{binary}$ and, *b)* Rényi-Jensen statistical complexity measure $\mathcal{C}^{(R)}_{\mathcal{J},q}\big|_{binary}$ for non-extensive parameter $q = 0.4$, 0.8, 1.0, 1.2, 1.6 evaluated for the logistic map (binary sequence) as function of parameter r ($3.5 \leq r \leq 4$, $\Delta r = 0.0001$). Figure take from Ref. [75]

for for $q \neq 1$. Here $\mathcal{D}_{\mathcal{K}^R_q}$ is the appropriate distance to the uniform distribution P_e defined in terms of Kullback-Rényi's entropy

$$\mathcal{D}_{\mathcal{K}^R_q}[P, P_e] \;=\; \mathcal{K}^{(R)}_q[P|P_e] \;=\; \mathrm{S}^{(R)}_q[P_e] - \mathrm{S}^{(R)}_q[P] \;. \tag{22}$$

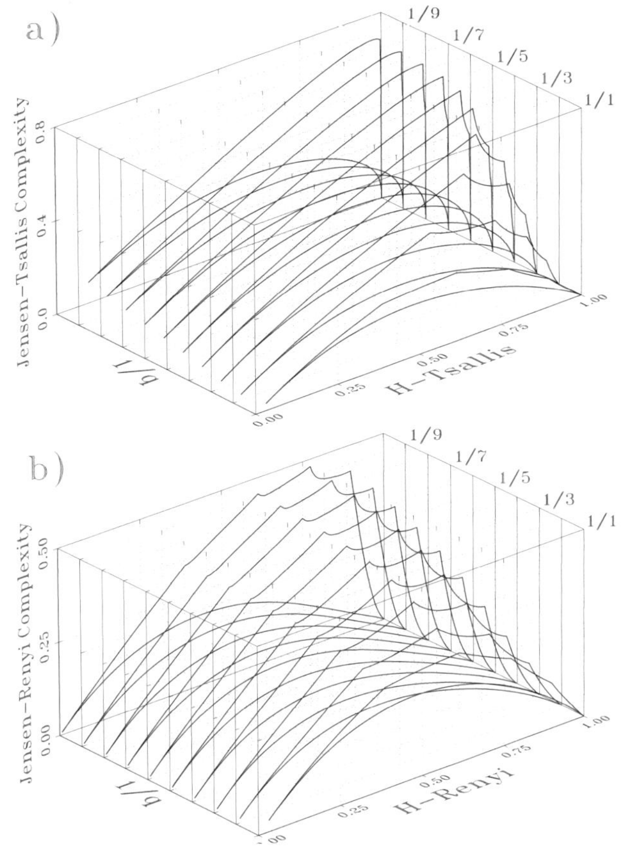

Figure 10: Variation of the extreme values for statistical complexity measure $\mathcal{C}_{min}$ and $\mathcal{C}_{max}$ for entropy-complexity plane *(a)* $\mathcal{H}_q^{(T)} \times \mathcal{C}_{\mathcal{J},q}^{(T)}$ and, *(b)* $\mathcal{H}_q^{(R)} \times \mathcal{C}_{\mathcal{J},q}^{(R)}$ with the non-extensive parameter q. In present case, $N = 6$ and the parameter q was varied between 1 and 9 with step 1. Figure take from Ref. [75]

In general, the entropic difference $\mathrm{S}[P_1] - \mathrm{S}[P_2]$ does not define an information gain (or divergence) because the difference is not necessarily positive definite. Something else is needed.

d) Jensen divergence $\mathcal{D}_J$ [15, 23]:

An important example is provided by Jensen's divergence, $J_S^{\beta}[P_1, P_2]$ with $0 \leq \beta \leq 1$ [58, 71]

$$J_S^{\beta}[P_1, P_2] = \mathrm{S}[\beta P_1 + (1-\beta) P_2] - \beta\ \mathrm{S}[P_1] - (1-\beta)\ \mathrm{S}[P_2]\ , \tag{23}$$

whose positivity depends on the concavity of entropy S. In the following we consider only the case $\beta = 1/2$ and the corresponding supra-index will be omitted for simplicity of the notation.

If $\mathrm{S} \equiv \mathrm{S}_1^{(S)}$ is the Shannon entropy measure, we have

$$J_{\mathrm{S}_1^{(S)}}[P_1, P_2] = \mathrm{S}_1^{(S)}\left[\frac{(P_1+P_2)}{2}\right] - \frac{1}{2}\,\mathrm{S}_1^{(S)}[P_1] - \frac{1}{2}\,\mathrm{S}_1^{(S)}[P_2]\ , \tag{24}$$

that may also be expressed in terms of relative entropies

$$J_{\mathrm{S}_1^{(S)}}[P_1, P_2] = \frac{1}{2}\,K_1^{(S)}\left[P_1\left|\frac{(P_1+P_2)}{2}\right.\right] + \frac{1}{2}\,K_1^{(S)}\left[P_2\left|\frac{(P_1+P_2)}{2}\right.\right]\ , \tag{25}$$

with $K_1^{(S)}$ the relative Kullback-Leibler Shannon measure given by Eq. (15). $K_1^{(S)}$ is a concave function and the Jensen-Shannon divergence, $J_{\mathrm{S}_1^{(S)}}$, verifies the following properties

$$(i) \qquad J_{\mathrm{S}_1^{(S)}}[P_1, P_2] \geq 0\ , \tag{26}$$

$$(ii) \qquad J_{\mathrm{S}_1^{(S)}}[P_1, P_2] = J_{\mathrm{S}_1^{(S)}}[P_2, P_1]\ , \tag{27}$$

$$(iii) \qquad J_{\mathrm{S}_1^{(S)}}[P_1, P_2] = 0 \Leftrightarrow P_2 = P_1\ . \tag{28}$$

Moreover, it square root satisfies the triangle inequality

$$(iv) \quad \{J_{\mathrm{S}_1^{(S)}}[P_1, P_2]\}^{1/2} + (\{J_{\mathrm{S}_1^{(S)}}[P_2, P_3]\}^{1/2} = \{J_{\mathrm{S}_1^{(S)}}[P_1, P_3]\}^{1/2}\ , \tag{29}$$

which implies that the square root of the Jensen-Shannon divergence is a metric [64, 71]. Note also that it is defined in terms of entropies and, in consequence, is an extensive quantity in the thermodynamical sense. Thus, the corresponding statistical complexity will be an intensive quantity as well.

The Jensen divergence (given by Eq. (23)) associated with a measure for which definite concavity properties cannot be established, the Jensen measure itself may lose its positive character. The relation between Eqs. (24) and (25) allows one to conceive of a generalization of Jensen's divergence by simply replacing the relative entropy in Eq. (25), using for the purpose either Tsallis, escort-Tsallis or Rényi relative entropies (among several possibilities). One finds

$$\mathcal{J}_{\mathrm{S}_q^{(T)}}[P_1, P_2] = \frac{1}{2}\,K_q^{(T)}\left[P_1\left|\frac{(P_1+P_2)}{2}\right.\right] + \frac{1}{2}\,K_q^{(T)}\left[P_2\left|\frac{(P_1+P_2)}{2}\right.\right]\ , \tag{30}$$

$$\mathcal{J}_{\mathrm{S}_q^{(G)}}[P_1, P_2] = \frac{1}{2}\, K_q^{(G)}\left[P_1\left|\frac{(P_1+P_2)}{2}\right.\right] + \frac{1}{2}\, K_q^{(G)}\left[P_2\left|\frac{(P_1+P_2)}{2}\right.\right] , \tag{31}$$

$$\mathcal{J}_{\mathrm{S}_q^{(R)}}[P_1, P_2] = \frac{1}{2}\, K_q^{(R)}\left[P_1\left|\frac{(P_1+P_2)}{2}\right.\right] + \frac{1}{2}\, K_q^{(R)}\left[P_2\left|\frac{(P_1+P_2)}{2}\right.\right] . \tag{32}$$

We will call $\mathcal{D}_J$ the distance to the uniform distribution associated with Jensen-Shannon divergence

$$\mathcal{D}_{J_1^S} = J_{\mathrm{S}_1^{(S)}}[P, P_e] = \mathrm{S}_1^{(S)}\left[\frac{(P+P_e)}{2}\right] - \frac{1}{2}\,\mathrm{S}_1^{(S)}[P] - \frac{1}{2}\,\mathrm{S}_1^{(S)}[P_e] . \tag{33}$$

Likewise, $\mathcal{D}_{\mathcal{J}_q^T}$, $\mathcal{D}_{\mathcal{J}_q^G}$ and $\mathcal{D}_{\mathcal{J}_q^R}$ would be the distance to the uniform distribution associated with the Jensen-Tsallis, Jensen-escort-Tsallis and Jensen-Rényi divergence (q-Jensen divergences), respectively,

$$\mathcal{D}_{\mathcal{J}_q^T} = \mathcal{J}_{\mathrm{S}_q^{(T)}}[P, P_e] , \tag{34}$$

$$\mathcal{D}_{\mathcal{J}_q^G} = \mathcal{J}_{\mathrm{S}_q^{(G)}}[P, P_e] , \tag{35}$$

$$\mathcal{D}_{\mathcal{J}_q^R} = \mathcal{J}_{\mathrm{S}_q^{(R)}}[P, P_e] . \tag{36}$$

Let $\mathcal{Q}_q^{(\nu)}$ be the generalized disequilibrium defined as

$$\mathcal{Q}_q^{(\nu)}[P, P_e] = Q_0^{(\nu)}\, \mathcal{D}_\nu[P, P_e] , \tag{37}$$

where ν stands for the different measures of distance $\mathcal{D}$ advanced above. Thus, $\nu = E$, W, K, $\mathcal{K}_q$, J, $\mathcal{J}_q$ (Euclidean, Wootters, Kullback, q-Kullback, Jensen, q-Jensen). $Q_0^{(\nu)}$ is the concomitant normalization constant such that $0 \leq \mathcal{Q}_q^{(\nu)} \leq 1$. We will have:

$$Q_0^{(E)} = \frac{N}{N-1} , \tag{38}$$

$$Q_0^{(W)} = 1/\cos^{-1}\left\{\left(\frac{1}{N}\right)^{1/2}\right\} , \tag{39}$$

$$Q_0^{(K^S)} = Q_0^{(K_q^R)} = 1/\ln N , \tag{40}$$

$$Q_0^{(\mathcal{K}_q^T)} = Q_0^{(\mathcal{K}_q^G)} = \frac{(q-1)}{N^{(q-1)}-1} , \tag{41}$$

$$Q_0^{(J^S)} = -2\left\{\left(\frac{N+1}{N}\right)\ln(N+1) - \ln(2N) + \ln N\right\}^{-1} , \tag{42}$$

$$Q_0^{(\mathcal{J}_q^T)} = Q_0^{(\mathcal{J}_q^G)} = \tag{43}$$
$$= (1-q) \cdot \left\{ 1 - \left[\frac{(1+N^q)(1+N)^{(1-q)} + (N-1)}{2^{(2-q)} N} \right] \right\}^{-1} ,$$

$$Q_0^{(\mathcal{J}_q^R)} = 2\,(1-q) \cdot \tag{44}$$
$$\left\{ \ln \left(\frac{(N+1)^{(1-q)} + (N-1)}{2^{(1-q)} N} \right) + (1-q) \ln \left(\frac{N+1}{2N} \right) \right\}^{-1} .$$

1.5 Generalized Statistical Complexity measures

On the basis of the functional product form $\mathcal{C} = \mathcal{Q} \cdot \mathcal{H}$ we obtain a family of statistical complexity measures $\mathcal{C}_{\nu,q}^{(\kappa)}$, for each one of four distinct disorder measures $\mathcal{H}_q^{(\kappa)}$ and disequilibriums $\mathcal{Q}_q^{(\nu)}$ just enumerated, namely,

$$\mathcal{C}_{\nu,q}^{(\kappa)}[P] = \mathcal{H}_q^{(\kappa)}[P] \cdot \mathcal{Q}_q^{(\nu)}[P] . \tag{45}$$

with $\kappa = S, T, G, R$ (Shannon, Tsallis, escort-Tsallis, Rényi) for fixed q. In Shannon's instance ($\kappa = S$) we have, of course, $q = 1$. The index $\nu = E, W, K, \mathcal{K}_q, J, \mathcal{J}_q$ tell us that the disequilibrium-distance is to be evaluated with the appropriate distance measure (Euclidean, Wootters, Kullback, q-Kullback, Jensen, q-Jensen), respectively.

It is interesting to note that for $\nu = K, \mathcal{K}_q$ the SCM family becomes

$$\mathcal{C}_{K,q}^{(\kappa)}[P] = \mathcal{H}_q^{(\kappa)}[P] \cdot (1 - \mathcal{H}_q^{(\kappa)}[P]) . \tag{46}$$

Such is the generalized functional form advanced by Shiner, Davison, and Landsberg [13] for the SCM. For $\kappa = S$ and $q = 1$ we get the original SDL statistical complexity $C_{SDL} \equiv \mathcal{C}_{K,1}^{(S)}$. Note that the generalized SDL complexity measure is evaluated from the normalized information measure $\mathcal{H}_1^{(S)}$. One could raise the objection that this SCM is just a simple function of the entropy. As a consequence, it might not contain new information vis-a-vis the measure of order. Such an objection is discussed at length in Refs. [72, 73, 74].

We stress the fact that the remaining members of the family $\mathcal{C}_{\nu,q}^{(\kappa)}$, with $\nu = E, W, J, \mathcal{J}_q$ are not trivial functions of the entropy because they depend on two different probabilities distributions, the one associated with the system under analysis, P, and the uniform distribution P_e. Furthermore, it has been shown that for a given $\mathcal{H}$ value, there exists a range of possible SCM values, from a minimum one $\mathcal{C}_{min}$ up to a maximal value $\mathcal{C}_{max}$. Evaluation

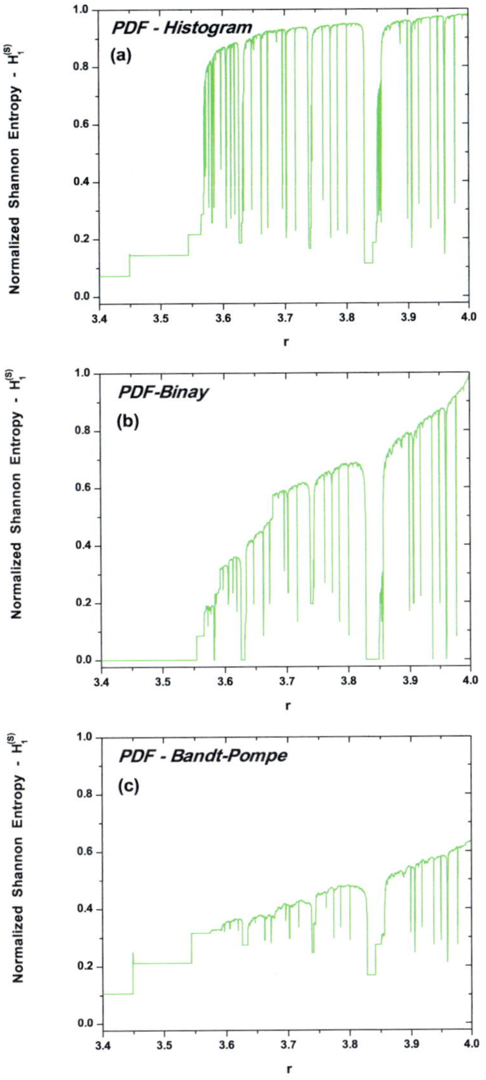

Figure 11: Normalized Shannon entropy, $\mathcal{H}_1^{(S)}$ evaluated for the logistic map as function of the parameter r ($3.4 \leq r \leq 4$, $\Delta r = 0.0003$) considering either *a)* an histogram-determined PDF, $\mathcal{H}_{hist} \equiv \mathcal{H}_1^{(S)}\Big|_{histogram}$; *b)* a binary PDF, $\mathcal{H}_{bin} \equiv \mathcal{H}_1^{(S)}\Big|_{binary}$ one or; *c)* a Bandt-Pompe PDF, $\mathcal{H}_{BP} \equiv \mathcal{H}_1^{(S)}\Big|_{Bandt-Pompe}$.

of $\mathcal{C}_{\nu,q}^{(\kappa)}$ yields, consequently, new information according to the peculiarities of the pertinent probability distribution. A general procedure to obtain the bounds $\mathcal{C}_{min}$ and $\mathcal{C}_{max}$ corresponding to the generalized $\mathcal{C}_{\nu,q}^{(\kappa)}$-family is given in [75]. Thus, it is clear that important additional information related to the correlational structure between the components of the

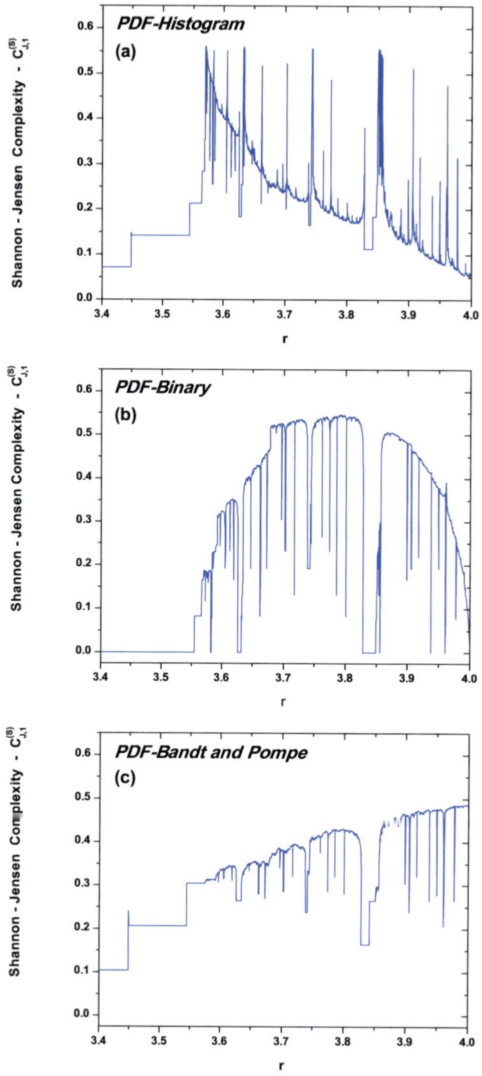

Figure 12: Shannon-Jensen generalized statistical complexity, $\mathcal{C}^{(S)}_{J,1}$ evaluated for the logistic map as function of the parameter r ($3.4 \leq r \leq 4$, $\Delta r = 0.0003$) considering either *a)* an histogram-determined PDF, $\mathcal{C}_{hist} \equiv \mathcal{C}^{(S)}_{J,1}\big|_{histogram}$; *b)* a binary PDF, $\mathcal{C}_{bin} \equiv \mathcal{C}^{(S)}_{J,1}\big|_{binary}$ one or; *c)* a Bandt-Pompe PDF, $\mathcal{C}_{BP} \equiv \mathcal{C}^{(S)}_{J,1}\big|_{Bandt-Pompe}$.

physical system is provided by evaluating the statistical complexity.

1.6 Time evolution

In statistical mechanics one is often interested in isolated systems characterized by an initial, arbitrary, and discrete probability distribution. Evolution towards equilibrium is to be described, as the overriding goal. At equilibrium, we can think, without loss of generality, that this state is given by the uniform distribution P_e. In order to study the time evolution of the SCM a diagram of $\mathcal{C}$ *versus* time t can then be used. But, as we know, the second law of thermodynamics states that in isolated system entropy grows monotonically with time ($d\mathcal{H}/dt \geq 0$) [76]. This implies that $\mathcal{H}$ can be regarded as an arrow of time, so that an equivalent way to study the time evolution of the SCM is to plot $\mathcal{C}$ *versus* $\mathcal{H}$. In this way, the normalized entropy-axis substitutes for the time-axis. This kind of diagram $\mathcal{H} \times \mathcal{C}$ has been used also to study changes in the dynamics of a system originated by modifications of some characteristic parameters (see, for instance, Refs. [11, 23, 55, 75, 77], and references therein).

1.7 Additional issues

As we mentioned before, if $\kappa = S$ and $\nu = E$ we recover the statistical complexity measure definition given originally by López-Ruiz, Mancini and Calbet (LMC) ($\mathcal{C}_{LMC} \equiv \mathcal{C}^{(S)}_{E,1}$) [11]. It has been pointed out by Crutchfield and co-workers [12] that the LMC measure is marred by some troublesome characteristics that we list below:

(1) it is neither an intensive nor an extensive quantity.

(2) it vanishes exponentially in the thermodynamic limit for all one-dimensional, finite-range systems.

The above authors forcefully argue that a reasonable SCM should

(3) be able to distinguish among different degrees of periodicity;

(4) vanish only for unity periodicity.

Finally, and with reference to the ability of the LMC measure to adequately capture essential dynamical aspects, some difficulties have also been encountered [78].

The product functional form for the generalized SCM makes it impossible to overcome the second deficiency mentioned above. In previous works we have shown that, after performing some suitable changes in the definition of the disequilibrium-distance, by means of utilization of either Wootters distance [14] or Jensen's divergence [15], one is in a position to obtain a generalized SCM that is:

(i) able to grasp essential details of the dynamics (*i.e.*, chaos, intermitency, *etc.*)

(ii) capable of discerning between different degrees of periodicity, and

(iii) an intensive quantity if Jensen's divergence is used.

2 Methodologies for selecting PDFs

An important point for the evaluation of the previous information quantifiers (entropy and generalized statistical complexity measures) is the proper determination of the underlying probability distribution function (PDF) P, associated with a given dynamical system or time series $\mathcal{S}(t) = \{x_t; t = 1, \cdots, M\}$. This often neglected issue deserves serious consideration as the probability distribution P and the sample space Ω are inextricably linked.

Many schemes have been proposed for a proper selection of the probability space (Ω, P). We can mention, among others: *(a)* frequency count [79] *(b)* procedures based on amplitude statistics (histograms) [31], *(c)* binary symbolic dynamics [80], *(d)* Fourier analysis [81] and, *(e)* wavelet transform [82, 83]. Their applicability depends on particular characteristics of the data such as stationarity, length of the time series, variation of the parameters, level of noise contamination, *etc.* In all these cases the global aspects of the dynamics can be somehow captured, but the different approaches are not equivalent in their ability to discern all the relevant physical details. One must also acknowledge the fact that the above techniques are introduced in a *rather ad hoc fashion and are not directly derived from dynamical properties of the system under study*. This can be adequately effected, for instance, by recourse to the Bandt-Pompe methodology [9].

In the present work we compare the results of using Information Theory quantifiers (Entropy and Generalized Statistical Complexity) for the logistic map (see next section) after a proper evaluation of the all important probability distribution function (PDF) is made by recourse to one of the three distinct approaches in coordinate space, namely,

(i) histograms of the x_t−values,

(ii) binary representations, and

(iii) the Bandt-Pompe technique.

The temporal information-content of each being discussed. The main idea is to shown the importance of inclusion of time causality in the corresponding PDF construction for an accurate description of the dynamics under study. Note also, that if the PDF evaluation is made in the frequency space similar results are obtained (not shown). If the PDF is obtained by recourse of the Fast Fourier Transform (FFT) no temporal information is included. Contrary, in the case of Discrete Wavelet Transform is used for determination of the PDF, frequency and temporal information are included in average way (for details see [82, 83]).

2.1 PDF based on histograms

In order to extract a PDF *via* amplitude-statistics, the interval $[a, b]$ (with a and b the minumum and maximum of the time series $\mathcal{S}(t) = \{x_t; t = 1, \cdots, M\}$) is first divided into a finite number N_{bin} of non overlapping subintervals A_i: $[a, b] = \bigcup_{i=1}^{N_{bin}} A_i$ and $A_i \bigcap A_j = \emptyset$, $\forall i \neq j$. One then employs the usual histogram-method, based on counting the relative frequencies of the time series values within each subinterval.

It should be clear that the resulting PDF lacks any information regarding temporal ordering (temporal causality). The only pieces of information we have here are the x_t-values that allow one to assign inclusion within a given bin, ignoring their temporal location (this is, the subindex i). Note that N in equations (4) to (46) is equal to N_{bin}. Let us also point out that it is relevant to consider a judiciously chosen optimal value for N_{bin} (see *i.e.*, De Micco *et al.* [31]).

2.2 PDF based on binary representation:

Following the original work of Lopez-Ruiz, Mancini and Calbet [11], for each parameter value, r, the dynamics of the logistic map was reduced to a binary sequence (0 if $x \leq \frac{1}{2}$; 1 if $x > \frac{1}{2}$) and binary strings of length $L = 12$ without overlap were considered as states of the system. In this case some temporal information is, in average, retained. Why? Because we face a two-steps procedure in this approach. In making the binary assignment (1st. step), part of the temporal information is lost by averaging, but, in considering words of length L (without overlap) in a second step we would are indeed respecting such "averaged" temporal information.

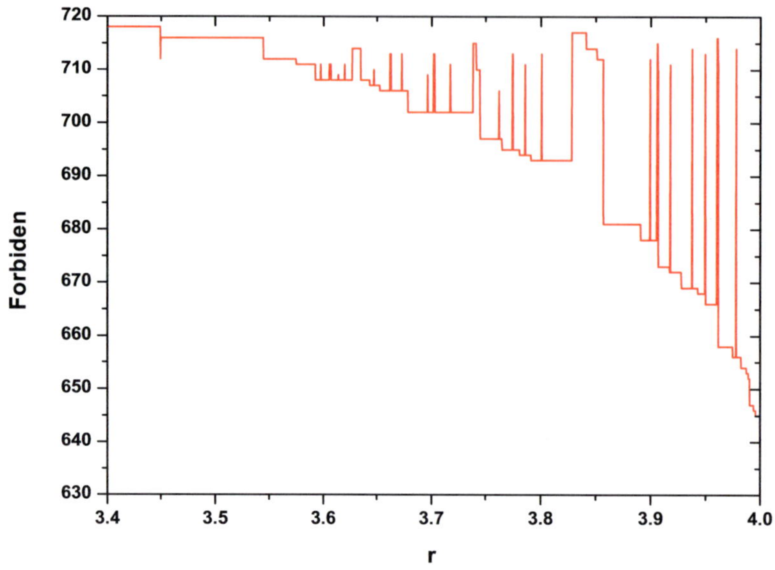

Figure 13: Forbidden patterns for the logistic map as function of parameter r ($3.4 \leq r \leq 4$, $\Delta r = 0.0003$). Embedding dimension $D = 6$.

2.3 PDF based on Bandt and Pompe's methodology

To use the Bandt and Pompe [9] methodology for evaluating the probability distribution P associated with the time series (dynamical system) under study, one starts by considering partitions of the pertinent D-dimensional space that will hopefully "reveal" relevant details of the ordinal-structure of a given one-dimensional time series $\mathcal{S}(t) = \{x_t; t = 1, \cdots, M\}$ with embedding dimension $D > 1$ and time delay τ. In the following we take $\tau = 1$ [9]. We are interested in "ordinal patterns" of order D [9, 84] generated by

$$(s) \mapsto \left(x_{s-(D-1)}, x_{s-(D-2)}, \cdots, x_{s-1}, x_s \right) , \tag{47}$$

which assigns to each time s the D-dimensional vector of values at times $s, s-1, \cdots, s-(D-1)$. Clearly, the greater the $D-$value, the more information on the past is incorporated into our vectors. By "ordinal pattern" related to the time (s) we mean the permutation $\pi = (r_{D-1}, r_{D-2}, \cdots, r_1, r_0)$ of $[D-1, D-2, \cdots, 1, 0]$ defined by

$$x_{s-r_{D-1}} \leq x_{s-r_{D-2}} \leq \cdots \leq x_{s-r_1} \leq x_{s-r_0} . \tag{48}$$

In order to get a unique result we set $r_i < r_{i-1}$ if $x_{s-r_i} = x_{s-r_{i-1}}$. This is justified if the values of x_t have a continuous distribution so that equal values are very unusual. Otherwise, it is possible to break these equalities by adding small random perturbations.

Thus, for all the $D!$ possible permutations π of order D, their associated relative frequencies can be naturally computed by the number of times this particular order sequence is found in the time series divided by the total number of sequences. The probability distribution $P = \{p(\pi)\}$ is defined by

$$p(\pi) \; = \; \frac{\sharp\{s|s \leq M - D + 1;\ (s),\ \texttt{has type}\ \pi\}}{M - D + 1} \ . \tag{49}$$

In this expression, the symbol $\sharp$ stands for "number".

The procedure can be better illustrated with a simple example; let us assume that we start with the time series $\{1, 3, 5, 4, 2, 5, \ldots\}$, and we set the embedding dimension $D = 4$. In this case the state space is divided into 4! partitions and 24 mutually exclusive permutation symbols are considered. The first 4-dimensional vector is $(1, 3, 5, 4)$. According to Eq. (47) this vector corresponds with $(x_{s-3}, x_{s-2}, x_{s-1}, x_s)$. Following Eq. (48) we find that $x_{s-3} \leq x_{s-2} \leq x_s \leq x_{s-1}$. Then, the ordinal pattern which allows us to fulfill Eq. (48) will be $[3, 2, 0, 1]$. The second 4-dimensional vector is $(3, 5, 4, 2)$, and $[0, 3, 1, 2]$ will be its associated permutation, and so on. For the computation of the Bandt and Pompe PDF we follow the very fast algorithm described by Keller and Sinn [84], in which the different ordinal patterns are generated in lexicographic ordering.

The Bandt-Pompe methodology is not restricted to time series representative of low dimensional dynamical systems but can be applied to any type of time series (regular, chaotic, noisy, or reality based), with a very weak stationary assumption (for $k = D$, the probability for $x_t < x_{t+k}$ should not depend on t [9]). It also assumes that enough data are available for a correct embedding procedure. Of course, the embedding dimension D plays an important role in the evaluation of the appropriate probability distribution because D determines the number of accessible states $D!$. Also, it conditions the minimum acceptable length $M \gg D!$ of the time series that one needs in order to extract reliable statistics.

The main difference between entropy and complexity measures evaluated with a PDF-histogram or a PDF-Bandt and Pompe is that, unlike those quantifiers based on a PDF-histogram, those obtained à la Bandt and Pompe are invariant with respect to strictly monotonic distortions of the data. This point will be the very important in the case of empiric natural time series (*i.e.*,, sedimentary data are related to rainfall by an unknown nonlinear function [85]). However, due to their invariance properties, quantifiers based on Bandt and Pompe's perspective, when applied to any other data-series that is strictly monotonic in the given data will yield the same results. Therefore, in general, ordinal data processing tools based on ranked numbers would lead to results that could be more useful for data analysis than those obtained using tools based on metric properties.

3 The Logistic Map

The logistic map constitutes a paradigmatic example, often employed as a testing-ground in order to illustrate new concepts in the treatment of dynamical systems. In such a vein we discuss here the application of the generalized statistical complexity measures (see Eq. (45)).

We deal then with the map $F : x_n \to x_{n+1}$ [3, 4], described by the ecologically motivated, dissipative system described by the first order difference equation

$$x_{n+1} = r \cdot x_n \cdot (1 - x_n) \tag{50}$$

with $0 \leq x_n \leq 1$ and $0 < r \leq 4$.

Figure 1.a depicts the well known bifurcation diagram for the logistic map for $0 < r \leq 4$. In Fig. 1.b a zoom in the range of $3.4 \leq r \leq 4.0$ is presented and in Fig. 2, the corresponding Lyapunov exponent Λ is also shown. Let us briefly review, with reference to these two figures (Figs. 1 and 2), some exceedingly well known results for this map that we need in order to put into an appropriate perspective the properties of our family of generalized statistical complexity measures.

If $0 < r \leq 1$ only the value $x_n = 0$ is observed. For values of the control parameter $1 < r \leq 3$ there exists only a single steady-state solution. Increasing the control parameter past $r = 3$ forces the system to undergo a period-doubling bifurcation. Cycles of period 8, 16, 32, $\cdots$ occur and, if r_n denotes the value of r where a 2^n cycle first appears, the r_n converge to a limiting value $r_\infty \cong 3.5699456$ [3, 4]. As r grows still more, a quite rich, and well-known structure arises. In order to be in a position to better appreciate at once the long-term behavior for all values of r lying between 3.4 and 4.0, we plot the pertinent bifurcation diagram in Fig. 1.b. We immediately note there the cascade of further period-doubling that occurs as r increases, until, at r_∞, the maps become chaotic and the attractors change from comprising a finite set of points to becoming an infinite set. For $r > r_\infty$ the orbit-diagram reveals an "strange" mixture of order and chaos. The large window beginning near $r = 3.8284$ contains a stable period-3 cycle.

In Fig. 2 we see that the non-zero Lyapunov characteristic exponent Λ remains negative for $r < r_\infty$. We notice that Λ approaches zero at the period-doubling bifurcation. The onset of chaos is apparent near $r \cong 3.5699$, where Λ first becomes positive. As stated above, for $r > r_\infty$ the Lyapunov exponent increases globally, except for the dips one sees in the windows of periodic behavior. Notice the particularly large dip due to the period-3 window near $r = 3.8284$.

4 Results and Discussion

Time series with total length $M = 10^6$ data were generated for the logistic map for parameter value r in the range $3.4 \leq r \leq 4.0$ with step of $\Delta r = 0.0003$. The concomitant probabilities are assigned according to the frequency of occurrence after running over at least 10^5 iterations, in order that all the transient died. For the evaluation of PDF-histogram we consider $N_{bin} = N = 2^{14}$ bins in the interval $[0, 1]$. In the case of PDF-binary time series the word length was fixed to $L = 12$ bits ($N = 2^{12}$) and for the case of PDF-Bandt and Pompe, the embedding dimension was fixed to $D = 6$ ($N = 6! = 720$). Note that, in the case of binary representation, similar results were obtained when the binary strings' length was taken to be $L = 16$.

The behavior ("degree of chaoticity") of the Lyapunov exponent Λ as a function of the parameter r is displayed in Fig. 2. From this figure, we see that Λ and, as a result, the associated degree of chaoticity grows globally with r (since there are many periodic windows where Λ drops to negative values), reaching a maximum at $r = 4$. One would expect that a sensible statistical complexity measure should accompany such a globally growth. In other words, a reasonable statistical complexity measure should take very small values for $r < r_\infty$ and then grow globally together with the degree of chaoticity.

We present and discuss first the roll of the different distances entering in the definition of the disequilibrium, as well as the different entropic forms. In order to do that, we exemplified with the use of PDF-binary.

Figures 3.a and 4.a display the Euclidean and Kullback-Leibler disequilibrium, $\mathcal{Q}^{(E)}\Big|_{binary}$ and $\mathcal{Q}^{(K)}\Big|_{binary}$, respectively, for the logistic map (PDF-binary representation). We depict in Figs. 3.b and 4.b the SCM corresponding to $\kappa = S$ (Shannon entropy) with $\nu = E$ (Euclidean distance) and $\nu = K$ (Kullback-Leibler Shannon relative entropy) respectively.

These figures correspond to the López-Ruiz, Mancini, Calbet (LMC), $\mathcal{C}_{LMC} \equiv \mathcal{C}^{(S)}_{E,1}\Big|_{binary}$, and Shiner, Davison, Landsberg (SDL), $\mathcal{C}_{SDL} \equiv \mathcal{C}^{(S)}_{K,1}\Big|_{binary}$, statistical complexity measures, evaluated for the logistic map (binary sequence) as functions of the logistic parameter $3.4 \leq r \leq 4.0$.

We observe, in both instances, an abrupt statistical complexity growth around $r > r_\infty \cong 3.57$. After we pass this point, the two measures behave in quite different fashions. Notwithstanding the fact that both measures almost vanish within the large stability windows, in the inter-windows region they noticeably differ. In the latter zone, the LMC measure globally decreases, reaching almost null values. Consider, for example, $3.58 \leq r \leq 3.62$ in Fig. 3. The many peaks indicate a local complexity growth. Comparison with the orbit-diagram (see Fig. 1.b) indicates that the peaks coincide with the periodic windows. The original LMC measure ($\kappa = S$ and $\nu = E$) regards this periodic motion as having a stronger statistical complexity-character than that pertaining to the neighboring chaotic zone. Note, instead, that the SDL measure ($\kappa = S$ and $\nu = K$) does grow in the inter-windows region and rapidly falls within the periodic windows (see Fig. 4.b).

Plots of $\mathcal{Q}^{(W)}\Big|_{binary}$ and $\mathcal{Q}^{(J)}\Big|_{binary}$, Wootters and Jensen-Shannon disequilimbrium respectively, for the logistic map (PDF-binary representation) are shown in Figs. 5.a and 6.a. The Shannon-Wootters ($\kappa = S$ and $\nu = W$) and Shannon-Jensen ($\kappa = S$ and $\nu = J$) statistical complexity measures are denoted by $\mathcal{C}_W \equiv \mathcal{C}^{(S)}_{W,1}\Big|_{binary}$ and $\mathcal{C}_{JS} \equiv \mathcal{C}^{(S)}_{J,1}\Big|_{binary}$ respectively. They are evaluated for the logistic map (PDF-binary representation) as functions of the parameter r ($3.4 \leq r \leq 4.0$) and we shown in Figs. 5.b and 6.b. As in the previous cases of LMC and SDL complexity, for both complexities $\mathcal{C}_W$ and $\mathcal{C}_{JS}$ an abrupt complexity growth around $r \sim r_\infty$ is observed. We also note that the behavior of both measures, $\mathcal{C}_W$ and $\mathcal{C}_{JS}$, as the logistic parameter r varies, is similar. This similarity between both measures is due to the fact, that the Jensen-Shannon divergence and the the Wootters distance are coincident at third order in a series expansion and its difference at fourth order is about $1/200$ [86].

Globally, the decay of $\mathcal{C}_{JS}$ is swifter the closer r becomes to the special value $r = 4$. Note that for the region $r_\infty \leq r \leq 4$ both statistical complexity measures do grow in the inter-windows region and rapidly fall within the periodic windows. More importantly, we clearly appreciate here the fact the both statistical complexity measures do distinguish among different periodicities, as rightly demanded in Ref. [12]. The "Lyapunov"-based criterium mentioned above is satisfied by the two measures as well. For a physicist, $\mathcal{C}_{J,1}^{(S)}$, the Shannon-Jensen statistical complexity measure, is the best one because it is an intensive quantity, which is not the case for $\mathcal{C}_{W,1}^{(S)}$.

We present in Fig. 7 plots of a similar character to the interesting ones given by Crutchfield and Young in Ref. [2] (see their Fig. 2), while in (our) Fig. 7.a, we depict the Lyapunov exponent Λ *versus* the normalized Shannon entropy $\mathcal{H}_1^{(S)}$. Figures 7.b and 7.c exhibit the entropy-complexity plane $\mathcal{H} \times \mathcal{C}$ for $\mathcal{C}_{E,1}^{(S)}\Big|_{binary}$ and $\mathcal{C}_{J,1}^{(S)}\Big|_{binary}$ *versus* $\mathcal{H}_1^{(S)}\Big|_{binary}$, respectively (we consider the r-range $3.4 \leq r \leq 4.0$, although the control parameter does not explicitly appear, of course). The two continuous curves represent the corresponding minimum, $\mathcal{C}_{min}$, and maximum, $\mathcal{C}_{max}$, complexity values evaluated with the general procedure descripted in Ref. [75]. Note that, for the case of periodic windows, if $\mathcal{H}_1^{(S)}\Big|_{binary} < H^* \approx 0.3$, we can ascertain that $\Lambda < 0$, while for $\mathcal{H}_1^{(S)}\Big|_{binary} > H^*$ we see that $\Lambda > 0$, which entails chaotic behavior. As illustrated by Fig. 7.b, the LMC statistical complexity $\mathcal{C}_{E,1}^{(S)}\Big|_{binary}$ is larger for periodic than for chaotic motion, which is wrong!. As illustrated by Fig.7.c, the Shannon-Jensen statistical complexity measure, $\mathcal{C}_{J,1}^{(S)}\Big|_{binary}$, on the other hand, behaves in opposite manner, and is also different for distinct degrees of periodicity.

Figure 8 amplifies the interesting zone $3.83 \leq r \leq 3.87$, which include the window of period 3 in the bifurcation diagram of the logistic map, in order to illustrate further details. Again, the LMC-complexity $\mathcal{C}_{E,1}^{(S)}\Big|_{binary}$ is *(a)* unable to distinguish among different periodic windows mayor and *(b)* assigns greater complexity to periodic regions than to neighboring chaotic ones. The Shannon-Jensen complexity, $\mathcal{C}_{J,1}^{(S)}\Big|_{binary}$, instead, behaves in a quite different (and seemingly more correct) fashion. It does indeed differentiate among distinct degrees of periodicity (see the bifurcation diagram of Fig. 8.a). Summing up: the generalized statistical complexity measure with $\kappa = S$ (Shannon) and $\nu = J$ (Jensen divergence) *(i)* becomes intensive, *(ii)* is able to distinguish among distinct degrees of periodicity, and *(iii)* yields a better description of dynamical features (a better grasp of dynamical details).

Extensions of the generalized statistical complexity measures to non-extensive statistics are given by $\mathcal{C}_{JT} \equiv \mathcal{C}_{\mathcal{J},q}^{(T)}\Big|_{binary}$ (Tsallis-Jensen) and, $\mathcal{C}_{JR} \equiv \mathcal{C}_{\mathcal{J},q}^{(R)}\Big|_{binary}$ (Rényi-Jensen). Figures 9.a and 9.b plots the $\mathcal{C}_{JT}$ and $\mathcal{C}_{JR}$ measures for q = 0.4, 0.8, 1.0, 1.2, 1.6 for the logistic map (PDF-binary) as a function of r. The global behavior of these statistical complexity measures is similar for the different q-values. Nevertheless, as q grows, a better (more detailed) picture ensues, that amplifies differences between different behaviors (both periodic and chaotic).

Changes in $\mathcal{C}_{min}$ and $\mathcal{C}_{max}$ *versus* q are also of interest (for a detail analysis see Ref. [75]) if one like to visualize the dynamics in the corresponding entropy-complexity plane. We show the behavior of this two curves for $N = 6$ and for different values of the parameter q in the cases of the statistical complexity measures Tsallis-Jensen and Rényi-Jensen type in Figs. 10.a and 10.b, respectively. Note that in the case of Tsallis-Jensen, the typical like "half-moon" area for $q = 1$ move to a like "triangular" area for increasing values of q. However, for the Rényi-Jensen case, no "triangular" area ensues. The original picture obtained for $q = 1$ remains globally invariant as q changes.

We discuss now the influence of different selection of PDF over the Information Theory quantifiers, entropy and complexity. In Fig. 11 we depict the normalized Shannon entropy, $\mathcal{H}_1^{(S)}$ evaluated for the logistic map as function of the parameter r considering either *i)* an histogram-determined PDF, $\mathcal{H}_{hist} \equiv \mathcal{H}_1^{(S)}\Big|_{histogram}$ (Fig. 11.a); *ii)* a binary, $\mathcal{H}_{bin} \equiv \mathcal{H}_1^{(S)}\Big|_{binary}$ (Fig. 11.b) one or; *iii)* a Bandt-Pompe PDF, $\mathcal{H}_{BP} \equiv \mathcal{H}_1^{(S)}\Big|_{Bandt-Pompe}$ (Fig. 11.c). We observe in all instances an abrupt entropy growth around $r > r_\infty \cong 3.5699$. After we pass this point, the entropy displays a trend to increase, taking its maximum value at $r = 4$. The several "drops" in entropic values in the interval $r_\infty < r < 4$ correspond to the periodic windows, as can easily be confirmed comparing with the bifurcation diagram and with the Lyapunov exponent depicted in Figs. 1 and 2 respectively.

It is interesting to observe that, in the PDF-histogram-instance, the normalized entropy for $r = 4$ is almost unity ($\mathcal{H}_{hist} \simeq 0.983$). The reason is that the invariant measure of the logistic map is in this case given by an almost constant function, namely,

$$\rho(x) = \begin{cases} \frac{1}{\pi\sqrt{x(1-x)}} & 0 < x < 1 \\ 0 & other \quad x \end{cases} . \tag{51}$$

Results with a PDF-binary are even worse ($H_{bin} = 1$), because the invariant measure of Eq. (51) is symmetric around $x = 0.5$. Only the PDF-Bandt and Pompe leads to the realistic value $\mathcal{H}_{BP} \simeq 0.6$. Furthermore, the doubling period-cascade is lost if a PDF-binary is used.

The generalized statistical complexity measure $\mathcal{C}_{J,1}^{(S)}$ evaluated with: *i)* an histogram determined PDF, $\mathcal{C}_{hist} \equiv \mathcal{C}_{J,1}^{(S)}\Big|_{histogram}$; *ii)* a binary, $\mathcal{C}_{bin} \equiv \mathcal{C}_{J,1}^{(S)}\Big|_{binary}$ one or; *ii)* a Bandt-Pompe PDF, $\mathcal{C}_{BP} \equiv \mathcal{C}_{J,1}^{(S)}\Big|_{Bandt-Pompe}$ are depicted in Fig. 12. In all cases we note an abrupt statistical complexity growth around $r > r_\infty$. After we pass this point, the corresponding generalized complexity measure behaves in quite different fashions. Notwithstanding the fact that in all instances the complexity measure decreases, it remains constant within the periodic windows (consider for example the period three window $r \in [3.8284,\ 3.8570]$). Also, for different periodic windows the complexity values differ according with the kind of periodicity-degree of that window.

For the statistical complexity evaluated using PDF-histograms, Fig. 12.a, one sees that, after $r > r_\infty$, an overall decreasing tendency becomes evident. A minimum value exists at $\mathcal{C}_{hist} \equiv \mathcal{C}_{J,1}^{(S)}\Big|_{histogram} \cong 0$ for $r = 4$. The several peaks observed in the region $r_\infty < r < 4$ signal a local complexity growth. The comparison with the bifurcation diagram (see Fig. 1.b) indicates that these peaks are correlated with periodic windows. In consequence, they signal the transition between different dynamical behaviors, *i.e.*, from chaotic to periodic ones.

As we mention before, the statistical complexity evaluated with a PDF-binary exhibits a parabolic behavior according to the variation of the parameter $r_\infty < r < 4$ (see Fig. 12.b), with a value $\mathcal{C}_{bin} \equiv \mathcal{C}_{J,1}^{(S)}\Big|_{binary} \cong 0$ for $r = 4$. Globally, the decay of $\mathcal{C}_{bin}$ is the more swifter the closer we approach the value $r = 4$. Note also that for this region the statistical complexity grows in the inter-windows region and rapidly falls within the periodic windows.

In the instance of a PDF evaluated with the Bandt and Pompe methodology, Fig.12.c, the statistical complexity measure $\mathcal{C}_{BP} \equiv \mathcal{C}_{J,1}^{(S)}\Big|_{Bandt-Pompe}$ exhibits overall increasing values as a function of the parameter r, adopting its maximum value for $r = 4$, which corresponds to the case of totally developed chaos. Also we observe drops in its values, that are associated with periodic windows. The behavior of the normalized Shannon entropy ($\mathcal{H}_{BP}$) and of the generalized statistical complexity measure ($\mathcal{C}_{BP}$) take into account in a natural way, as emphasized above, "time causality", which allows in fact for the ability to distinguish between chaotic and stochastic dynamics [55]. Indeed, if the time causality is not totally taken into account, as is the case of both the PDF-histogram and the PDF-binary, one has $\mathcal{H} \cong 1$ and $\mathcal{C} \cong 0$ for both kind of dynamics representation, in contradiction to what happens using the Bandt and Pompe methodology to determine the PDF, for which the situation $\mathcal{H} \cong 1$ together with $\mathcal{C} \cong 0$ is only reached for the uncorrelated stochastic dynamics (white noise) [55]. In plain words, "*chaos is not noise*" even if they share some common characteristics. In point of fact, chaos is representative of deterministic processes, and thus time-causality constitutes an important facet that must be taken into account for a proper characterization.

The above described special characteristics of the PDF obtained using Bandt and Pompe's methodology can be given deeper insight if we consider the emergence of the so-called "forbidden patterns" [87, 88]. As shown recently by Amigó *et al.* [87, 89, 90, 91], in the case of deterministic one-dimensional maps, not all the possible ordinal patterns can be effectively materialized into orbits, which in a sense makes these patterns "forbidden". Indeed, the existence of these *forbidden ordinal patterns* becomes a persistent fact that can be regarded as a "new" dynamical property. Thus, for a fixed pattern-length (embedding dimension D) the number of forbidden patterns of a time series (unobserved patterns) is independent of the series length M. Remark that this independence does not characterize other properties of the series such as proximity and correlation, which die out with time [87, 91]. For example, in the time series generated by the logistic map $x_{k+1} = 4 \cdot x_k \cdot (1 - x_k)$, if we consider patterns of length $D = 3$, the pattern $[2, 1, 0]$ is forbidden. That is, the pattern $x_{k+2} < x_{k+1} < x_k$ never appears [87].

In the case of time series generated by *unconstrained stochastic process* (uncorrelated process) every ordinal pattern has the same probability of appearance [87, 89, 90, 91]. That is, if the data set is long enough, all the ordinal patterns will eventually appear. In this case, when the number of time-series' observations becomes long enough the associated probability distribution function should be the uniform distribution, and the number of observed patterns will depend only on the length M of the time series under study. For correlated stochastic processes the probability of observing individual pattern depends not only on the time series length M but also on the correlation structure [92, 93]. The existence of a non-observed ordinal pattern does not qualify it as "forbidden", only as *"missing"*, and it could be due to the finite length of the time series.

In Fig. 13 we depict the number of forbidden patterns ($D = 6$) for the logistic map as function of the parameter r within the range $3.5 \leq r \leq 4$. See the decreasing tendency in the number of forbidden patterns as a function of r, with a minimum value at $r = 4$. The fluctuation in the values of forbidden patterns for the different periodic windows and chaotic zones can be associated with the numerical precision attained in the evaluation of Eq. (48), which defines not only the patterns but also the finite number of realization-data one considering.

Finally, we display in Fig. 14 the entropy-complexity plane, $\mathcal{H} \times \mathcal{C}$, for the logistic map with the three different instances of PDF-evaluation under consideration (Histogram, Binary and Bandt-Pompe), in the $r-$range $3.4 \leq r \leq 4$ (the control parameter does not explicitly appear in the graph, of course). The two continuous curves represent, respectively, the maximum, $\mathcal{C}_{max}$, and minimum $\mathcal{C}_{min}$, statistical complexity values evaluated as explained in [75] for the corresponding N (number of degree of freedom). Note that, for the case of periodic windows, if $\mathcal{H} < H^*$ ($H^*_{hist} \approx 0.4$, $H^*_{bin} \approx 0.3$ and $H^*_{BP} \approx 0.35$) we can ascertain that the Lyapunov exponent $\Lambda < 0$, while for $\mathcal{H} > \mathcal{H}^*$ we observed that $\Lambda > 0$, which entails chaotic behavior. As evidenced by Fig. 14, in all the instances of PDF-evaluation, periodic behaviors exhibit lower values than chaotic ones, as demanded in [12]. The "Lyapunov"-based criteria mentioned above is satisfied by the three PDF evaluation instances as well.

For a physicist, the Shannon-Jensen statistical complexity measure evaluated with the Bandt and Pompe PDF is the best one because it is an intensive quantity (which is not the case when the disequilibrium is evaluated in terms of Wootters distance) and also clearly distinguishes deterministic (chaos) from stochastic process [55]. The importance of the $\mathcal{H} \times \mathcal{C}$-plane [55] resides in the fact that this kind of diagrams yield information of a system independently of the values that the different control parameters may adopt. The bounds yield also information that depends on the particular characteristics of a given system (for instance, the existence of global extrema), or on the peculiarities of the system's internal configuration for which such extrema obtain. Similar results, to the previously described are obtained when the other functional forms of entropy and generalized statistical complexity are considered.

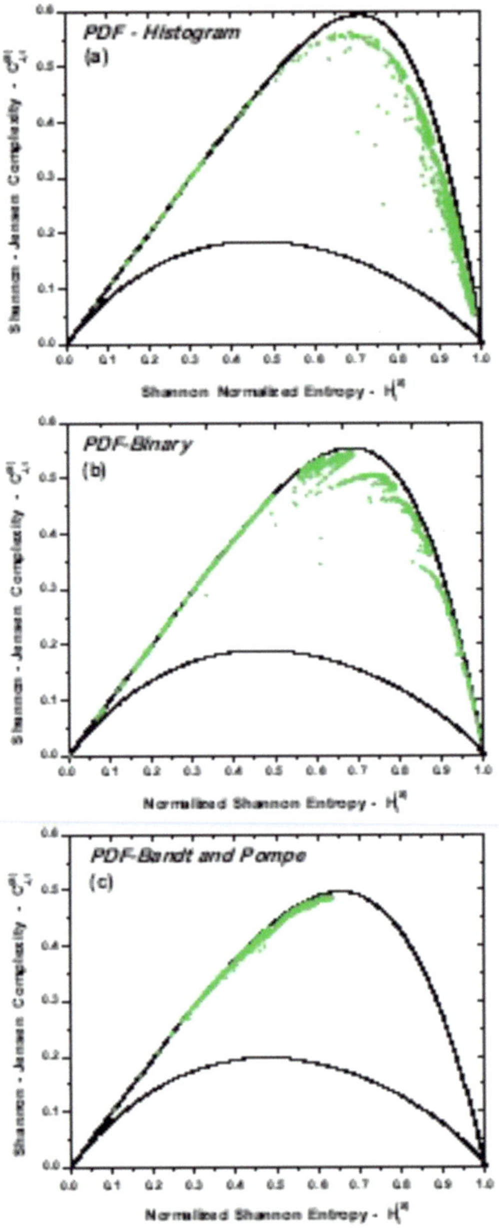

Figure 14: Entropy-Complexity plane for the logistic map (parameter $3.4 \leq r \leq 4$, $\Delta r = 0.0003$) for: *a)* PDF-histogram ($N = N_{bin} = 2^{14}$), *b)* PDF-binary ($L = 12$, $N = 2^{12}$), *c)* PDF-Bandt and Pompe, ($D = 6$, $N = 720$). We also display the maximum and minimum possible values of the generalized statistical complexity (continuous lines).

5 Conclusions

We conclude that the following remarks are in order when a generalized statistical complexity measure (SCM) is going to be used for characterizing a time series:

- A statistical complexity measure depends also on the selection of an appropriate disequilibrium $\mathcal{Q}$. The distance-choice is also critical. In this vein, see the text and for an illustrations of the defects associated with either the Euclidean, Kullback-Leibler or the Wootters distances. The choice of the Jensen divergence seems the most appropriate due to it has the best properties.
- Special care is required in the selection of the probability distribution function (PDF). If determinism is an important feature to be taken into account, the Bandt and Pompe prescription has advantages over other choices. For the statistical complexity this choice is not sufficient by itself.
- Chaotic time series are located near the $\mathcal{C}_{max}$ curve in any of the $\mathcal{H} \times \mathcal{C}$ representations planes (if Jensen divergence based disequilibrium is used). But the PDF Bandt and Pompe prescription locates chaos very near from the top, with high complexity and $\mathcal{H}-$values near 0.5. The other "non-causal" PDF prescriptions place deterministic systems close to truly random ones (near $\mathcal{C} = 0$ and $\mathcal{H} = 1$).
- The number of forbidden ordering patterns is another quantifier that yields deterministic behavior in the full chaos case $r = 4$. However, it does not exhibit the bifurcation diagram in as good a fashion as $\mathcal{C}_{BP}$ does.

Conflict of interest: None.

Acknowledgment

This work was partially supported by Universidad Nacional de Mar del Plata, ANPCyT (PICT 2010-2335) and CONICET (PIP 112-200801-1420 CCTMDP). OAR acknowledges partial support from CONICET, Argentina, and CAPES, PVE fellowship, Brazil.

References

[1] Lempel, A.; Ziv, J. Complexity of finite sequences. *IEEE Trans. Inf. Theory* **22** 75–81 (1976).

[2] Crutchfield, J.P.; Young, K. Inferring statistical complexity. *Phys. Rev. Lett.* **63** 105–108 (1989).

[3] Sprott, J.C. *Chaos and Time Series Analysis*, Oxford University Press, Oxford, 2004.

[4] Ott, E.; Sauer, T.; Yorke, J.A. *Coping with Chaos*, Wiley, New York, 1994.

[5] Solomonoff, R.J. A formal theory of inductive inference, Part 1 and Part 2. *Information and Control* **7** 1–22, 224–254 (1964).

[6] Kolmogorov, A.N. Three approaches to the quantitative definition of information. *Problems in Information Transmission* **1** 1–7 (1965).

[7] Chaitin, G.J. On the length of programs for Computing finite binary sequences. *J. Assoc. Comput. Mach.* **13** 547–569 (1966).

[8] Marwan, N.; Wessel, N.; Meyerfeldt, U.; Schirdewan, A.; Kurths, J. Recurrence-plot based measures of complexity and their application to heart-rate-variability data. *Phys. Rev. E* **66** 026702 (2002).

[9] Bandt, C.; Pompe, B. Permutation entropy: A natural complexity measure for time series. *Phys. Rev. Lett.* **88** 174102 (2002).

[10] Wackerbauer, R., Witt, R.A.; Atmanspacher, H.; Kurths, J.; Scheingraber, H. A comparative classification of complexity-measures. *Chaos, Solitons & Fractals* **4** 133–173 (1994).

[11] López-Ruiz, R.; Mancini, H.; Calbet, X. A statistical measure of complexity. *Phys. Lett. A* **209** 321–326 (1995).

[12] Feldman, D.P.; Crutchfield, J.P. Measures of statistical complexity: Why?. *Phys. Lett. A* **238** 244–252 (1998).

[13] Shiner, J.S.; Davison, M.; Landsberg, P.T. Simple measure for complexity, *Phys. Rev. E* **59** 1459–1464 (1999).

[14] Martín, M.T.; Plastino, A.; Rosso, O.A. Statistical complexity and disequilibrium, *Phys. Lett. A* **311** 126–132 (2003).

[15] Lamberti, P.W.; Martín, M.T.; Plastino, A.; Rosso, O.A. Intensive entropic nontriviality measure. *Physica A* **334** 119–131 (2004).

[16] Grassberger, P. Toward a qualitative theory of self-generated complexity. *Int. J. Theor. Phys.* **25** 907–938 (1988).

[17] Kantz, H.; Kurths, J.; Meyer-Kress, G. *Nonlinear Analysis of Physiological Data.* Springer, Berlin, 1998.

[18] Rosso, O.A.; Martín, M.T.; Plastino, A. Brain electrical activity analysis using wavelet based informational tools. *Physica A* **313** 587–608 (2002).

[19] Rosso, O.A.; Martín, M.T.; Plastino, A. Brain electrical activity analysis using wavelet based informational tools (II): Tsallis non-extensivity and complexity measurements. *Physica A* **320** 497–511 (2003).

[20] Plastino, A.; Martín, M.T.; Rosso, O.A. Generalized information measures and the analysis of brain electrical signals. In *Nonextesive Entropy - Interdisciplinary applications.* Edited by: Gell-Mann, M.; Tsallis; C. Santa Fe Institute Studies in the Sciences of Complexity, 261–293, 2004.

[21] Rosso, O.A.; Martín, M.T.; Plastino, A. Evidence of self-organization in brain electrical activity using wavelet based informational tools. *Physica A* **347** 444–464 (2005).

[22] Plastino, A.; Rosso, O.A. Entropy and statistical complexity in brain activity. *European Physical Society - Europhysics News 36* 224–228 (2005). Special Number on: "Nonextensive Statistical Mechanics - Theory and Applications", Boon, J.P.; Tsallis, C. (Guest Editors).

[23] Rosso, O.A.; Martín, M.T.; Figliola, A.; Keller, K.; Plastino, A. EEG analysis using wavelet-based informational tools. *J. Neurosc. Meth.* **153** 163–182 (2006).

[24] Kowalski, A.M.; Martín, M.T.; Plastino, A.; Proto, A.N.; Rosso, O.A. Wavelet statistical complexity analysis of classical limit. *Phys. Lett. A* **311** 180–191 (2003).

[25] Kowalski, A.M.; Martín, M.T.; Plastino, A.; Rosso, O.A. Entropic non-triviality, the classical limit, and geometry-dynamics correlations. *Int. J. of Modern Phys. B* **14** 2273–2285 (2005).

[26] Kowalski, A.M.; Martín, M.T.; Plastino, A.; Rosso, O.A. Bandt-Pompe approach to the classical-quantum transition. *Physica D* **233** 21–31 (2007).

[27] Kowalski, A.M.; Martín, M.T.; Plastino, A.; Rosso, O.A. Chaos and complexity in the classical-quantum transition *Int. J. Applied Mathematics & Statistics* (2011) in press.

[28] González, C.M.; Larrondo, H.A.; Rosso, O.A. Statistical complexity measure of pseudorandom bit generators. *Physica A* **354** 281–300 (2005).

[29] Larrondo, H.A.; González, C.M.; Martín, M.T.; Plastino, A.; Rosso, O.A. Intensive statistical complexity measure of pseudorandom number generators. *Physica A* **356** 133–138 (2005).

[30] Larrondo, H.A.; González, C.M.; Martín, M.T.; Plastino, A.; Rosso, O.A. Random number generators and causality. *Phys. Lett. A* **352** 421–425 (2006).

[31] De Micco, L.; González, C.M.; Larrondo, H.A.; Martín, M.T.; Plastino, A.; Rosso, O.A. Randomizing nonlinear maps via symbolic dynamics. *Physica A* **387** 3373–3383 (2008).

[32] De Micco, L.; Larrondo, H.A.; Plastino, A.; Rosso, O.A. Quantifiers for randomness of chaotic pseudo-random number generators. *Philosophical Transactions of the Royal Society A* **367** 3281–3296 (2009).

[33] Saco, P.M.; Carpi, L.C.; Figliola, A.; Serrano, E.; Rosso, O.A. "Entropy Analysis of the Dynamics of EL Niño/Southern Oscillation during the Holocene", *Physica A* **389** 5022–5027 (2010).

[34] Korol, A.M.; Foresto, P.; Pérez, S.; Martín, M.T.; Rosso, O.A. Preliminary characterization of erythrocytes deformability on the entropy–complexity plane. *Open Medical Informatics Journal* **4** 164–170 (2010).

[35] Korol, A.M.; Rosso, O.A.; Martín, M.T.; D'Arrigo, M.; Riquelme B.D. Impairment of erythrocytes incubated in glucose medium: a wavelet information theory analysis. *Cell Biochemistry and Biophysics* **60** 329–334 (2011).

[36] Rosso, O.A.; Masoller, C. Detecting and quantifying stochastic and coherence resonances via Information Theory complexity measurements *Phys. Rev. E* **79** 040106(R) (2009).

[37] Rosso, O.A.; Masoller, C. Detecting and quantifying temporal correlations in stochastic resonances by Information Theory measures. *European Physics Journal B* **69** 37–43 (2009).

[38] Zunino, L.; Soriano, M.C.; Fischer, I.; Rosso, O.A.; Mirasso C. Permutation information theory approach to unveil delay dynamics from time series analysis. *Phys. Rev. E* **82** 046212 (2010).

[39] Soriano, M.C.; Zunino, L.; Rosso, O.A.; Mirasso, C. Quantifying complexity of the chaotic regime of a semiconductor laser subject to feedback via Information Theory measures. In Semiconductor Lasers and Laser Dynamics IV, Panajotov, K; Sciamanna, M.; Valle, A.A.; Michalzik, R. (Editors). *Proc. of SPIE* **7720** 77202G (2010).

[40] Masoller, C.; Rosso, O. Quantifying the complexity of the delayed logistic map. *Philosophical Transactions of the Royal Society A* bf 369 425–438 (2011).

[41] Zunino, L.; Zanin, M.; Tabak, B.M.; Pérez, D.G.; Rosso, O.A. Forbidden patterns, permutation entropy and stock market inefficiency. *Physica A* **388** 2854–2864 (2009).

[42] Zunino, L.; Zanin, M.; Tabak, B.M.; Pérez, D.G.; Rosso, O.A. Complexity-entropy causality plane: a useful approach to quantify the stock market inefficiency. *Physica A* **389** 1891–1901 (2010).

[43] Zunino, L.; Tabak, B.M.; Serinaldi, F.; Zanin, M.; Pérez, D.G.; Rosso, O.A. Commodity predictability analysis with a permutation information theory approach. *Physica A* **390** 876-890 (2011).

[44] Rosso, O.A.; Vicente, R.; Mirasso C. Encryption test of pseudo-aleatory messages embedded on chaotic laser signals: an Information Theory approach. *Phys. Lett. A* **372** 1018–1023 (2008).

[45] Tiana-Alsina, J.; Torrent, M.C.; Rosso, O.A.; Masoller, C.; Garcia-Ojalvo, J. Quantifying the statistical complexity of low-frequency fluctuations in semiconductor lasers with optical feedback. *Physical Review A* **82** 013819 (2010).

[46] Soriano, M.C.; Zunino, L.; Rosso, O.A.; Fischer, I.; Mirasso C. Time scales of a chaotic semiconductor laser with Optical feedback under the lens of a permutation information analysis. *IEEE Journal of Quantum Electronic* **47** 256–261 (2011).

[47] Pérez, D.G.; Zunino, L.; Garavaglia, M.; Rosso, O.A. Wavelet entropy and fractional Brownian motion time series. *Physica A* **365** 282–288 (2006).

[48] Pérez, D.G.; Zunino, L.; Martín, M.T.; Garavaglia, M.; Plastino, A.; Rosso, O.A. Model-free stochastic processes studied with q-wavelet-based informational tools. *Phys. Lett. A* **364** 259–266 (2007).

[49] Zunino, L.; Pérez, D.G.; Garavaglia, M.; Rosso, O.A. Wavelet entropy of stochastic processes. *Physica A* **379** 503–512 (2007).

[50] Zunino, L.; Pérez, D.G.; Martín, M.T.; Plastino, A.; Garavaglia, M.;Rosso, O.A. Characterization of gaussian self-similar stochastic processes using wavelet-based informational tools. *Phys. Rev. E* **75** 021115 (2007).

[51] Rosso, O.A.; Zunino, L.; Pérez, D.G.; Figliola, A.; Larrondo, H.A.; Garavaglia, M.; Martín, M.T.; Plastino, A. Extracting features of Gaussian self-similar stochastic processes via the Bandt and Pompe approach. *Phys. Rev. E* **76** 061114 (2007).

[52] Zunino, L.; Pérez, D.G.; Martín, M.T.; Garavaglia, M.; Plastino, A.; Rosso, O.A. Permutation entropy of fractional Brownian motion and fractional Gaussian noise. *Phys. Lett. A* **372** 4768–4774 (2008).

[53] Zunino, L.; Pérez, D.G.; Kowalski, A.; Martín, M.T.; Garavaglia, M; Plastino, A.; Rosso, O.A. Fractional Brownian motion, fractional Gaussian noise, and Tsallis permutation entropy. *Physica A* **387** 6057–6068 (2008).

[54] Carpi, L.C.; Rosso, O.A.; Saco, P.M.; Goméz-Ravetti M. Analyzing complex networks evolution through Information Theory Quantifiers. *Phys. Lett. A* **375** 801–804 (2011).

[55] Rosso, O.A.; Larrondo, H.A.; Martín, M.T.; Plastino, A.; Fuentes, M.A. Distinguishing noise from chaos. *Phys. Rev. Lett.* **99** 154102 (2007).

[56] Shannon, C.E. A mathematical theory of communication. *Bell Syst. Technol. J.* **27** 379–423, 623–656 (1948).

[57] Aczél, J.; Daròczy Z. On measures of information and their characterization. *Mathematics in Science and Engineering* **115**. Academic Press, London, 1975.

[58] Basseville, M. Information: entropies, divergences et mayennes. *Institut de Recherche en Informatique et Systèmes Aléatoires*, **Publication Interne 1020**. Rennes Cedex: IRISA, 1996 .

[59] Kowalski, A.M.; Martín, M.T.; Plastino, A.; Rosso, O.A.; Casas, M. Distances in probability space and the Generalized Statistical Complexity setup. *Entropy* **13** 1055–1075 (2011).

[60] Wootters, K.W. Statistical distance and Hilbert space. *Phys. Rev. D* **23** 357–362 (1981).

[61] Cover, T.M.; J.A. Thomas, J.A. *Elements of Information Theory*, Wiley, New York, 1991.

[62] Grandy Jr., W.T; Milonni, P.W. (Eds.) *Physics and probability: Essays in Honor of Edwin T. Jaynes*, Cambridge University Press, New York, 1993.

[63] Kullback, S.; Leibler, R.A. On Information and Sufficiency. *Annals of Mathematical Statistics* **22**, 79–86 (1951).

[64] Briët, J.; Harremoës, P. Properties of classical and quantum Jensen-Shannon divergence. *Phys. Rev. A* **79** 052311 (2009).

[65] Kolmogorov, A.N. A new metric invariant of transitive dynamic system and automorphysms in Lebesgue spaces. *Dokl. Akad. Nauk. SSSR* **119** 861–864 (1958).

[66] Sinai, Y.G. On the concept of entropy of dynamical system. *Dokl. Akad. Nauk. SSSR* **124** 768–771 (1959).

[67] Beck, C.; Schlögl, F. *Thermodynamic of chaotic systems.* New York: Cambridge University Press; 1993.

[68] Tsallis, C. Possible generalization of Boltzmann-Gibbs statistics. *J. Stat. Phys.* **52** 479–487 (1988).

[69] Martín, M.T.; Plastino, A.R.; Plastino, A. Tsallis-like information measures and the analysis of complex signals. *Physica A* **275** 262–271 (2000).

[70] Rényi, A. *Probability theory.* North-Holland, Amsterdam, 1970.

[71] Lamberti, P.W.; Majtley, A.P. Nonlogaritmic Jensen-Shannon divergence. *Physica A* **329** 81–90 (2003).

[72] Crutchfield J.P.; Feldman D.P.; Shalizi C.R. Comment I on: simple measure of complexity. *Phys Rev E* **62** 2996–2997 (2000).

[73] Binder, P.M.; Perry N. Comment II on: simple measure of complexity. *Phys. Rev. E* **62** 2998–2999 (2000).

[74] Shiner J.S.; Davison M.; Landsberg P.T. Replay to comments on: simple measure for complexity. *Phys Rev E* **62** 3000–3003 (2000).

[75] Martín, M.T.; Plastino, A.; Rosso, O.A. Generalized statistical complexity measures: Geometrical and analytical properties. *Physica A* **369** 439–462 (2006).

[76] Plastino, A.R.; Plastino, A. Symmetries of the Fokker-Plank equation and Fisher-Frieden arrow of time. *Phys. Rev. E* **54** 4423–4326 (1996).

[77] Rosso, O.A.; De Micco, L.; Larrondo, H.A.; Martín, M.T.; Plastino, A. Generalized statistical complexity measure. *Int. J. Bif. and Chaos* **20** 775–785 (2010).

[78] Anteneodo, C.; Plastino, A.R. Some features of the López-Ruiz-Mancini-Calbet (LMC) statistical measure of complexity. *Phys. Lett. A* **223** 348–354 (1996).

[79] Rosso, O.A.; Craig, H.; Moscato, P. Shakespeare and other english renaissance authors as characterized by Information Theory complexity quantifiers. *Physica A* **388** 916–926 (2009).

[80] Mischaikow, K.; Mrozek, M.; Reiss, J.; Szymczak, A. Construction of symbolic dynamics from experimental time series. *Phys. Rev. Lett.* **82** 1114–1147 (1999).

[81] Powell, G.E.; Percival, I. A spectral entropy method for distinguishing regular and irregular motion of hamiltonian systems. *J. Phys. A: Math. Gen.* **12** 2053–2071 (1979).

[82] Rosso, O.A.; Blanco, S.; Jordanova, J.; Kolev, V.; Figliola, A.; Schürmann, M.; Başar, E. Wavelet entropy: a new tool for analysis of short duration brain electrical signals. *J. Neurosc. Meth.* **105** 65–75 (2001).

[83] Rosso, O.A.; Mairal, M.L. [2002] Characterization of time dynamical evolution of electroencephalographic records. *Physica A* **312** 469–504 (2002).

[84] Keller, K.; Sinn, M. Ordinal analysis of time series. *Physica A* **356** 114–120 (2005).

[85] Saco, P.M.; Carpi, L.C.; Figliola, A.; Serrano, E.; Rosso, O.A. Entropy analysis of the dynamics of el Niño/southern oscillation during the holocene. *Physica A* **389** 5022–5027 (2010).

[86] Majtey, A.; Lamberti, P.W.; Martín, M.T.; Plastino, A. Wootters' distance revisited: a new distinguishability criterium. *European Phys. Journal D* **32** 413–419 (2005).

[87] Amigó, J.M.; Zambrano, S.; Sanjúan, M.A.F. True and false forbidden patterns in deterministic and random dynamics. *Europhys. Lett.* (2007) 50001 **79**.

[88] Zanin, M. [2008] Forbidden patterns in financial time series. *Chaos* **18** 013119 (2008).

[89] Amigó, J.M.; Kocarev, L.; Szczepanski, L.J. Order patterns and chaos. *Phys. Lett. A* **355** 27-31 (2006) .

[90] Amigó, J.M.; Zambrano, S.; Sanjuán, M.A.F. Combinatorial detection of determinism in noisy time series. *Europhys. Lett.* **83** 60005 (2008).

[91] Amigó, J.M. *Permutation complexity in dynamical systems.* Springer-Verlag, Berlin, Germany, 2010 .

[92] Carpi, L.C.; Saco, P.M.; Rosso, O.A. Missing Ordinal Patterns in Correlated Noises. *Physica A* **389** 2020–2029 (2010).

[93] Rosso, O.A.; Carpi, L.C.; Saco, P.M.; Gómez Ravetti, M.; Plastino, A., Larrondo, H.A. Causality and the entropy-complexity plane: robustness and missing ordinal patterns. *Physica A* **391** 42–55 (2012).

Send Orders for Reprints to reprints@benthamscience.net

CHAPTER 9

The Fisher Information: Properties and Physico-Chemical Applications

Jesus S. Dehesa [a,b,*], Rodolfo O. Esquivel [a,c],
Angel Ricardo Plastino [a,b,e] and Pablo Sánchez-Moreno [a,d]

[a] Institute "Carlos 1" for Computational and Theoretical Physics, University of Granada, Granada, Spain

[b] Department of Atomic, Molecular and Nuclear Physics, University of Granada, Granada, Spain

[c] Department of Chemistry, Metropolitan Autonomous University, Mexico D.F., Mexico

[d] Department of Applied Mathematics, University of Granada, Granada, Spain

[e] CeBio y Secretaría de Investigación, Universidad Nacional Buenos Aires-Noroeste, UNNOBA, Junin, Argentina

Abstract

The Fisher information functional occupies a very special position among the uncertainty measures of quantum systems of any dimensionality. In contrast with (global) entropic measures of Shannon, Rényi or Tsallis types, it contitutes a local measure in the sense that it is sensitive to the oscillatory character of the probability density distribution. Moreover, the Fisher information is closely related to various macroscopic properties of the system described by different density functionals. Here, we review its main analytical properties such as, inequality-based relationships with some radial expectation values and its uncertainty relation. Tighter versions of these inequalities are shown for those systems subject to an arbitrary central potential. Moreover, the Fisher information is computed for D-dimensional hydrogenic systems explicitly in terms of the radial and angular hyperquantum numbers characterizing the quantum states of the system in both position and momentum spaces. Finally, the utility of this quantity is discussed for various physico-chemical processes of recent interest: Abstraction and nucleophilic substitution reactions.

* Corresponding author. E-mail: dehesa@ugr.es

Andres M. Kowalski, Raul D. Rossignoli and Evaldo M. F. Curado (Eds)

Keywords: Information theory, Fisher information, D-dimensional hydrogenic systems, chemical reactions, abstraction and nucleophilic substitution reactions.

1 Introduction

The parameter-dependent Fisher information was firstly introduced by Ronald A. Fisher [1] in 1925 to quantify the statistical information about the parameters which is contained in data samples obeying a given parametric model. This quantity has been shown extremely useful in a great variety of subjects, ranging from applied mathematics and statistics to classical and quantum physics, as it is described in the 2004 monograph of Frieden [2]. Let us only highlight here that it is the main theoretic tool of the extreme physical information method, a general variational principle which allows one to derive the most fundamental equations of physics (*e.g.*, Maxwell equations, Einstein field equations, and Schrödinger, Klein-Gordon and Dirac equations), various laws of statistical physics and some laws governing nearly incompressible turbulent fluid flows.

Applications of this quantity to a large amount of problems in theoretical physics have received a strong impulse when it was realized that the spatial distribution of the single-particle probability density $\rho(\mathbf{r})$, which is the basic variable of the density functional methods [3], can be quantitatively measured by its translationally-invariant (t.i.) version $F[\rho]$ in a different and complementary manner as the well-known Shannon entropy [4]. These two quantities characterize in a different and complementary way the information-theoretic content of the density $\rho(\mathbf{r})$ which describes the physical state under consideration. One may think of these two quantities as flip sides of the same coin: the Shannon entropy giving a measure of the extent to which the density is spread, and the t.i. Fisher information a measure of its narrowness.

In this chapter we will focus on the t.i. Fisher information of $\rho(\mathbf{r})$ defined as

$$F[\rho] = \int \frac{\|\nabla\rho(\mathbf{r})\|^2}{\rho(\mathbf{r})} d\mathbf{r},$$

which is called differently in various contexts: Fisher information for locality, intrinsic accuracy or, simply, Fisher information. The higher this quantity is, the more localized is the density, the smaller is the uncertainty and the higher is the accuracy in estimating the localization of the particle [2, 5].

The Fisher information is very sensitive to the fluctuations or irregularities of the density because it is a functional of the D-dimensional gradient ∇ [6]. This allows it to have a unique position among the measures of the internal disorder of a physical system as expressed by the spreading of the density $\rho(\mathbf{r})$ all over the space. The variance and the information-theoretic measures of Shannon, Rényi [7] and Tsallis [8] types have a global character in the sense that they are practically insensitive to local rearrangements of the position variable $\mathbf{r}$. The Fisher information controls the localization of the density around its nodes, appropriately grasping the oscillatory nature of the wavefunctions of the quantum-mechanical states; this confers it a relevant role in the characterization of numerous scientific phenomena of standard ($D = 3$) and non-standard D-dimensional systems. Indeed, the Fisher information has been shown (i) to be closely related to various density functionals describing physical and chemical observables of the system, particularly the kinetic and Weiszäcker energies [3,9–12], (ii) to characterize correlation properties in atomic systems [13, 14], (iii) to identify the most distinctive nonlinear spectral phenomena (the avoided crossings) of atomic systems in strong external fields [15], (iv) to explain the periodicity and shell structure in the periodic table, from hydrogen to lawrencium [16,17], (v) to predict the stationary points and bond breaking/forming regions in chemical reactions [18], and (vi) to build up uncertainty relations stronger than the variance-based Heisenberg relation [5,19–21], (vi) to variationally characterize various quantum equation of motion [22] and (vii) to rederive the classical thermodynamics without requiring the usual concept of Boltzmann entropy [23].

In this chapter we review in Section 2 the known analytical properties of the Fisher information of quantum systems with arbitrary dimensionality, highlighting the inequality-type relationships with various radial expectation values as well as the exact expressions in terms of $\langle r^2 \rangle$ and $\langle p^2 \rangle$, the uncertainty relation and its improvement for central potentials. Then, in Section 3, we collect a few recent applications of this concept to various physico-chemical processes.

2 Analytical properties of the Fisher information

Let us start pointing out that the Fisher information is known to be related with the Shannon entropy and the variance by means of the well-known deBruijn and Cramér-Rao relations [24, 25], respectively, the latter being expressed as

$$F[\rho] \times V[\rho] \geq D^2,$$

where $V[\rho]$ denotes the variance of the density $\rho(\mathbf{r})$. In the rest of the Section we will consider the inequality-type relations of the Fisher information with various radial expectation values, the uncertainty relation fulfilled by the position and momentum Fisher informations, and the exact expressions of these quantities as well as the improvement of position-momentum uncertainty relation for quantum systems subject to arbitrary central potentials. Finally, the explicit expressions of the Fisher informations of both ground and excited states of D-dimensional hydrogenic systems will be given in terms of the hyperquantum numbers characterizing the state, not only in position space but also in momentum space.

2.1 Relation with radial expectation values

The use of the classical Redheffer-Weyl inequalities have been used to find [16] the following inequalities between the Fisher information and the first few radial expectation values:

$$F[\rho] \geq (\alpha + D - 1)^2 \frac{\langle r^{\alpha-1}\rangle}{\langle r^{2\alpha}\rangle}; \quad \alpha \geq \max\{-D+1, -1\},$$

and

$$F[\rho] \geq \langle r^{-2}\rangle \left[(D-2)^2 + \frac{(\alpha+1)^2 \langle r^{\alpha-1}\rangle^2}{\langle r^{2\alpha}\rangle\langle r^{-2}\rangle - \langle r^{\alpha-1}\rangle^2}\right]; \quad \alpha \geq -1,$$

which extend to D dimensions some upper bounds to the Fisher information previously obtained in a variational way for three-dimensional systems [12, 17]. Moreover, the accuracy of these bounds is at times improved by taking into account some logarithmic expectation values; in this sense it is interesting to consider the following family of upper bounds

$$F[\rho] \geq \langle r^{-2}\rangle \left[(D-2)^2 + \frac{\langle r^{-2}\rangle^2}{\langle r^{-2}(\ln r)^2\rangle\langle r^{-2}\rangle - \langle r^{-2}\ln r\rangle^2}\right]$$

On the other hand, the variational Frieden's principle of extreme physical information [2] and, in particular, the minimization problem of the Fisher information have allowed us [26] to find that the D-dimensional density $\rho_F(\mathbf{r}) \equiv g(\mathbf{r})$ which minimizes the Fisher information subject to the m constraints $\langle r^k\rangle$, fulfills the differential equation

$$\left[\frac{\nabla g(\mathbf{r})}{g(\mathbf{r})}\right]^2 + 2\nabla\left[\frac{\nabla g(\mathbf{r})}{g(\mathbf{r})}\right] + \lambda_0 + \sum_{k=1}^{m} \lambda_k r^k = 0$$

The general solution of this equation has not yet been found. In the case of $D = 3$, this problem has been partially solved [12] for (i) a single constraint $\langle r^k \rangle$, with $k = 1$ and 2, and (ii) two constraints of the types $\{\langle r^{-2}\rangle, \langle r^{-1}\rangle\}$ and $\{\langle r^{-2}\rangle, \langle r^{2}\rangle\}$. Let us here collect the lower bound

$$F[\rho] \geq 4\langle r^{-1}\rangle$$

It is relevant to mention that this inequality becomes an equality for the following minimizer density

$$g(\mathbf{r}) = \frac{1}{\pi}\langle r^{-1}\rangle^3 \exp\left(-2\langle r^{-1}\rangle r\right),$$

because it has very interesting information-theoretic properties [18]. In particular, these properties have recently allowed us [26] to nicely interpret the periodicity and shell structure of the atomic systems.

2.2 Uncertainty relations

It is well known that the Heisenberg uncertainty principle has been mathematically formulated by means of various spreading measures of the density $\rho(\mathbf{r})$, initially with the variance giving rise to the Heisenberg-Kennard relation

$$\langle \mathbf{r}^2\rangle\langle \mathbf{p}^2\rangle \geq \frac{D^2}{4} \tag{2.1}$$

Later, since this inequality is very weak and, at times, even misleading, other uncertainty relations based on arbitrary moments $\langle r^k\rangle$ with $k \neq 2$ [27] and different information-theoretic measures (Shannon, Rényi and Tsallis entropies) have been proposed, as recently reviewed [28]. Let us now discuss the uncertainty relation

$$F[\rho] \times F[\gamma] \geq 4D^2, \tag{2.2}$$

which we have just proposed in terms of the Fisher informations $F[\rho]$ and $F[\gamma]$ of the position and momentum single-particle densities $\rho(\mathbf{r})$ and $\gamma(\mathbf{p})$, respectively. This relation has been variationally shown [21] to be valid for wavefunctions which are real either in position or in momentum space. Moreover we have observed that for general wavefunctions, the Fisher product $F[\rho] \times F[\gamma]$ can be made arbitrarily small. Indeed, let us take in the one-dimensional case the following multiparametric family of Hermite wavefunctions

$$\psi_n(x) = \pi^{-\frac{1}{4}} e^{-\frac{x^2}{2}} \sum_{k=0}^{n} a_k \frac{1}{\sqrt{2^k k!}} H_k(x),$$

where a_k are complex coefficients such that $\sum_{k=0}^{n} |a_k|^2 = 1$. These functions span a linear subspace of the general space of normalized wavefunctions $\psi(x)$, $x \in \mathbb{R}$. Now, we use the numerical Monte Carlo approach to count the percentage number of states f_n which violate the relation $F[\rho] \times F[\gamma] \geq 4$ in the following way. First, we generate a large number A_n of random states $\psi_n(x)$ in the subspace, uniformly distributed according to the Haar measure, which is done by yielding A_n n-tuples $(a_1, a_2, \ldots, a_n)$. Then, we determine the number B_n of the A_n states which violate the inequality and we compute the ratio $f_n = B_n/A_n$ of states violating the relation $F[\rho] \times F[\gamma] \geq 4$. We found that f_n is equal to 0.33, 0.092, 0.03, 0.0093 and 0.0025 for $n = 2, 3, 4, 5$ and 6, respectively. In summary, we observe that as n increases both A_n and B_n increase in such a way that its proportion f_n decreases very fast. This allows us to conjecture that the proportion of states which violate the uncertainty relation $F[\rho] \times F[\gamma] \geq 4$ tends rapidly to zero when n goes up to infinite. Indeed, our numerical evidence suggests that f_n decreases exponentially with n. In other terms, there are actually plenty of complex wavefunctions satisfying this uncertainty relation. Thus, we feel that the latter relation is not the most universal Fisher-information-based uncertainty relation.

2.3 Central potentials: Exact values and improved uncertainty relation

Here we will show two issues valid for quantum systems subject to central potentials: Firstly, the position and momentum Fisher information are exactly expressed in terms of the two second-order expectation values $\{\langle r^{-2}\rangle, \langle p^2\rangle\}$ and $\{\langle p^{-2}\rangle, \langle r^2\rangle\}$, respectively. And secondly, we substantially improve the previous uncertainty relation for general solutions of arbitrary central potentials. To do that we begin with the Schrödinger equation of a D-dimensional particle moving in the central potential $V_D(r)$, $r \equiv \|\mathbf{r}\|$, is

$$\left[-\frac{1}{2}\nabla^2 + V_D(r)\right]\psi(\mathbf{r}) = E_D\psi(\mathbf{r}).$$

The physical solutions of this equation admit the following separable form

$$\psi(\mathbf{r}) = R_{E,l}(r)\mathcal{Y}_{\{\mu\}}(\Omega_{D-1}),$$

in the hyperspherical coordinates $\mathbf{r} = (r, \theta_1, \theta_2, \ldots, \theta_{D-1}) = (r, \Omega_{D-1})$, where E_D is the energy and $\{\mu\} = (l \equiv \mu_1, \mu_2, \ldots, \mu_{D1} \equiv |m|)$ denote the integer hyperquantum numbers characterizing the physical states of the particle with the values

$$l = 0, 1, 2, \ldots, \qquad l \geq \mu_2 \geq \cdots \geq |\mu_{D-1}|.$$

The symbols $\mathcal{Y}_{\{\mu\}}(\Omega_{D-1})$ and $R_{E,l}(r)$ stand, respectively, for the known hyperspherical harmonics [6] and the radial eigenfunctions which are the solutions of the radial Schrödinger equation

$$\left[-\frac{1}{2}\frac{d^2}{dr^2} - \frac{D-1}{r}\frac{d}{dr} + \frac{l(l+D-2)}{2r^2} + V_D(r)\right] R_{E,l}(r) = E_D R_{E,l}(r).$$

This equation cannot be solved unless we know the explicit analytical form of the D-dimensional potential $V_D(r)$. However, taking into account the numerous algebraic properties of the hyperspherical harmonics, making profit of the fact that the Fisher information is a gradient functional of the density and that the momentum operator of the particle has a gradient representation, we have proved [29] that the Fisher information of the particle has the following exact expressions

$$F[\rho] = 4\langle \mathbf{p}^2\rangle - 2(2l + D - 2)|m|\langle \mathbf{r}^{-2}\rangle, \tag{2.3}$$

$$F[\gamma] = 4\langle \mathbf{r}^2\rangle - 2(2l + D - 2)|m|\langle \mathbf{p}^{-2}\rangle, \tag{2.4}$$

in the position and momentum spaces, respectively. Moreover, by use of the nonnegativity of the functional

$$G(\lambda) = \int_0^{+\infty} \left(u'(r) - \lambda\frac{u(r)}{r}\right)^2 dr,$$

where $u(r) = r^{(D-1)/2}R_{E,l}(r)$, we have found [20] that the members of the couples of expectation values $(\{\langle r^{-2}\rangle, \langle p^2\rangle\}, \{\langle p^{-2}\rangle, \langle r^2\rangle\})$ are mutually related by means of the inequalities

$$\langle \mathbf{p}^2\rangle \geq \left(l + \frac{D-2}{2}\right)^2 \langle \mathbf{r}^{-2}\rangle,$$

$$\langle \mathbf{r}^2\rangle \geq \left(l + \frac{D-2}{2}\right)^2 \langle \mathbf{p}^{-2}\rangle.$$

Taking into account these expressions into Eqs. (2.3) and (2.4), one obtains the following lower bounds

$$F[\rho] \geq 4\left(1 - \frac{|m|}{2l + D - 2}\right)\langle \mathbf{p}^2\rangle$$

$$F[\gamma] \geq 4\left(1 - \frac{|m|}{2l + D - 2}\right)\langle \mathbf{r}^2\rangle$$

on the position and momentum Fisher information. The multiplication of the last two inequalities yields the following formula

$$F[\rho]F[\gamma] \geq 16\left(1 - \frac{2|m|}{2l + D - 2}\right)^2 \langle \mathbf{r}^2\rangle\langle \mathbf{p}^2\rangle,$$

which connects the position-momentum Fisher product with the Heisenberg-Kennard uncertainty product, manifesting the uncertainty character of the former product in a clear way.

Moreover, taking into account the general Heisenberg-Kennard relation (2.1) or better its improvement for central potentials [20] given by

$$\langle \mathbf{r}^2 \rangle \langle \mathbf{p}^2 \rangle \geq \left(l + \frac{D}{2} \right)^2 ,$$

one finds the following position-momentum Fisher-information-based uncertainty relation

$$F[\rho]F[\gamma] \geq 16 \left(1 - \frac{2|m|}{2l + D - 2} \right)^2 \left(l + \frac{D}{2} \right)^2 \tag{2.5}$$

which is valid for all wavefunctions of arbitrary central potentials. Remark that for $l = m = 0$, one recovers the relation $F[\rho] \times F[\gamma] \geq 4D^2$ previously shown for wavefunctions real in either position or momentum space.

2.4 Hydrogenic systems: Explicit values

Let us now examine the position and momentum Fisher informations of the ground and excited states of the D-dimensional hydrogenic systems which are subject to the Coulomb potential $V_D(r) = -\frac{Z}{r}$, $r \equiv \|\mathbf{r}\|$, with $\mathbf{r} = (r, \theta_1, \theta_2, \ldots, \theta_{D-1}) \equiv (r, \Omega_{D-1})$. They include a large diversity of physical systems such as hydrogenic atoms and ions, exotic atoms, antimatter atoms, Rydberg atoms, ions and molecules, among others. To achieve a deeper knowledge of these systems is basic not only to gain full insight into the intimate structure of matter but also for the understanding of diverse phenomena in quantum cosmology, nanotechnology (*e.g.*, for low dimensional semiconductor nanostructures such as quantum wells, wires and dots), quantum computation (*e.g.*, one and two-dimensional hydrogenic atoms have been proposed as qubits), D-dimensional physics and quantum field theory. The existence of hydrogenic systems with dimensionality other than 3 has been observed for $D < 3$ and suggested for $D > 3$. See [30, 31] for further details.

The stationary states of these systems have the energies

$$E = -\frac{Z^2}{2\eta^2}; \quad \eta = n + \frac{D-3}{2},$$

and the associated electronic probability densities in position and momentum spaces are given [30, 31] by

$$\rho(\mathbf{r}) = N_{n,l}^2 \tilde{r}^{2l} e^{-\tilde{r}} \left[L_{n-l-1}^{(2l+D-2)}(\tilde{r}) \right]^2 \left| \mathcal{Y}_{\{\mu\}}(\Omega_{D-1}) \right|^2 ,$$

with $\tilde{r} = \frac{2Zr}{\eta}$, and

$$\gamma(\mathbf{p}) = K_{n,l}^2 \frac{(\eta p)^{2l}}{(1+\eta^2 p^2)^{2l+D+1}} \left[C_{n-l-1}^{(l+\frac{D-1}{2})} \left(\frac{1-\eta^2 p^2}{1+\eta^2 p^2} \right) \right]^2 |\mathcal{Y}_{\{\mu\}}(\Omega_{D-1})|^2$$

respectively. The symbols $N_{n,l}$ and $K_{n,l}$ are the corresponding normalized factors, and $L_k^{(\alpha)}(x)$ and $C_k^{(\alpha)}(x)$ denote the Laguerre and Gegenbauer polynomials of degree k and parameter α, respectively.

From Eqs. (2.3) and (2.4) and the known expectation values $\{\langle r^{-2}\rangle, \langle p^2\rangle, \langle p^{-2}\rangle, \langle r^2\rangle\}$ [30,31], we have found the value

$$F[\rho] = \frac{4Z^2}{\eta^3}(\eta - \mu_{D-1}),$$

for the Fisher information in position space, and

$$F[\gamma] = \frac{2\eta^2}{Z^2}\left[5\eta^2 - 3L(L+1) - (8\eta - 6L - 3)\mu_{D-1} + 1\right],$$

for the Fisher information in momentum space, with the grand principal quantum number $L = l + \frac{D-3}{2}$. Remark that the position (momentum) quantity is directly (inversely) proportional to the quadratic nuclear charge, so that the position-momentum Fisher product does not depend on the Coulomb strength; it has the form

$$F[\rho]F[\gamma] = \frac{8(\eta - \mu_{D-1})}{\eta}\left[5\eta^2 - 3L(L+1) - (8\eta - 6L - 3)\mu_{D-1} + 1\right],$$

which clearly fulfils the uncertainty relation expressed not only by the general inequality (2.2) but also that given by the improved inequality (2.5) for any central potential.

Finally, it is interesting to give the three-dimensional values

$$F[\rho] = \frac{4Z^2}{\eta^3}(n - m),$$

and

$$F[\gamma] = \frac{2n^2}{Z^2}\left[5n^2 - 3l(l+1) - (8n - 6l - 3)m + 1\right],$$

for the position and momentum Fisher information of real hydrogenic systems, respectively. Notice that they are given in terms of the three quantum numbers (n, l, m) which characterize the quantum states. We observe that for the states with $l = m = 0$, we obtain $F[\rho] \times F[\gamma] = 40n^2 + 8$ which is bigger than $4D^2 = 36$ since $n \geq 1$.

3 Application to some physico-chemical processes

In this section we show that the Fisher information constitutes a valuable tool for detecting the transition state, the stationary points of a chemical reaction and the bond breaking/forming regions of elementary reactions such as the simplest hydrogen abstraction and the identity S_N2 exchange ones [18]. These results support the concept of a continuum of transient of Zewail and Polanyi for the transition state rather than a single state [32], which is also in agreement with reaction force analyses [33]. We followed the IRC (Internal Rection Coordinate) path of some selected elementary chemical reactions in order to analyze the course of the reactions by use of the Fisher information in position and momentum spaces as well as other charge density descriptors such as the Molecular Electrostatic Potential (MEP). Two processes were chosen: the chemical probes under study are the simplest hydrogen abstraction reaction $H^\bullet + H_2 \rightarrow H_2 + H^\bullet$ and the identity S_N2 reaction $H^- + CH_4 \rightarrow CH_4 + H^-$

The central quantities in this study are the Fisher information in position and momentum spaces which will be here denoted for convenience as $I_r \equiv F[\rho]$ and $I_p \equiv F[\gamma]$, respectively.

3.1 Abstraction reaction

The reaction $H_a^\bullet + H_2 \rightarrow H_2 + H_b^\bullet$ is the simplest radical abstraction reaction involving a free radical (atomic hydrogen) as a reactive intermediate. This kind of reaction involves at least two steps (S_N1 reaction type): in the first step a new radical (atomic hydrogen in this case) is created by homolysis and in the second one the new radical recombines with another radical species. Such homolytic bond cleavage occurs when the bond involved is not polar and there is no electrophile or nucleophile at hand to promote heterolytic patterns. When the bond is made, the product has a lower energy than the reactants and it follows that breaking the bond requires energy.

In Figure 1 the Fisher information in both position and momentum spaces are depicted. At a first glance, we observe that both quantities behave in a similar way toward the reactive complexes ($H_a^\bullet \cdots H - H$ and $H - H \cdots H_b^\bullet$) and tends to decrease toward the TS region, but with a very important difference that we analyze below. It is worth mentioning

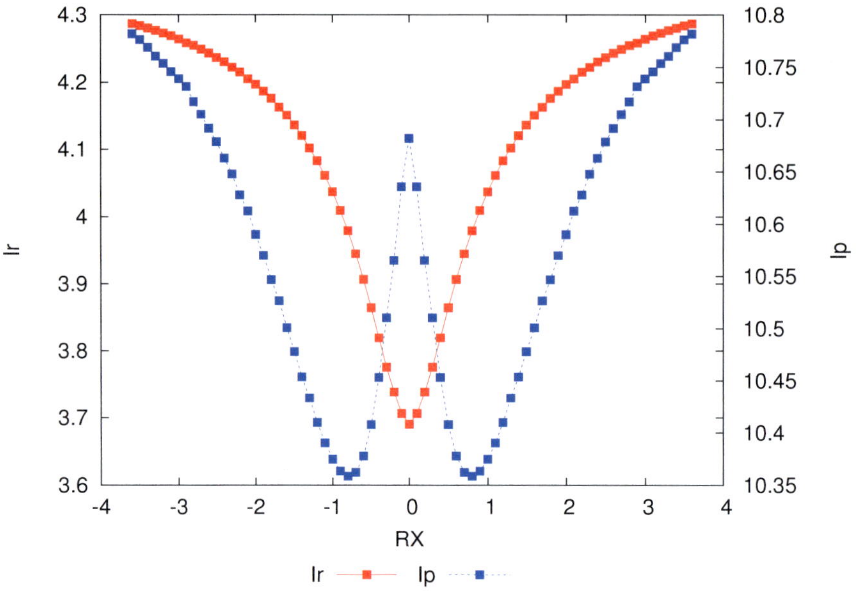

Figure 1: Fisher entropies in position (red line) and momentum (blue line) spaces for the IRC of $H_a^\bullet + H_2 \rightarrow H_2 + H_b^\bullet$.

that according to previous studies [34, 35], we have insight into the structural features of the distributions in both spaces, *i.e.*, concerning the spreading (localization/delocalization) of the densities. However, the behavior of the densities about their local changes (uniformity/irregularity) can only be provided by a local measure such as the Fisher information. Both position and momentum Fisher information measures possess more structure at the vicinity of the TS as it may be observed from Fig. 1. It is worthy to remark that this phenomenon is not present in the energy profile. By closer inspection, we note that the position Fisher information I_r possesses a global minima at the TS, whereas the momentum one, I_p, possesses a local maxima and two local minimum at the vicinity of the TS (approximately $R_X \simeq 0.9$). This is interesting on chemical grounds since the structure observed for the Fisher information in momentum space at the vicinity of the TS can be associated with a process of bond breaking/forming (depending on the reaction direction) followed by stabilization of the structure at the TS [34, 35].

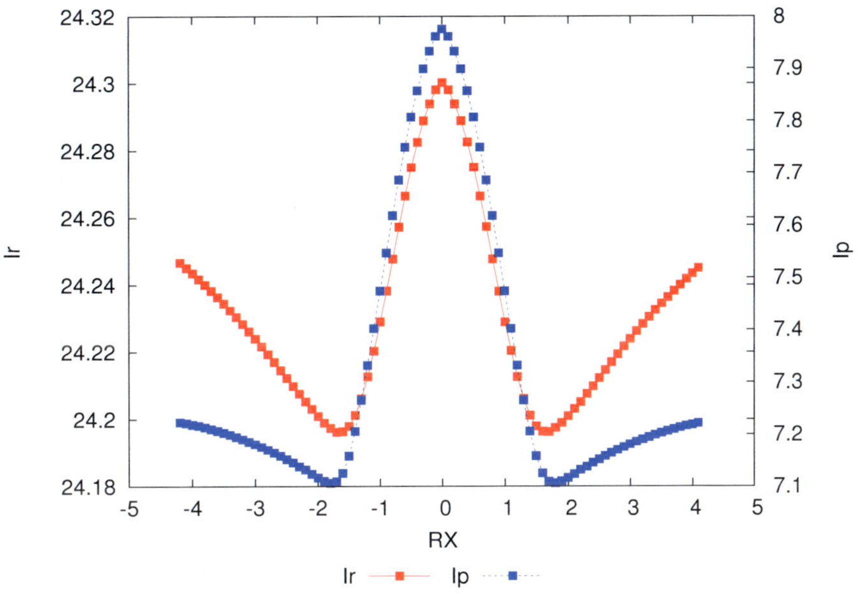

Figure 2: Fisher information in position (red line) and momentum (blue line) spaces for the IRC of $H_a^- + CH_4 \rightarrow CH_4 + H_b^-$.

3.2 Nucleophilic substitution reaction

Next we analyze a typical nucleophilic substitution (S_N2) reaction whose chemical process involves only one step in contrast with the abstraction reaction studied before which involves two steps. In the anionic form, the S_N2 mechanism is characterized by being kinetically of second order (first order in each of the reactants: the nucleophile Y^- and the substrate R_X, where X^- is the nucleofuge or leaving atom). For identity S_N2 reactions $X = Y$. It was postulated that the observed second order kinetics is the result of passage through the well-known Walden inversion transition state where the nucleophile displaces the nucleofuge (leaving group) from the backside in a single concerted reaction step.

If we represent both Fisher informations, I_r and I_p in the same picture, Figure 2, we can observe that they show a similar structure, both possessing a maximum at the TS and minimal regions at its vicinity. This behavior is significantly different than the abstraction reaction analyzed before in that the position Fisher information shows the opposite behavior as compared with the momentum Fisher information at the TS region. From previous studies [34, 35] with Shannon entropies, we found a delocalized position density and a localized momentum density in the TS region, *i.e.*, corresponding with a chemically relaxed structure (structurally less distorted with low kinetic energy and null dipole moment, see below). In contrast, the reactive complexes toward reactants/products show more localized densities with less delocalized momentum densities, *i.e.*, the chemical structures at these regions are structurally distorted and posses more kinetic energy as compared with the TS. In the vicinity of the TS, at around $R_X \simeq |1.7|$, critical points for these measures are observed; they correspond to ionic complexes that characterize position densities which are highly localized and with highly delocalized momentum densities and high kinetic energies. At a first glance, it seems likely that these regions correspond with BCER where bond breaking may start occurring.

One of the principal differences between the S_N2 reaction with respect to the abstraction one is that for the former the course of the reaction occurs by an heterolytic rupture with an exchange of charge, whereas for the latter the mechanism is homolytic, *i.e.*, a spin coupling process occurs. In this reaction as the incoming hydrogen approaches the molecule, it transfers charge, through the carbon bonding, to the leaving hydrogen so as to reach an equally charged distribution among the incoming/leaving hydrogens. As this process evolves, the gradient of the distribution involved in the position Fisher information I_r increases so as to reach a maximum at the TS.

4 Conclusions

The Fisher information, which measures the gradient content of the quantum-mechanical wavefunctions of physical systems, is a basic variable to interpret numerous quantum phenomena in physico-chemical processes. This is mainly due to its close connection to various energetic quantities, specially the kinetic energy, and its ability to grasp the oscillatory character of the single-particle density of the system. In this work we review some analytical properties of Fisher's measure that are relevant in quantum mechanics, with emphasis on its inequality-based relations with the radial expectation values and the uncertainty relation. We show the improvement of these properties when the average potential of the system is spherically symmetric. Moreover, the Fisher information of the D-dimensional hydrogenic systems is explicitly computed in terms of the hyperquantum numbers of both ground and excited states for position and momentum spaces. Finally, we illustrate the applicability and usefulness of the position and momentum Fisher informations in various physico-chemical processes of practical interest.

Conflict of interest: None.

Acknowledgments

This work has partially been supported by the MICINN grant FIS2011-24540, the grants FQM-7276 and FQM-4643 of the Junta de Andalucía, and the Mexican grants 08226 CONACyT and PIFI 3.3 PROMEP-SEP. JSD, ARP and PSM belong to the research group FQM-207.

References

[1] R. A. Fisher, Theory of statistical estimation, Proc. Cambridge Phil. Soc. 22, 700–725 (1925), Reprinted in Collected Papers of R.A. Fisher, edited by J.H. Bennet (University of Adelaide Press, South Australia), 1972, 15–40.

[2] B. R. Frieden, *Science from Fisher Information*, Cambridge University Press, Cambridge, 2004.

[3] R. G. Parr, W. Yang, *Density-Functional Theory of Atoms and Molecules*, Oxford Univ. Press, New York, 1989.

[4] C. E. Shannon, W. Weaver, *The Mathematical Theory of Communication*, University of Illinois Press, Urbana, 1949.

[5] A. Stam, Some inequalities satisfied by the quantities of information of Fisher and Shannon, Inform. Control 2, 101–112 (1959).

[6] J. S. Avery, *Hyperspherical Harmonics and Generalized Sturmians*, Kluwer, Dordrecht, 2000.

References

[1] R. A. Fisher, Theory of statistical estimation, Proc. Cambridge Phil. Soc. 22, 700–725 (1925), Reprinted in Collected Papers of R.A. Fisher, edited by J.H. Bennet (University of Adelaide Press, South Australia), 1972, 15–40.

[2] B. R. Frieden, *Science from Fisher Information*, Cambridge University Press, Cambridge, 2004.

[3] R. G. Parr, W. Yang, *Density-Functional Theory of Atoms and Molecules*, Oxford Univ. Press, New York, 1989.

[4] C. E. Shannon, W. Weaver, *The Mathematical Theory of Communication*, University of Illinois Press, Urbana, 1949.

[5] A. Stam, Some inequalities satisfied by the quantities of information of Fisher and Shannon, Inform. Control 2, 101–112 (1959).

[6] J. S. Avery, *Hyperspherical Harmonics and Generalized Sturmians*, Kluwer, Dordrecht, 2000.

[7] A. Rényi, On measures of entropy and information, Proceedings of the Fourth Berkeley Symp. Math. Stat. Probability, vol. 1, 1961, Reprinted in P. Durán (editor), Selected Papers of Alfréd Rényi, Akademia Kiado, Budapest, 1976, pp. 565-580, p. 547.

[8] C. Tsallis, Possible generalization of Boltzmann-Gibbs statistics, J. Stat. Phys. 52, 479–487 (1998).

[9] S.B. Sears, R.G. Parr, U. Dinur, On the quantum-mechanical kinetic-energy as a measure of the information in a distribution, Israel J. Chem. 19, 165–173 (1980).

[10] B. R. Frieden, A. Plastino, A. R. Plastino, B. H. Soffer, Schrödinger link between nonequilibrium thermodynamics and Fisher information, Phys. Rev. E 66, 046128 (2002).

[11] S. E. Massen, C. P. Panos, A link of information entropy and kinetic energy for quantum many-body systems, Phys. Lett. A 280, 65–69 (2011).

[12] E. Romera, J.S. Dehesa, Weizsäcker energy of many-electron systems, Phys. Rev. A 50, 256 (1994).

[13] E. Romera, J.S. Dehesa, The Fisher-Shannon information plane, an electron correlation tool, J. Chem. Phys. 120, 8906–8912 (2004).

[14] A. Nagy, Fisher information in a two-electron entangled artificial atom, Chem. Phys. Lett 425, 154–156 (2006).

[15] R. González-Férez, J.S. Dehesa, Characterization of atomic avoided crossings by means of Fisher's information, Eur. Phys. J. D 32, 39–43 (2005).

[16] E. Romera, Stam's principle, D-dimensional uncertainty-like relationships and some atomic properties, Mol. Phys. 100, 3325–3329 (2002).

[17] E. Romera, J.C. Angulo, J.S. Dehesa, Fisher entropy and uncertaintylike relationships in many-body systems, Phys. Rev. A 59, 4064–4067 (1999).

[18] S. López-Rosa, R. O. Esquivel, J. C. Angulo, J. Antolin, J. S. Dehesa, N. Flores-Gallegos, Fisher information study in position and momentum spaces for elementary chemical reactions, J. Chem. Theory Comput. 6, 145 (2010).

[19] R. N. Silver, Essays in honor to Edwin T. Jaynes (W. T. Grandy Jr., P. W. Milonni, eds.), Cambridge University Press, 1993.

[20] P. Sánchez-Moreno, R. González-Férez, J.S. Dehesa, Improvement of the Heisenberg and Fisher-information-based uncertainty relations for d-dimensional central potentials, New J. Phys. 8, 330 (2006).

[21] P. Sánchez-Moreno, A. R. Plastino, J. S. Dehesa, A quantum uncertainty relation based on Fisher's information, J. Phys. A: Math. Theor. 44, 065301 (2011).

[22] B. R. Frieden, A. Plastino, A. R. Plastino, B. H. Soffer, Fisher-based thermodynamics: Its Legendre transform and concavity properties, Phys. Rev. E 60, 48–53 (1999).

[23] A. Plastino, A. R. Plastino, B. H. Soffer, Fisher info and thermodynamics first law, Physica A 369, 432–438 (2006).

[24] T. M. Cover, J. A. Thomas, *Elements of Information Theory*, Wiley, N.Y., 1991.

[25] A. Dembo, T. M. Cover, J. A. Thomas, Information theoretic inequalities, IEEE Trans. Information Theory 37, 1501–1528 (1991).

[26] S. López-Rosa, J. C. Angulo, J. S. Dehesa, R. J. Yáñez, Existence conditions and spreading properties of extreme entropy D-dimensional distributions, Physica A 387, 2234–2255 (2008), Erratum: ibid 387, 4729-4730 (2008).

[27] S. Zozor, M. Portesi, P. Sánchez-Moreno, J. S. Dehesa, The position-momentum uncertainty relation based on power moments with arbitrary orders, Phys. Rev. A 83, 052107 (2011).

[28] J. S. Dehesa, S. López-Rosa, D. Manzano, Entropy and complexity analyses of d-dimensional quantum systems, Statistical Complexities: Application to Electronic Structure (K. D. Sen, ed.), Springer, Berlin, 2011.

[29] E. Romera, P. Sánchez-Moreno, J.S. Dehesa, Uncertainty relation for Fisher information of D-dimensional single-particle systems with central potentials, J. Math. Phys. 47, 103504 (2006).

[30] J. S. Dehesa, S. López-Rosa, B. Olmos, R. J. Yáñez, Fisher information of D-dimensional hydrogenic systems in position and momentum spaces, J. Math. Phys. 47, 052104 (2006).

[31] J. S. Dehesa, S. López-Rosa, A. Martínez-Finkelshtein, R. J. Yáñez, Information theory of D-dimensional hydrogenic systems: Application to circular and Rydberg states, Int. J. Quantum Chem. 110, 1529–1548 (2010).

[32] J. C. Polanyi, A. H. Zewail, Acc. Chem. Res. 28, 119 (1995).

[33] P. Jaque, A. Toro-Labbé, P. Geerlings, F. De Proft, J. Phys. Chem. A 113, 332 (2009).

[34] R. O. Esquivel, N. Flores-Gallegos, C. Iuga, E. Carrera, J. C. Angulo, J. Antolín, Phys. Lett. A 374, 948-951 (2010).

[35] R O. Esquivel, N. Flores-Gallegos, J. S. Dehesa, J. C. Angulo, J. Antolín, K. D. Sen, J. Phys. Chem. A 114, 1906-1916 (2010).

CHAPTER 10

Entanglement and Entropy

J. Batle [a], A. Plastino [b,c*], A. R. Plastino [d,e], and M. Casas [c†]

[a] Universitat de les Illes Balears Mallorca, Spain

[b] National University La Plata, IFLP-CONICET, C.C. 727, La Plata 1900, Argentina

[c] Physics Department and IFISC-CSIC, University of Balearic Islands, 07122, Palma de Mallorca, Spain

[d] Instituto Carlos I de Fisica Teorica y Computacional, Universidad de Granada, Spain

[e] CeBio y Secretaría de Investigación,
Universidad Nacional Buenos Aires-Noroeste, UNNOBA, Junin, Argentina

† Late

Abstract

We review links between the entanglement concept and the entropic-information one, as represented by different measures like the Shannon and Tsallis ones. This remarkable linkage illuminates several aspects of the physics of information-theoretic activity and it is thus of importance to understand the structure of the concomitant connections

Keywords: Entropy; Entanglement; Information Theory; Entropic measures.

1 Introduction

Entanglement is one of the most fundamental and non-classical features exhibited by quantum systems [1], that lies at the basis of some of the most important processes studied by quantum information theory [1, 2, 3, 4, 5] such as quantum cryptographic key distribution [6], quantum teleportation [7], superdense coding [8], and quantum computation [9, 10]. It is thus imperative to be in a position to ascertain whether the state ρ representing some physical system is appropriate for developing a given non-classical task by detecting its entanglement-degree. In this context one encounters the so-called "separability problem", which is of great importance in quantum information theory. Unless explicitly stated we

* Corresponding author. E-mail: plastino@fisica.unlp.edu.ar

Andres M. Kowalski, Raul D. Rossignoli and Evaldo M. F. Curado (Eds)

just consider entanglement between two parties, the common situation in quantum communication protocols.

Entropy is a macroscopic property that measures the microscopic disorder within the a system. It is moreover an important ingredient of the second law of thermodynamics, one of the pillars of our current understanding of Nature. Thermodynamic systems are made up of microscopic objects, *e.g.*, atoms or molecules, which "carry" energy. According to the above mentioned thermodynamics' law, entropy quantifies the amount of energy which can not be utilized during energy conversions. In applied thermodynamics, as a matter of convention, entropy is measured in units of energy per temperature (joules per kelvin). It is important to realize that, in technical applications, machines are basically energy conversion devices. Thus, such devices can only be driven by convertible energy and appeal to the entropic concept becomes unavoidable. The same applies to biological organisms. There are two related definitions of entropy: the thermodynamical and the statistical ones. The former definition was developed in the early 1850s by R. Clausius and essentially describes how to measure the entropy of an isolated system in thermal equilibrium. Significantly enough, no reference to the microscopic nature of matter is neither needed or desired. The statistical notion was concocted by L. Boltzmann in the 1870s by recourse to an ingenious analysis of the statistical behavior of atoms. Boltzmann went on to show that his definition is equivalent to Clausius' to within a constant number (Boltzmann's constant). Thus, the thermodynamic definition of entropy provides its empiric definition, while the statistical definition provides both explanations and a deeper understanding of its nature. The statistical concept of entropy admits, in turn, an information-theoretical interpretation. Roughly speaking, the entropy of a system can be construed as a measure of our lack of information about the precise state of the system. This was the point of view emphasized by Jaynes. This information-theoretical concept of entropy is applicable to any physical system, regardless of its size or number of internal constituents. In fact, it is also applicable to elementary, microscopic systems characterized by few degrees of freedom. In the case of a quantum system described by a density matrix ρ, appropriate entropic functionals $S[\rho]$ are closely related to the degree of mixedness of ρ. In the case of a pure quantum state, where we have the maximum possible knowledge about the state of the system allowed by the laws of Nature, we have $S[\rho] = 0$. A completely mixed state, corresponding to maximal ignorance about the state of the system, constitutes the opposite situation where $S[\rho]$ adopts its maximum value. In the present work we are going to refer to the concept of entropy in the information-theoretical sense, which is also related to the idea of entropy as a measure of the degree of mixedness of a density matrix ρ. In what follows we are going to consider only quantum systems described by Hilbert spaces of discrete and finite dimensions.

More to the point, what we wish here is to briefly review links between the entanglement concept and the entropic-information one. That such linkage exist is indeed remarkable and, moreover, illuminates several aspects of the physics of information-theoretic activity, so that a glance to the structure of the concomitant connections serves a quite useful purpose, as we hope to make clear below.

2 The separability problem

We begin by pointing out that it is of the essence to discriminate those micro-states that contain classical correlations only and thus no entanglement. This is referred to as the separability problem. Historically, the violation of the so called Bell's inequalities has become equivalent to the notion of entanglement. For every pure entangled state there is a Bell inequality that is violated and, in consequence, from a historic viewpoint, the first separability criterion is that of Bell (see [11] and references therein). It is not known, however, whether in the case of many entangled mixed states, violations exist. There exist states that, after subjected to a process denominated "distillation" of entanglement, that is effected by performing local operations and classical operations (LOCC), *i.e.*, operations performed on a party only of a bipartite one, eventually violate the Bell inequalities. However, some other systems do not.

Indeed, an entangled state does not necessarily violate a Bell-type inequality. If not, we say that it admits a local variable model. Werner [12] provided us with a family of mixed states (the *Werner* density matrix ρ_W) that does no violate the aforementioned inequalities. Werner also stated the current mathematical separability definition: a state of a composite quantum system (two subsystems A and B) is called "entangled" if it can not be represented as a convex linear combination of product states. In other words, the density matrix $\rho_{AB} \in \mathcal{H}_A \otimes \mathcal{H}_B$ represents an entangled state if it cannot be expressed as

$$\rho_{AB} = \sum_k p_k \, \rho_A^{(k)} \otimes \rho_B^{(k)}, \tag{1}$$

with $0 \leq p_k \leq 1$ and $\sum_k p_k = 1$. On the contrary, states of the form (1) are called separable. The above definition is physically meaningful because entangled states (unlike separable states) cannot be prepared locally by acting on each subsystem individually [13] (LOCC operations). An example of a LOCC operation is provided by

$$\rho\prime = (U_A \otimes U_B) \, \rho \, (U_A \otimes U_B)^{\dagger}, \tag{2}$$

where $U_A(U_B)$ represents a local action (unitary transformation) acting on subsystem $A(B)$. Equivalently, a separable state is a state that can be written as a mixture of factorizable pure states. Apparently this should be the end of the story. If one is able to write a given state as a convex combination of product states as in (1), then that state is separable. In practice, though, this is an impossible task because there are infinitely many ways of decomposing a state ρ, *i.e.*, a state can be prepared in many ways. Another intriguing feature is that a mixture of entangled states is not necessary entangled. The set of separable or unentangled states is convex, *viz.*, any linear convex combination of separable states gives another separable state. Thus we arrive at the *separability problem*: given a state ρ describing a quantum system, is it entangled or not? This (not generally solved) problem can be related to the challenging open questions of modern linear algebra: the characterization of positive maps. However, we require a criterion which could decide whether a state ρ is entangled or not. Such procedure can be cast following a simple algorithm or recipe (a *functional* criterion) or abstractly (a *non-functional* criterion).

3 Characterization of entanglement

Detecting entanglement may not be sufficient. Given a matrix ρ representing the state of the system, it is possible that interaction with environment destroys the entanglement present in ρ. We therefore require a means of *quantifying* entanglement. This feature, quantification, it is decisive whenever we must decide whether the entanglement provided by a physical set up (*e.g.*, a certain physical implementation for quantum computation) is sufficient or not in order to accomplish with the quantum information related tasks. The characterization of entanglement obviously implies a knowledge of the physical meaning of entanglement itself.

In point of fact entanglement should be regarded as a resource, like energy, which underlies most of the striking new applications of the newborn science of quantum information. This extremely non-classical feature finds lots of practical applications impossible to achieve without it.

As for the meaning of entanglement, *no one really knows what it actually is*, which also happens in trying to define the absolute meaning of the word "energy". One merely describes the spectra of capacities acquired *via* entanglement, a situation resembling the historical development of Thermodynamics, where entropy appeared in order to clarify the processes involving temperature, work and heat, and one had to find links to that newborn quantity through the specific heat, Joules and calories, that is, computable magnitudes.

In the context of quantum information, entanglement receives several definitions, *viz.*,

- for a physicist working in quantum cryptography, entanglement is an essential tool for absolute secure communication;
- for a physicist working in teleportation, entanglement is the basic tool for making teleportation of states possible;
- for a physicist in the field of quantum correlations/Bell inequalities, entanglement represents some sort of "extra" correlation that enables to refute local hidden variable theories trying to describe physical reality in local terms;
- for a computer scientist, entanglement is the basic ingredient for building new kinds of algorithms that solve problems exponentially faster that classical Turing machines.

Entanglement between parties, *i.e.*, between certain degrees of freedom of the parties, arises in the form of correlations between them *after* these parties have "spoken" to each other through some interaction. This fact is physically meaningful because particles which do not interact or have interacted in the past are not expected to shown any quantal correlation. Of course this is not the case for non-interacting particles with some associated statistics (*e.g.*, identical fermions or bosons), which clearly possess intrinsic correlations, though they are useless for quantum information purposes. Thus, the characterization of entanglement takes into account the distinguishability of the subsystems. To characterize entanglement one might choose a physical system being representative of maximal quantal correlation, and define it as an "entanglement ruler". Of course the situation is more involved, but intuitively valid. A feature to keep in mind is that entanglement cannot be enhanced by local operations acting on the subsystems individually or by classical communication between them (LOCC operations).

The so-called entanglement of formation for two-qubit systems is given by Wootters' expression [14]

$$E[\rho] = h\left(\frac{1+\sqrt{1-C^2}}{2}\right), \quad (3)$$

where

$$h(x) = -x\log_2 x - (1-x)\log_2(1-x), \quad (4)$$

and the concurrence C is given by

$$C = max(0, \lambda_1 - \lambda_2 - \lambda_3 - \lambda_4), \quad (5)$$

λ_i, $(i = 1, \ldots 4)$ being the square roots, in decreasing order, of the eigenvalues of the 4x4 matrix $\rho\tilde{\rho}$, with

$$\tilde{\rho} = (\sigma_y \otimes \sigma_y)\rho^*(\sigma_y \otimes \sigma_y), \quad (6)$$

while the σ's are Pauli matrices.

The above expression has to be evaluated by recourse to the matrix elements of $\hat{\rho}$ computed with respect to an appropriate basis. C^2 can be regarded as a proper measure of entanglement, as (3) is a monotonic increasing function of C^2. For pure states (3) reduces to the usual von Neumann entropy. As a matter of fact, $C^2 = 4\,\mathrm{det}\rho_A$ for pure two-qubits states. The general form for a bipartite pure state [15] $|\psi\rangle \in \mathcal{H}_N \otimes \mathcal{H}_K$ is ($N \leq K$ is assumed) $C(|\psi\rangle) = \sqrt{2[|\langle\psi|\psi\rangle|^2 - Tr(\rho_N^2)]}$, with $\rho_N = \mathrm{Tr}_K(|\psi\rangle\langle\psi|)$. Although there is no general expression for the concurrence of mixed bipartite states, F. Mintert *et al.* derive in [16] a lower bound for the concurrence of mixed bipartite quantum states, valid in arbitrary dimensions.

3.1 q-Entropies

The separability issue has interesting echoes in information theory and its associate information measures or entropies. When one deals with a classical composite system described by a suitable probability distribution defined over the concomitant phase space, the entropy of any of its subsystems is always equal or smaller than the entropy characterizing the whole system. This is also the case for separable states of a composite quantum system [17, 18]. In contrast, a subsystem of a quantum system described by an entangled state may have an entropy greater than the entropy of the whole system. Indeed, the von Neumann entropy of either of the subsystems of a bipartite quantum system described (as a whole) by a pure state provides a natural measure of the amount of entanglement of such state. Thus, a pure state (which has null entropy) is entangled if and only if its subsystems have an entropy larger than the one associated with the system as a whole.

Regrettably enough, the situation is more complex when the composite system is described by a mixed state. There are entangled mixed states such that the entropy of the complete system is smaller than the entropy of one of its subsystems. Alas, entangled mixed states such that the entropy of the system as a whole is larger than the entropy of either of its subsystems do exist as well. Consequently, the classical inequalities relating the entropy of the whole system with the entropy of its subsystems provide only necessary, but not sufficient, conditions for quantum separability. There are several entropic (or information) measures that can be used in order to implement these criteria for separability. Considerable attention has been paid, in this regard, to the q-entropies [11, 18, 19, 20, 21, 22, 23, 24, 25], which incorporate both Rényi's [26] and Tsallis' [27, 28, 29] families of information measures as special instances (both admitting, in turn, Shannon's measure as the particular case associated with the limit $q \to 1$). The "q-entropies" depend upon the eigenvalues p_i of the density matrix ρ of a quantum system through the quantity $\omega_q = \sum_i p_i^q$. More explicitly, we shall consider either the Rényi entropies [26],

$$S_q^{(R)} = \frac{1}{1-q} \ln(\omega_q), \tag{7}$$

or the Tsallis' entropies [27, 28, 29]

$$S_q^{(T)} = \frac{1}{q-1}(1-\omega_q), \tag{8}$$

which have found applications in many different fields of Physics. These entropic measures incorporate the important (because of its relationship with the standard thermodynamic entropy) instance of the von Neumann measure, as a particular limit ($q \to 1$) situation

$$S_1 \;=\; -Tr\left(\hat{\rho}\ln\hat{\rho}\right). \tag{9}$$

The concomitant *conditional* $q-$ *entropies* are defined as

$$S_q^{(T)}(A|B) \;=\; \frac{S_q^{(T)}(\rho_{AB}) - S_q^{(T)}(\rho_B)}{1+(1-q)S_q^{(T)}(\rho_B)} \tag{10}$$

for the Tsallis case, while its Rényi counterpart is

$$S_q^{(R)}(A|B) \;=\; S_q^{(R)}(\rho_{AB}) - S_q^{(R)}(\rho_B), \tag{11}$$

where $\rho_B = Tr_A(\rho_{AB})$ (the conditional q-entropy $S_q^{(T)}(B|A)$ is defined in a similar way as (10), replacing ρ_B by $\rho_A = Tr_B(\rho_{AB})$). The "classical q-entropic inequalities" finally read

$$\begin{aligned} S_q^{(T,R)}(A|B) \;&\geq\; 0, \\ S_q^{(T,R)}(B|A) \;&\geq\; 0, \end{aligned} \tag{12}$$

valid for all separable states for several q-values [30].

The early motivation for the studies reported in [18, 19, 20, 21, 22, 23, 24, 25] was the development of practical separability criteria for density matrices. The discovery by Peres of the partial transpose criteria, which for two-qubits and qubit-qutrit systems turned out to be both necessary and sufficient, rendered that original motivation somewhat outmoded. However, its study provides some physical insight into the issue of quantum separability.

4 Entanglement, q-entropies and mixedness

The amount of entanglement and the degree of mixture (mixedness) of a quantum states are important attributes for characterizing it. There is a quite interesting relationship between these two traits, because, as the degree of mixture increases, quantum states tend to be endowed with smaller entanglement-degrees. Even more, in the very important case of two-qubits systems, states with a large enough mixedness are always separable [31].

Thus, a detailed knowledge of the relation between the degree of mixture and the amount of entanglement becomes essential in order to understand the limitations that mixedness imposes on variegated quantum information processes such as teleportation or computing.

Of course, in order to study the relationship between entanglement and mixture we need appropriate quantitative measures for these two quantities. The entanglement of formation provides a natural quantitative measure of entanglement with a clear physical motivation. As for mixedness, there are several indicators that can be useful within the present context. The von Neumann measure

$$S_1 = -Tr\left(\hat{\rho}\ln\hat{\rho}\right), \tag{13}$$

is important because of its relationship with the thermodynamic entropy. Additionally, the so called participation ratio,

$$R(\hat{\rho}) = \frac{1}{Tr(\hat{\rho}^2)}, \tag{14}$$

is particularly convenient for calculations [31, 32]. The q-entropies, introduced below are functions of the quantity

$$\omega_q = Tr\left(\hat{\rho}^q\right), \tag{15}$$

and yield a whole family of measures for the degree of mixture. In particular, when $q = 2$ they are simply related to the participation ratio (14).

We will focus now our attention upon some aspects of the aforementioned relationship between entanglement and purity (the "inverse" of the mixture-degree), using the q-entropies as measures of mixedness [33]. We will center our interest in the probability (density) $F(R)$ of finding two qubit state with participation ratio R. We shall also discuss the limiting instance $q \to \infty$ and its connection with the use of the largest eigenvalue of $\hat{\rho}$ as a measure of the degree of mixture.

5 Distribution of two-qubits states according to their mixture

Before beginning with our task, some preliminary words are in order. The space $\mathcal{S}$ of all (pure and mixed) states of a quantum system described by an N-dimensional Hilbert space can be regarded as a cartesian product space $\mathcal{S} = \mathcal{P} \times \Delta$, where $\mathcal{P}$ stands for the family of all complete sets of ortonormal projectors $\{\hat{P}_i\}_{i=1}^{N}$, $\sum_i \hat{P}_i = I$ (I being the identity matrix), and Δ is the set of all real N-tuples $\{\lambda_1, \ldots, \lambda_N\}$, with $\lambda_i \geq 1$ and $\sum_i \lambda_i = 1$. The natural measure

$$\mu = \nu \times \mathcal{L}_{N-1} \tag{16}$$

is the one used for the random generation of two-qubits states ($N = 4$). Also, in order to study the distribution of two-qubit states according to their degree of mixture, we identify the simplex Δ with a regular tetrahedron T_Δ of side length 1, in $\mathcal{R}^3$, centred at the origin. Next we consider two special q-cases that are relevant in our study.

5.1 The case $q = 2$

In this case the degree of mixture is characterized by the quantity $\omega_2 = Tr(\hat{\rho}^2) = \sum_i p_i^2$. This quantity is easily seen to be related to the distance $r = \mid \mathbf{r} \mid$ to the centre of the tetrahedron T_Δ by

$$r^2 = -\frac{1}{8} + \frac{1}{2}\omega_2. \tag{17}$$

Thus, the states with a given degree of mixture lie on the surface of a sphere of radius r concentric with the tetrahedron T_Δ.

The volume associated with states endowed with a value of ω_2 lying within a small interval dw_2 is clearly associated with the volume dV of the subset of points in T_Δ whose distances to the centre of T_Δ are between r and $r + dr$, with $rdr = \omega_2\, d\omega_2$. Let Σ_r denote the sphere of radius r concentric with T_Δ. The volume dV is then proportional to the area $A(r)$ of the part of Σ_r which lies within T_Δ. In order to compute the aforementioned area, it is convenient to separately consider three different ranges for the radius r.

Let us first consider the range of values $r \in [0, h_1]$, where $h_1 = \frac{1}{4}\sqrt{\frac{2}{3}}$ is the radius of a sphere tangent to the faces of the tetrahedron T_Δ. In this case the sphere Σ_r lies completely within the tetrahedron T_Δ. Thus, the area we are interested in is just the area of the sphere,

$$A_I(r) = 4\pi r^2. \tag{18}$$

We now consider a second range of values of the radius, $r \in [h_1, h_2]$, where $h_2 = \frac{\sqrt{2}}{4}$ denotes the radius of a sphere which is tangent to the sides of the tetrahedron T_Δ. In this case, the area of the portion of Σ_r which lies within T_Δ is

$$A_{II}(r) = 4\pi\left[r^2 - 2r(r-h_1)\right]. \tag{19}$$

Finally, we consider the range of values $r \in [h_2, h_3]$, where $h_3 = \frac{\sqrt{6}}{4}$ is the radius of a sphere passing through the vertices of T_Δ. In this case the area A_{III} of the part of the sphere Σ_r lying within T_Δ is

$$A_{III}(r) = 4(S_A - 3S_B), \tag{20}$$

where

$$\begin{aligned} S_A &= r^2(3\alpha - \pi), \\ S_B &= r^2\left[h(-\pi + 2\sin^{-1}(C_1C_2)) + 2\sin^{-1}\sqrt{\frac{1-C_1^2C_2^2}{1+C_2^2}}\right]. \end{aligned} \tag{21}$$

The quantities appearing in the right hand sides of the above expressions are defined by

$$\alpha = \cos^{-1}\left[\frac{\cos A - \cos^2 A}{\sin^2 A}\right]; \quad A = 2\sin^{-1}(D_1/r); \quad D_1 = \frac{1}{2}\left(\frac{1}{2} - \sqrt{r^2 - \frac{1}{8}}\right), \tag{22}$$

and

$$\begin{aligned} h &= h_1/r; \quad C_1 = \frac{h}{\sqrt{1-h^2}}; \quad C_2 = \frac{C_B}{\sqrt{1-C_B^2}}; \\ C_B &= \sqrt{\frac{D_2^2 - D_1^2}{r^2 - D_1^2}}; \quad D_2 = r\sqrt{1-h^2}. \end{aligned} \tag{23}$$

Using the relation between r and the participation rate $R = 1/Tr(\rho^2)$,

$$r^2 = -\frac{1}{8} + \frac{1}{2R}, \tag{24}$$

we analytically obtain the probability $F(R)$ of finding a quantum state with a participation rate R,

$$F(R) = f(r)\left|\frac{dr}{dR}\right|. \tag{25}$$

In Eq. (25) $f(r) = A(r)/(\text{Volume}[T_\Delta])$, while $A(r)$ is given by equations (18-20). The

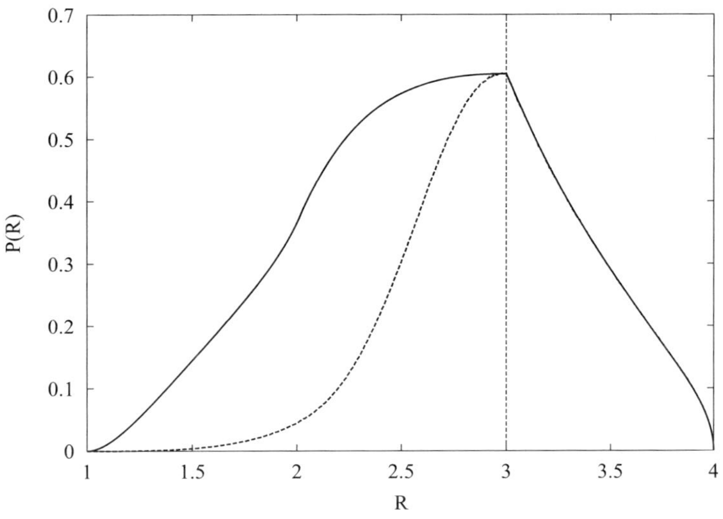

Figure 1: Probability (density) $P(R)$ of finding two-qubits states with a given amount of mixedness R. The solid line corresponds to all states and the dashed line to those ones which are separable *only*, given by $P_{sep}(R)$. This curve is normalized to the overall probability of finding a separable state so that both $P(R)$ and $P_{sep}(R)$ can be compared at the same scale. The maximum occurs at the separability border $R = 3$.

behavior of $F(R)$ is shown in Fig.1. We note that this distribution $F(R)$ was first determined, in numerical fashion, by Zyczkowski *et al.* in [31]. Instead, here we compute $F(R)$ analytically [33]. As expected, these calculations coincide with the concomitant numerical results and also with the ones reported in [31].

5.2 Info-theoretical approach to entanglement and R

According to the $R-$value, two zones are clearly to be distinguished. In Fig.1, entanglement is to be encountered only in the zone lying at the "west" ($R < 3$) of the graph. As matter of fact, Fig. 2 depicts, for several values of the participation ratio R, the probability $P_R(C^2)$ of finding in our 15-dimensional space $\mathcal{S}$ of two-qubits a bipartite state of concurrence squared C^2 (5). The inset is a $\langle C^2 \rangle$ *vs.* R plot. It is well known that i) no entanglement exists for $R > 3$ and ii) $R = 1$ only for pure states. It is clear that R plays an important role in predetermining the possibility of finding entanglement. Obviously, to each curve $P_R(C^2)$ *vs.* C^2

one can assign a Shannon information measure

$$S_R = -\int_0^1 dC^2 \, P_R(C^2) \ln [P_R(C^2)], \tag{26}$$

that measures the informational amount contained in the distribution $P_R(C^2)$.

We can establish a correlation between the participation ratio and our knowledge regarding the entanglement-distribution (ED) in $\mathcal{S}$ by plotting S_R *vs.* R, as done in Fig.3, a logarithmic plot. The vertical axis corresponds to the logarithm of the information gain $G = [S_R - H_{max}]/S_{R=1}$, gain evaluated with reference to the uniform distribution, that always entails maximum ignorance. For pure states ($R = 1$) $S_{R=1}$ is large, but does not quite reach the uniform value H_{max}. We reiterate that the information (or ignorance) units are calibrated so that $G = 1$ corresponds to the ignorance concerning the ED for bipartite pure states. It is clearly appreciated that, as R grows, G steadily increases. We can assert, for instance, that if $R = 1.5$, our information gain becomes ~ 10 times the one that one has at $R = 1$. The upper, dashed horizontal line indicates the information gain corresponding to the knowledge, with reference to our 15-dimensional space, of the probability $P(C^2)$ for all states, mixed or not (See for instance [33]). It is clear that we acquire information because the number of entangled states decreases with R.

We remind that for a continuous probability distribution (cpd) the entropy is defined only up to an arbitrary additive constant. Only entropic *differences* make sense (the information gain is obtained with respect to H_{max}). Additionally, for cpd's, these differences can become infinite, because one needs an infinite amount of information to select a point out of the continuum.

5.3 The case $q = \infty$

The quantity ω_q is not appropriately suited to discuss the limit $q \to \infty$. However, $\omega_q^{1/q}$ does exhibit a nice behavior when $q \to \infty$. Indeed, we have

$$\lim_{q\to\infty} (Tr\rho^q)^{1/q} = \lim_{q\to\infty} \left(\sum_i p_i^q\right)^{1/q} = \lambda_m, \tag{27}$$

where

$$\lambda_m = \max_i \{p_i\} \tag{28}$$

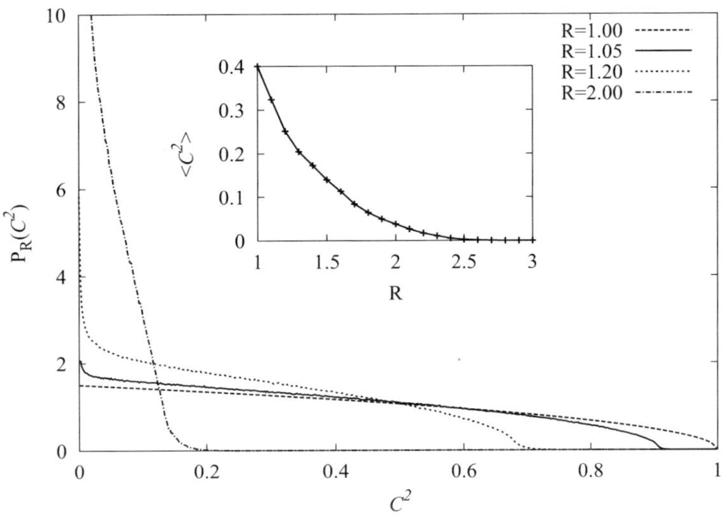

Figure 2: Probability (density) $P_R(C^2)$ of finding two-qubits states (generated according to the μ-measure) with fixed degree of mixture R, and endowed with a given value of the concurrence squared C^2. The range of available values for C^2 continuously decreases as we approach the limiting value $R = 3$. The inset shows the evolution of C^2 *vs.* R.

is the maximum eigenvalue of the density matrix ρ. Hence, in the limit $q \to \infty$, the q-entropies (when properly behaving) depend only on the largest eigenvalue of the density matrix. For example, in the limit $q \to \infty$, the Rényi entropy reduces to

$$S_{\infty}^{(R)} = -\ln(\lambda_m). \tag{29}$$

It is worth realizing that the largest eigenvalue itself constitutes a legitimate measure of mixture. Its extreme values correspond to (i) pure states ($\lambda_m = 1$) and (ii) the identity matrix ($\lambda_m = 1/4$). It is also interesting to mention that, for states diagonal in the so called Bell basis, the entanglement of formation is completely determined by λ_m (This is not the case, however, for general states of two-qubits systems).

In terms of the geometric representation of the simplex Δ, the set of states with a given value λ_m of their maximum eigenvalue is represented by the tetrahedron determined by the four planes

$$\lambda_m = 2(\mathbf{r} \cdot \mathbf{r}_i) + \frac{1}{4}, \quad i = 1, \ldots, 4. \tag{30}$$

The four vertices of this tetrahedron are given by the intersection points of each one of the

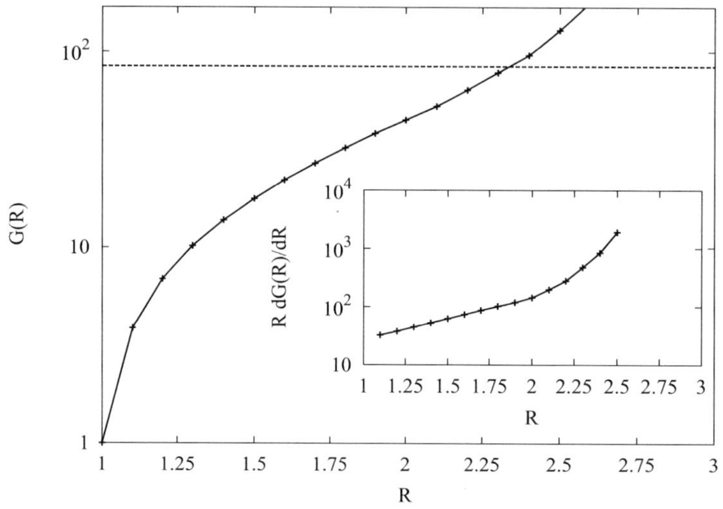

Figure 3: $\ln G$ *vs.* R–plot, with $G = [S_R - H_{max}]/S_{R=1}$ the information gain with respect to the uniform distribution H_{max} and S_R the information-amount contained in the distribution $P_R(C^2)$ of Fig.2. The horizontal dashed line indicates the information gain corresponding to the knowledge of the probability $P(C^2)$ for all states, mixed or not. The inset shows (again in a logarithmic plot) the evolution of the analogue of a specific heat $C_R = R\frac{dG}{dR}$ *vs.* R. A quadratic evolution is observed in the range $\in [1, 2]$, which becomes steeper in $\in [2, 3]$.

four possible triplets of planes that can be selected among the four alluded to planes.
For $q \to \infty$ the accessible states with a given degree of mixture are on the surface of a small tetrahedron T_l concentric with the tetrahedron T_Δ. We are going to characterize each tetrahedron T_l (representing those states with a given value of λ_m) by the distance l between (i) the common centre of T_Δ and T_l and (ii) each vertex of T_l. The volume associated with states with a value of λ_m belonging to a given interval λ_m is proportional to the area $A(l)$ of the portion of T_l lying within T_Δ.
Following a similar line of reasoning as the one pursued in the case $q = 2$, we consider three ranges of values for l. The first range of l-values is given by $l \in [0, h_1/3]$. The particular value $l = h_1/3$ corresponds to a tetrahedron T_l whose vertices are located at the centres of the faces of T_Δ. Within the aforementioned range of l-values, T_l is lies completely within T_Δ. Consequently, $A(l)$ coincides with the area of T_l,

$$A_I(l) = 24\sqrt{3}l^2. \tag{31}$$

The second range of l-values corresponds to $l \in [h_1/3, h_1]$. The area of the part of T_l lying within T_Δ is now

$$A_{II}(l) = 3\sqrt{3}\left[8l^2 - \frac{3}{2}(3l - h_1)^2\right] \tag{32}$$

Finally, the third range of l-values we are going to consider is $l \in [h_1, 3h_1]$. In this case we have

$$A_{III}(l) = \frac{3}{2}\sqrt{3}(3h_1 - l)^2 \tag{33}$$

In a similar way as in the $q = 2$ case, the above expressions for $A(l)$ lead to the analytical form of the probability (density) $F(\lambda_m)$ of finding a two-qubits state with a given value of its greatest eigenvalue,

$$F(\lambda_m) = \frac{A(l)}{\text{Volume}[T_\Delta]}\left|\frac{dl}{d\lambda_m}\right|. \tag{34}$$

Regarding $\omega_q^{1/q}$ as a measure of mixture, the probability of finding states with given degrees of mixture is depicted in Fig.4, for different values of q. In the limit $q \to \infty$, this probability distribution is given by $F(\lambda_m)$ (which is included in Fig.4). Remarkably enough, as q

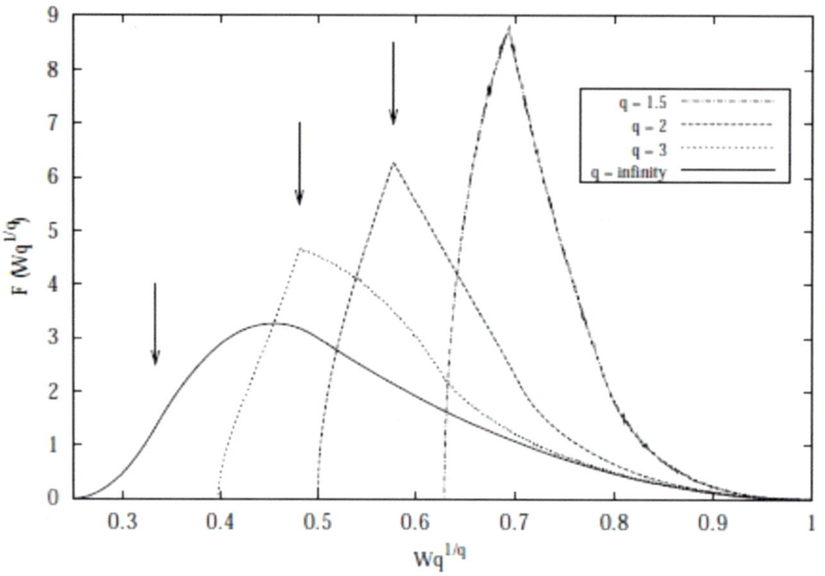

Figure 4: Probability $F(w_q^{1/q})$ *versus* $w_q^{1/q} = (\sum_i p_i^q)^{1/q}$ for several q values. The value of $w_q^{1/q}$ up to which all the states are separable is also indicated. This Fig. is taken from J. Batle *et al.* in Physics Letters A 296 (2002) 251258.

tends to infinity all discontinuities in the derivative of $F(\lambda_m)$ disappear. Is seems as in the λ_m-domain the distribution is completely smooth, as opposed to the R-domain.

6 q-Entropies and the separability threshold

Quantum states of two-qubits are always separable (that is, their entanglement is equal to zero) if the degree of mixture is high enough. This fact was first demonstrated in [31]. With the aid of q-entropies it is possible to give this property of two-qubits states a quantitative expression [33]. For values of the parameter q belonging to the interval $[2, \infty)$, a special value of the q-entropy exists such that states with an entropy greater than that particular value are always separable. Let us consider the subset $\Lambda(q; Z)$ of the simplex Δ consisting of all the points of Δ such that

$$\omega_q = \sum_{i=1}^{4} p_i^q \leq z. \tag{35}$$

Note that $\Lambda(q; z)$ is a convex set. In the geometric simplex picture the set $\Lambda(q; z)$ constitutes a subset of the tetrahedron. We shall design this subset with the same symbol $\Lambda(q; z)$. It should be clear that

$$q > 2 \implies \Lambda(q;z) \subset \Lambda(2;z). \tag{36}$$

Let $z(q)$ be that particular $z-$value for which the set $\Lambda(q;z)$ becomes tangent to the faces of the tetrahedron. The points where $\Lambda(q;z(q))$ "kisses" these faces are precisely the point within our faces at which the quantity ω_q takes its minimum value. These minima of ω_q correspond to a a q-entropy maximum (for example, to a maximum of Renyi's entropy $S_q^{(R)}$). Now, within a face of the tetrahedron one of the p_i's vanishes. More specifically, the maximum of a q-entropy corresponds to the centre of the face, where the remaining three non vanishing p_i's are equal to 1/3. Thus,

$$z(q) \;=\; 3\,(1/3)^q. \tag{37}$$

For $q = 2$, one has $z(2) \;=\; \frac{1}{3}$. This value of ω_2 corresponds to a participation ration $R = 3$. Consequently, the points belonging to $\Lambda(2;z(2))$ are precisely those with $R \leq 3$, which are always separable. Consequently,

$$\begin{aligned} q > 2 \quad &\Longrightarrow \quad z(q) \;<\; z(2) \\ &\Longrightarrow \quad \Lambda(q;z(q)) \;\subset\; \Lambda(q;z(2)), \end{aligned} \tag{38}$$

which combined with the inclusion relation (36) lead to

$$q > 2 \implies \Lambda(q;z(q)) \;\subset\; \Lambda(2;z(2)). \tag{39}$$

The above equation implies that all states belonging to $\Lambda(q;z(q))$ are separable. We have thus proved that, for $q \geq 2$, states with

$$w_q \;=\; Tr(\rho^q) \leq \; 3\,(1/3)^q \tag{40}$$

are always separable. In terms of the Renyi entropies we then can state that, for $q \geq 2$, states with

$$S_q^{(R)}[\hat{\rho}] \;\geq \ln 3, \tag{41}$$

are always separable. In the limit $q \to 1$ (see Eq.(29)), the separability criterium given by equation (41) implies that all states whose largest eigenvalue λ_m is less than $1/3$ are separable.

Note that, in terms of the Renyi entropies, the separability threshold does not depend (for $q \geq 2$) on the value of the parameter q. The problem of determining the limit values of the Renyi entropies $S_q^{(R)}$ such that states with entropies above them are always separable was first discussed by Zyczkowski *et al.* in [31], *via* a numerical study. On the basis of their numerical results, it was then conjectured that the limit value of $S_q^{(R)}$ is a decreasing function of q. In our considerations we are able to further assert that, for $q \geq 2$, *this limit value is not dependent on* q *and is equal to* $\ln 3$.

7 Conclusions

We have shown that intimate links between the concepts of entropy and entanglement exist and have explored some of the facets of these linkages. From our considerations one gathers that there is much to learn from the careful study of these connections. A glimpse of the issues involved therein has been provided as, hopefully, an interesting foretaste.

Conflict of interest: None.

Acknowledgements

We acknowledge partial financial support from the programs FQM-2445 and FQM-207 of the Junta de Andalucia, Spain. This work is also funded by the Spain Ministry of Science and Innovation (Project FIS2008-00781) and by FEDER funds (EU).

References

[1] Hoi-Kwong Lo, S. Popescu and T. Spiller (Editors), *Introduction to Quantum Computation and Information* (World Scientific, River-Edge, 1998).

[2] A. Galindo and M. A. Martín-Delgado, Rev. Mod. Phys. **74**, 347 (2002).

[3] M. A. Nielsen and I. L. Chuang, *Quantum Computation and Quantum Information* (Cambridge University Press, Cambridge, 2000).

[4] C. P. Williams and S.H. Clearwater, *Explorations in Quantum Computing* (Springer, New York, 1997).

[5] C. P. Williams (Editor), *Quantum Computing and Quantum Communications* (Springer, Berlin, 1998).

[6] A. Ekert, Phys. Rev. Lett. **67**, 661 (1991).

[7] C.H. Bennett, G. Brassard, C. Crepeau, R. Jozsa, A. Peres and W. K. Wootters, Phys. Rev. Lett. **70**, 1895 (1993).

[8] C. H. Bennett and S. J. Wiesner, Phys. Rev. Lett. **69**, 2881 (1993).

[9] A. Ekert and R. Jozsa, Rev. Mod. Phys. **68**, 773 (1996).

[10] G. P. Berman, G. D. Doolen, R. Mainieri, and V. I. Tsifrinovich, *Introduction to Quantum Computers* (World Scientific, Sinagapore, 1998).

[11] B. M. Terhal, Theor. Comp. Sci. **287**, 313 (2002).

[12] R. F. Werner, Phys. Rev. A **40**, 4277 (1989).

[13] A. Peres, *Quantum Theory: Concepts and Methods* (Kluwer, Dordrecht, 1993).

[14] W. K. Wootters, Phys. Rev. Lett. **80**, 2245 (1998).

[15] P. Rungta, V. Buzek , C. M. Caves, M. Hillery, and G. J. Milburn, Phys. Rev. A **64** 042315 (2001).

[16] F. Mintert, M. Kus, and A. Buchleitner, Phys. Rev. Lett. **92**, 167902 (2004).

[17] M. A. Nielsen and J. Kempe, Phys. Rev. Lett. **86**, 5184 (2001).

[18] K. G. H. Vollbrecht and M. M. Wolf, J. Math. Phys. **43**, 4299 (2002).

[19] R. Horodecki, P. Horodecki, M. Horodecki, Phys. Lett. A **210**, 377 (1996).

[20] R. Horodecki, M. Horodecki, Phys. Rev. A **54**, 1838 (1996).

[21] N. Cerf, C. Adami Phys. Rev. Lett. **79**, 5194 (1997).

[22] A. Vidiella-Barranco, Phys. Lett. A **260**, 335 (1999).

[23] C. Tsallis, S. Lloyd, M. Baranger, Phys. Rev. A **63**, 042104 (2001).

[24] C. Tsallis, P.W. Lamberti, D. Prato, Physica A **295**, 158 (2001).

[25] S. Abe, Phys. Rev. A **65**, 052323 (2002).

[26] C. Beck and F. Schlogl, *Thermodynamics of Chaotic Systems* (Cambridge University Press, Cambridge, 1993).

[27] C. Tsallis, J. Stat. Phys. **52**, 479 (1988).

[28] P. T. Landsberg and V. Vedral, Phys. Lett. A **247**, 211 (1998).

[29] J. A. S. Lima, R. Silva, and A. R. Plastino, Phys. Rev. Lett. **86**, 2938 (2001).

[30] M. Horodecki and P. Horodecki, Phys. Rev. A **59**, 4206 (1999)

[31] K. Zyczkowski K, P. Horodecki, A. Sanpera and M. Lewenstein, Phys. Rev. A **58**, 883 (1998).

[32] W.J. Munro, D.F.V. James, A.G. White, and P.G. Kwiat, *Phys. Rev. A* **64** (2001) 030302.

[33] J. Batle, M. Casas, A. R. Plastino and A. Plastino A, Phys. Lett. A **296**, 251 (2002).

Send Orders for Reprints to reprints@benthamscience.net

CHAPTER 11

Semiclassical Treatments and Information Theory

Flavia Pennini [a,*], Angelo Plastino [b], Gustavo Luis Ferri [c]

[a] Departamento de Física, Universidad Católica del Norte, Av. Angamos 0610, Antofagasta, Chile

[b] Instituto de Física La Plata–CCT-CONICET, Fac. de Ciencias Exactas, Universidad Nacional de La Plata, C.C. 67, 1900, La Plata, Argentina

[c] Facultad de Ciencias Exactas y Naturales, Universidad Nacional de La Pampa, Uruguay 151, Santa Rosa, La Pampa, Argentina

Abstract

We review here the difference between quantum statistical treatments and semiclassical ones, using as the main concomitant tool a semiclassical, shift-invariant Fisher information measure built up with Husimi distributions. Its semiclassical character notwithstanding, this measure also contains abundant information of a purely quantal nature. Such a tool allows us to refine the celebrated Lieb bound for Wehrl entropies and to discover thermodynamic-like relations that involve the degree of delocalization. Fisher-related thermal uncertainty relations are developed and the degree of purity of canonical distributions, regarded as mixed states, is connected to this Fisher measure as well.

Keywords: Information theory; Semiclassical information; Fisher measures, Phase space; Quantum distributions.

1 Introduction

R.A. Fisher advanced already in 1925 a method to measure the information content of continuous (rather than digital) inputs replacing binary computer codes by statistical distributions of classical probability theory [1]. In 1980 Wootters noticed that Fisher's information measure (FIM) and quantum mechanics share a common formalism since both relate probabilities to the squares of continuous functions [2].

Here we review materials from interesting work devoted to the physical applications of Fisher's information measure (see, for example, [1, 3, 4, 5, 6]). The first point to be

*Corresponding author. E-mail: fpennini@ucn.cl

Andres M. Kowalski, Raul D. Rossignoli and Evaldo M. F. Curado (Eds)

underlined is that Frieden and Soffer [3] have demonstrated just how Fisher's information measure (FIM) provides one with a powerful variational principle, that they call the extreme physical information one, from which most of the canonical Lagrangians of theoretical physics can be obtained [1, 3].

Additionally, FIM provides an interesting characterization of the "arrow of time", better than the one associated with Boltzmann's entropy [7, 8] because it works in more general situations. Thus, discussing the multiple FIM facets and their links to physics is a task of general interest. Abounding a bit, the Legendre transform structure of thermodynamics can be FIM-replicated as well, without changes, by replacing Boltzmann–Gibbs–Shannon's entropy S by Fisher's information measure I. The later possesses the all important concavity property [5], and using it allows for the development of a thermodynamics that treats both equilibrium and non-equilibrium situations on the same footing [5].

In this review we focus attention, following [9], on the thermal description of harmonic oscillator (HO), which illustrates in adequate fashion the salient points to be addressed in general.

The semiclassical approximation (SC) had a long and distinguished history and still remains a very important weapon in the physics arsenal, being indeed indispensable in many areas of science. The SC is important also because it usually facilitates an intuitive understanding of the underlying physics that frequently remains hidden in extensive numerical solutions of Schrödinger's equation. Although the SC-approach is as old as quantum mechanics itself, it constitutes and active field, as reported, for example, in [10] and [11].

We will here underline the differences between (i) statistical treatments of a purely quantal nature, on the one hand, and (ii) semiclassical ones, on the other, and show that they can be expressed entirely in terms of a special version, to be called I_τ, of Fisher's information measure: the so-called shift-invariant (translational invariance) FIM [1], associated to the classical phase space. I_τ is a functional of a semiclassical distribution function, namely, the Husimi function $\mu(x,p)$. We will show that the phase space measure I_τ is of help to (1) refine the so-called Lieb-bound [12] and (2) connect this refinement with the delocalization in phase space, that can be visualized as information loss. I_τ will turn out to be related to a well-known semiclassical measure, early introduced to characterize the same phenomenon: *Wehrl's entropy* W [12],

$$W = -k_B \langle \ln \mu \rangle_\mu, \tag{1}$$

for which Lieb established the above cited lower bound $W \geq 1$, which is a manifestation of the uncertainty principle [13]. k_B is the Boltzmann's constant. Henceforth we will set $k_B = 1$, for the sake of simplicity.

In the following section we describe some fundamental aspects of the HO canonical-ensemble description from a coherent states' viewpoint [9], the Husimi probability distribution function, and the Wehrl information measure.

2 Basic notions

2.1 Coherent states of the harmonic oscillator

In [9] the authors discuss quantum-mechanical phase space distributions expressed in terms of the celebrated coherent states $|z\rangle$ of the harmonic oscillator, eigenstates of the annihilation operator $\hat{a}$ [14, 15], *i.e.*,

$$\hat{a}|z\rangle = z|z\rangle, \tag{2}$$

with z a complex combination of the phase space coordinates x, p

$$z = x/(2\sigma_x) + \mathrm{i}p/(2\sigma_p), \tag{3}$$

where $\sigma_x = (\hbar/2m\omega)^{1/2}$, $\sigma_p = (\hbar m\omega/2)^{1/2}$, and $\sigma_x\sigma_p = \hbar/2$.

Coherent states span Hilbert's space, constitute an over-complete basis and obey resolution of unity [15]

$$\int \frac{\mathrm{d}^2 z}{\pi}\,|z\rangle\langle z| = \int \frac{\mathrm{d}x\,\mathrm{d}p}{2\pi\hbar}\,|x,p\rangle\langle x,p| = 1, \tag{4}$$

where the differential element of area in the $z-$plane is $\mathrm{d}^2 z = \mathrm{d}x\mathrm{d}p/2\pi\hbar$ and the integration is carried out over the whole complex plane.

The coherent state $|z\rangle$ can be expanded in terms of the states of the HO as follows

$$|z\rangle = \sum_{n=0}^{\infty} |\langle z|n\rangle|^2\,|n\rangle, \tag{5}$$

where $|n\rangle$ are eigenstates of the HO Hamiltonian whose form is $\hat{H} = \hbar\omega\,[\hat{a}^\dagger\hat{a} + 1/2]$ and we have

$$|\langle z|n\rangle|^2 = \frac{|z|^{2n}}{n!} e^{-|z|^2}. \tag{6}$$

2.2 Useful expressions

We write down now, for future reference, well-known quantal HO-expressions for, respectively, the partition function Z, the entropy S, the mean energy U, the mean excitation energy E, the free energy $F = U - TS$, and the specific heat C [16]

$$Z = \frac{e^{-\beta\hbar\omega/2}}{1 - e^{-\beta\hbar\omega}}, \tag{7}$$

$$S = \beta \frac{\hbar\omega}{e^{\beta\hbar\omega} - 1} - \ln(1 - e^{-\beta\hbar\omega}), \tag{8}$$

$$U = \frac{\hbar\omega}{2} + E, \tag{9}$$

$$E = \frac{\hbar\omega}{e^{\beta\hbar\omega} - 1}, \tag{10}$$

$$F = \frac{\hbar\omega}{2} + T\ln(1 - e^{-\beta\hbar\omega}), \tag{11}$$

$$C = \left(\frac{\beta\hbar\omega}{e^{\beta\hbar\omega} - 1}\right)^2 e^{\beta\hbar\omega}. \tag{12}$$

2.3 The semiclassical Husimi distribution

In the wake of a discussion advanced in [17], we will be mainly concerned with building "Husimi–Fisher" bridges. It is well-known that the oldest and most elaborate phase space (PS) formulation of quantum mechanics is that of Wigner [18, 19]. To every quantum state a PS function (the Wigner one) can be assigned. This PS function can, regrettably enough, assume negative values so that a probabilistic interpretation becomes questionable. Such limitation was overcome, among others, by Husimi [20]. In terms of the concomitant Husimi probability distributions, quantum mechanics can be completely reformulated [21, 22, 23, 24]. This phase space distribution has the form of

$$\mu(x, p) \equiv \mu(z) = \langle z|\hat{\rho}|z\rangle, \tag{13}$$

where $\hat{\rho}$ is the density operator of the system and $|z\rangle$ are the coherent states (see, for instance, [25] and references therein). The function $\mu(x, p)$ is normalized in the fashion

$$\int (\mathrm{d}x\,\mathrm{d}p/2\pi\hbar)\,\mu(x,p) = 1. \tag{14}$$

For a thermal equilibrium case $\hat{\rho} = Z^{-1}e^{-\beta\hat{H}}$, $Z = \mathrm{Tr}(e^{-\beta\hat{H}})$ is the partition function, $\beta = 1/T$, with T being the temperature. Specializing things for the HO of frequency ω, with eigenstates $|n\rangle$ associated to the eigenenergies $E_n = \hbar\omega\,(n + 1/2)$, one has

$$\langle z|\hat{\rho}|z\rangle = \frac{1}{Z}\sum_n e^{-\beta\hat{H}}|\langle z|n\rangle|^2, \tag{15}$$

with $|\langle z|n\rangle|^2$ given by Equation (6), and the normalized Husimi probability distribution is

$$\mu(z) = (1 - e^{-\beta\hbar\omega})\,e^{-(1-e^{-\beta\hbar\omega})|z|^2}. \tag{16}$$

2.4 The semiclassical entropy of Wehrl's

The Wehrl entropy is defined as [12]

$$W = -\int \frac{\mathrm{d}x\,\mathrm{d}p}{2\pi\hbar}\,\mu(x,p)\,\ln\mu(x,p), \tag{17}$$

where $\mu(x,p)$ is the "classical" distribution function (13) associated to the density operator $\hat{\rho}$ of the system. The uncertainty principle manifests itself through the inequality $W \geq 1$, which was first conjectured by Wehrl [12] and later proved by Lieb [13]. Equality holds if $\hat{\rho}$ is a coherent state. After integration over all phase space, turns out to be [9]

$$W = 1 - \ln(1 - e^{-\beta\hbar\omega}). \tag{18}$$

3 FIM's fundamentals

Let us consider a system that is specified by a physical parameter θ, while $\mathbf{x}$ is a real stochastic variable and $f_\theta(\mathbf{x})$, which in turn depends on the parameter θ, is the probability density for $\mathbf{x}$. An observer makes a measurement of $\mathbf{x}$ and estimates θ from this measurement, represented by $\tilde{\theta} = \tilde{\theta}(\mathbf{x})$. One wonders how well θ can be determined. Estimation theory [26] asserts that the best possible estimator $\tilde{\theta}(\mathbf{x})$, after a very large number of $\mathbf{x}$-samples is examined, suffers a mean-square error e^2 from θ that obeys a relationship involving Fisher's I, namely, $Ie^2 = 1$, where the Fisher information measure I is of the form

$$I(\theta) = \int \mathrm{d}\mathbf{x}\, f_\theta(\mathbf{x}) \left\{ \frac{\partial \ln f_\theta(\mathbf{x})}{\partial \theta} \right\}^2 . \tag{19}$$

This "best" estimator is called the *efficient* estimator. Any other estimator must have a larger mean-square error. The only proviso to the above result is that all estimators be unbiased, *i.e.*, satisfy $\langle \tilde{\theta}(\mathbf{x}) \rangle = \theta$. Thus, Fisher's information measure has a lower bound, in the sense that, no matter what parameter of the system we choose to measure, I has to be larger or equal than the inverse of the mean-square error associated with the concomitant experiment. This result, $I\, e^2 \geq 1$, is referred to as the Cramer–Rao bound [1, 27] (see section 7 in chapter 2). A particular I-case is of great importance: that of translation families [1, 4], *i.e.*, distribution functions (DF) whose form does not change under θ-displacements. These DF are shift-invariant (à la Mach, no absolute origin for θ), and for them Fisher's information measure adopts the somewhat simpler appearance [1]

$$I = \int \mathrm{d}\mathbf{x}\, f(\mathbf{x}) \left\{ \frac{\partial \ln f(\mathbf{x})}{\partial \mathbf{x}} \right\}^2 . \tag{20}$$

Fisher's measure is additive [1]: If x and p are independent, variables, $I(x+p) = I(x) + I(p)$. Notice that, for $\tau \equiv (x,p)$ (a point in phase-space), we face a shift-invariance situation. Since in defining z in terms of the variables x and p, these are scaled by their respective variances, the Fisher measure associated to the probability distribution $\mu(x,p)$ will be of the form [4]

$$I_\tau = \int \frac{\mathrm{d}x\, \mathrm{d}p}{2\pi\hbar}\, \mu(x,p)\, \mathcal{A}, \tag{21}$$

with

$$\mathcal{A} = \sigma_x^2 \left[\frac{\partial \ln \mu(x,p)}{\partial x} \right]^2 + \sigma_p^2 \left[\frac{\partial \ln \mu(x,p)}{\partial p} \right]^2 . \tag{22}$$

Given the μ-expression (16), I_τ becomes

$$I_\tau = 1 - e^{-\beta\hbar\omega}, \tag{23}$$

which, immediately yields

$$I_\tau\, e^2_{|z|}(\beta, \omega) = 1; \text{ (CR bound reached)}. \tag{24}$$

We realize at this point that the Fisher measure built up with Husimi distributions is to be best employed to estimate "phase space position" $|z|$. Further, *efficient estimation is possible for all temperatures*, a rather significant result. Comparison with Equation (18) allows one now to write

$$W = 1 - \ln(I_\tau) \Rightarrow W + \ln(I_\tau) = 1. \tag{25}$$

Since both W and I_τ are positive-definite quantities, (25) tells us that *they are complementary informational quantities*, thus if one of them gains, the other loses. Following Anderson *et al.* [9] let us now analyze the high and low temperatures limits. Given the form (23), when the temperature goes to zero ($\beta \to \infty$), $I_\tau \approx 1$, its maximum possible value, since we know that the ground state will be the only one to be populated. If, on the other hand, the temperature tends to infinity ($\beta \to 0$), then $I_\tau \approx \beta\hbar\omega$ and tends to zero, because we know beforehand that, in the limit, all energy levels will be populated in uniform. The uniform distribution is that of maximum ignorance [28, 29, 30, 31]. The range of I_τ is $0 \leq I_\tau \leq 1$, that of W is $1 \leq W \leq \infty$, as we can see in Figure 1. Using I_τ together with Equation (7) we notice that

$$I_\tau = \frac{e^{-\beta\hbar\omega/2}}{Z}, \tag{26}$$

so that it coincides with the canonical ensemble probability for finding the system in its ground state.

4 FIM and thermodynamics

Consider now the general definition (19) of Fisher's information measure in terms of the DF $\mu(x,p)$

$$I_\beta = \int \frac{\mathrm{d}x\,\mathrm{d}p}{2\pi\hbar}\,\mu(x,p)\left(\frac{\partial \ln \mu(x,p)}{\partial \beta}\right)^2, \tag{27}$$

with $\theta \equiv \beta$ is the parameter to be estimated. Since

$$\frac{\partial \ln \mu(x,p)}{\partial \beta} = \frac{\hbar\omega}{e^{\beta\hbar\omega} - 1}\,[1 - (1 - e^{-\beta\hbar\omega})\,|z|^2], \tag{28}$$

one readily ascertains that (i) the μ-mean value of (28) vanishes and (ii)

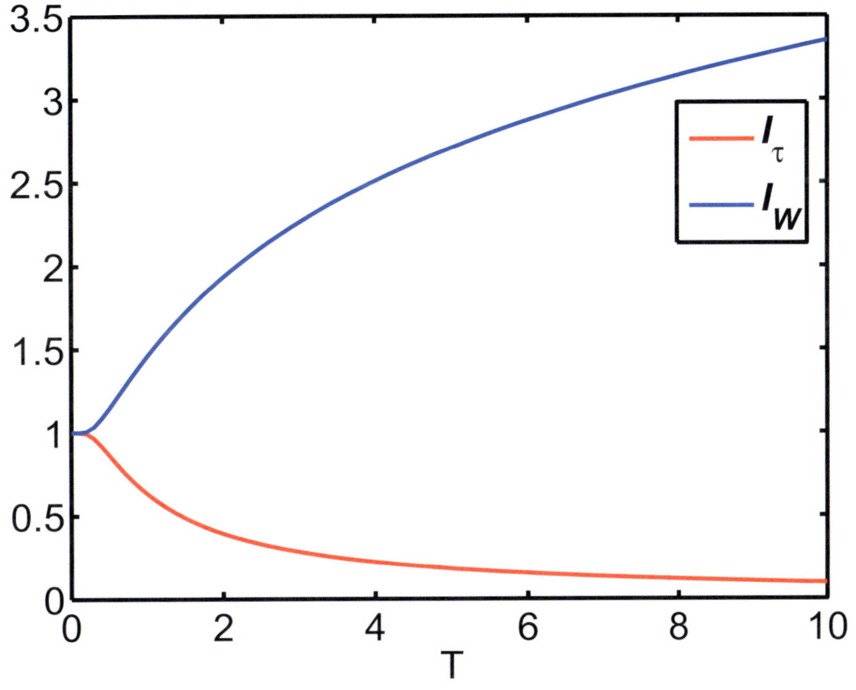

Figure 1: Fisher (I_τ) and Wehrl (W) information measures *vs.* T (in $\hbar\omega$ units) for HO-Husimi distribution.

$$I_\beta = \left(\frac{\hbar\omega}{e^{\beta\hbar\omega}-1}\right)^2 \quad (T=[0,\infty] \rightarrow I_\beta = [0,\infty])\,, \tag{29}$$

which, in view of (12), entails

$$I_\beta = \frac{e^{-\beta\hbar\omega}}{\beta^2}\, C. \tag{30}$$

Reflection upon the I_β-range (29) might led one to conclude that it constitutes a *Fisher manifestation of thermodynamics' third law.* Not only Shannon's measure but also Fisher's (for the HO, at least) vanishes at zero temperature. Replacing now (23) and (29) into the entropy expression (8) we immediately arrive at the relation

$$S = \beta\sqrt{I_\beta} - \ln I_\tau. \tag{31}$$

The HO entropy can be expressed as the sum of two terms: one associated with the Fisher information I_β and the other with the Fisher information for translation families I_τ corresponding to the phase space variables (x,p). Using Equation (7) we also have

$$\ln I_\tau = -\beta\frac{\hbar\omega}{2} - \ln Z = -(\beta E_{gs} + \ln Z) \tag{32}$$

with E_{gs} denoting the ground state energy. Thus,

$$S = \beta\left(\frac{\hbar\omega}{2} + \sqrt{I_\beta}\right) + \ln Z, \tag{33}$$

which is to be compared to the well known canonical ensemble general expression connecting S and the mean energy U [16]

$$S = \ln Z + \beta U, \tag{34}$$

we see that I_β is related to the excited spectrum contribution to U while I_τ is to be linked to the partition function. We will look now for a new connection between Fisher's measures I_τ and I_β. From (29) it is possible to rewrite I_β in the form

$$I_\beta \equiv \left(\frac{\hbar\omega\, e^{-\beta\hbar\omega}}{1-e^{-\beta\hbar\omega}}\right)^2, \tag{35}$$

and therefore

$$I_\tau\sqrt{I_\beta} = \hbar\,\omega\, e^{-\beta\hbar\omega} = -\frac{\partial}{\partial\beta}(e^{-\beta\hbar\omega}), \tag{36}$$

i.e., the product on the left hand side is the β-derivative of the Boltzmann factor (constant energy-wise) at the inverse temperature β. In other words, $I_\tau \sqrt{I_\beta}$ measures the β-gradient of the Boltzmann factor.

Equation (23) implies, *via* Equations (7) to (12), that the *quantal* HO expressions for the most important thermodynamic quantities can be expressed in terms of the *semiclassical* measure I_τ. For this end we define the semiclassical free energy

$$F_{sc} = T \ln I_\tau, \tag{37}$$

which is the semiclassical contribution to the HO free-energy $F = \hbar\omega/2 + F_{sc}$. Therefore, the thermodynamic quantities can be expressed as follows

$$\begin{aligned} Z &= \frac{e^{-\beta\hbar\omega/2}}{I_\tau}, & (38)\\ E &= \hbar\omega\,\frac{1 - I_\tau}{I_\tau}, & (39)\\ S &= \beta\hbar\omega\,\frac{1 - I_\tau}{I_\tau} - \frac{F_{sc}}{T}, & (40)\\ C &= (\beta\hbar\omega)^2\,\frac{1 - I_\tau}{I_\tau^2}, & (41) \end{aligned}$$

which shows that the semiclassical, Husimi-based I_τ−information measure does contain purely quantum statistical information. Furthermore, since from Helmholtz' free energy F, we can derive all of the HO quantal thermodynamics [16], we see that the the HO-quantum thermostatistics is, as far as information is considered, entirely of semiclassical nature, as it is completely expressed in terms of a semiclassical measure. We emphasize thus the fact that the semiclassical quantity I_τ contains all the relevant HO-statistical quantum information.

5 HO-illustrations

5.1 The maximum entropy technique

All the previous results are exact. No reference whatsoever needs to be made to Jaynes' Maximum Entropy Principle (MaxEnt) [31] up to this point. We wish now to consider a MaxEnt viewpoint. It is shown in [14] that the HO-energy can be cast as the sum of the ground-state energy $\hbar\omega/2$ plus the expectation value of the HO-Hamiltonian with respect to the coherent state $|z\rangle$, which is the sum of the ground-state energy plus a semiclassical excitation energy $\mathcal{E}$. One has for the semiclassical excitation HO-energy $e(z)$ at z [14]

$$e(z) = \langle z|\mathcal{H}|z\rangle - \hbar\omega/2 = \hbar\omega|z|^2, \;\; i.e.,$$
$$\mathcal{E}_\nu = \hbar\omega\langle|z|^2\rangle_\nu, \tag{42}$$

where the last expectation value is computed using the distribution $\nu(z)$. This semiclassical excitation energy $\mathcal{E}_\mu$ is given, for a Husimi distribution μ, by [25]

$$\mathcal{E}_\mu = \langle e(z)\rangle_\mu = \frac{\hbar\omega}{I_\tau}. \tag{43}$$

Note now that, from Equation (16), we can conveniently recast the HO-expression for μ into the Gaussian fashion

$$\mu(z) = I_\tau\, e^{-I_\tau\,|z|^2}, \tag{44}$$

peaked at the origin. The probability density μ of Equation (44) is clearly of the maximum entropy [31].

As a consequence, it proves convenient, at this stage, to view I_τ in the following light. The semiclassical form of the entropy S has exhaustively been studied by Wehrl. It is the (cf. 1) Shannon's information measure evaluated with Husimi distributions [12]. Assume we know *a priori* the value $\mathcal{E}_\nu = \hbar\omega\langle|z|^2\rangle_\nu$. We wish to determine the distribution $\nu(z)$ that maximizes the Wehrl entropy under this $\mathcal{E}_\nu-$value constraint. Accordingly, the MaxEnt distribution will be [31]

$$\nu(z) = e^{-\lambda_o}e^{-\eta\,E(z)}, \tag{45}$$

with λ_o the normalization Lagrange multiplier and η the one associated to $\mathcal{E}_\nu$. According to MaxEnt tenets we have [31]

$$\lambda_o = \lambda_o(\eta) = \ln\int\frac{\mathrm{d}^2z}{\pi}\,e^{-\eta\;\hbar\omega|z|^2} = -\ln\left(\eta\;\hbar\omega\right). \tag{46}$$

Now, the $\eta-$multiplier is determined by the relation [31]

$$-\mathcal{E}_\nu = \frac{\partial\lambda_o}{\partial\eta} = -\frac{1}{\eta}. \tag{47}$$

If we choose the Fisher-Husimi constraint given by Equation (43), this results in $\eta = I_\tau/\hbar\omega$ and from Equation (46) we get $\lambda_o = -\ln I_\tau$, *i.e.*, $e^{-\lambda_o} = I_\tau$, and we consequently arrive to the desired outcome

$$\nu(z) = I_\tau\, e^{-I_\tau\,|z|^2} \equiv \mu(z). \tag{48}$$

We have thus shown that the HO-Husimi distributions are MaxEnt-ones with the semi-classical excitation energy (43) as a constraint. It is clear from Equation (48) that I_τ plays there the role of an "inverse temperature".

The preceding argument suggests that we are tacitly envisioning the existence of a quantity T_W (the inverse of η) associated to the Wehrl measure that we here extend to extreme. This Wehrl temperature T_W governs the width of our Gaussian Husimi distribution. On account of

$$\mu(z) = I_\tau\, e^{-(I_\tau/\hbar\omega)\,\hbar\omega|z|^2} = I_\tau\, e^{-\mathcal{E}_\mu/T_W}, \tag{49}$$

which entails

$$T_W = \frac{\hbar\omega}{I_\tau}, \tag{50}$$

and it is easy to see from the range of I_τ that the range of values of T_W is then $\hbar\omega \leq T_W \leq \infty$. Due to the semiclassical nature of both W and μ, T_W has a lower bound greater than zero.

5.2 The phenomenon of delocalization

The two quantities W and I_τ have been shown to be related, for the HO, according to Equation (25). Since the Wehrl temperature T_W yields the width of our Gaussian Husimi distribution, we can conceive of introducing a "delocalization factor" D

$$D = \frac{T_W}{\hbar\omega}. \tag{51}$$

The above definition leads to the relation

$$W = 1 + \ln T_W - \ln \hbar\omega = 1 + \ln D. \tag{52}$$

As stressed above, W has been constructed as a delocalization measure [12]. The preceding considerations clearly motivate one to regard the Fisher measure built up with Husimi distributions as a "localization estimator" in phase space. The HO-Gaussian expression for μ (44) illuminates the fact that the Fisher measure controls both the height and the spread (which is $\sim [2I_\tau]^{-1}$). Obviously, *spread is here a "phase-space delocalization indicator"*. This fact is reflected by the quantity D introduced above.

Thus, an original physical interpretation of Fisher's measure emerges: *localization control.* The inverse of the Fisher measure, D, *turns out to be a delocalization-indicator.* Differentiating Fisher's measure (23) with respect T, notice also that

$$\frac{\mathrm{d}I_\tau}{\mathrm{d}T} = -\frac{\hbar\omega}{T^2}\, e^{-\beta\hbar\omega}, \tag{53}$$

so that Fisher's information decreases exponentially as the temperature grows. Our Gaussian distribution loses phase-space "localization" as energy and/or temperature are injected into our system, as reflected *via* T_W or D. Notice that (52) complements the Lieb bound $W \geq 1$. It tells us by just how much W exceeds unity. We see that it does it by virtue of delocalization effects. Moreover, this fact can be expressed using the shift-invariant Fisher measure. We will now show that D is proportional to the system's energy fluctuations.

5.3 Husimi distribution's moments

The second moment of the Husimi distribution $\mu(z)$ is an important measure to ascertain the "degree of complexity" of a quantum state (see below). It is a measure of the delocalization-degree of the Husimi distribution in phase space (see Ref. [32] for details and discussions). It is defined as

$$M_2 = \int \frac{\mathrm{d}^2 z}{\pi}\, \mu^2(z), \tag{54}$$

that, after explicit evaluation of M_2 from Equation (44) reads

$$M_2 = \frac{I_\tau}{2}. \tag{55}$$

Using now (52) we conclude that

$$M_2(D) = \frac{1}{2\,D}. \tag{56}$$

Thus, our energy-fluctuations turn out to be

$$\Delta_\mu e = \frac{\hbar\omega}{I_\tau} = \hbar\omega\, D, \tag{57}$$

with $(\Delta_\mu e)^2 = (\mathcal{E}^2)_\mu - \mathcal{E}_\mu^2$. As a consequence, we get

$$D = \frac{\Delta_\mu e}{\hbar\omega}. \tag{58}$$

An important result is thus obtained: the delocalization factor D represents energy-fluctuations expressed in $\hbar\omega-$terms. *Delocalization is clearly seen to be the counterpart of energy fluctuations.*

6 Thermal relations

Let us now go back to Equation (37) and revisit the entropic expression. It is immediately realized that we can recast the entropy S in terms of the quantal mean excitation energy E and the delocalization factor D as

$$\frac{E}{T} = S - \ln D, \tag{59}$$

i.e., if one injects into the system some excitation energy E, expressed in "natural" T units, it is apportioned partly as heat dissipation *via* S and partly *via* delocalization. More precisely, the part of this energy not dissipated is that employed to delocalize the system in phase space. Now, since $W = 1 - \ln I_\tau = 1 + \ln D$, the above equation can be recast in alternative forms, as

$$S = \frac{E}{T} + \ln D = \frac{E}{T} - \ln I_\tau; \quad \text{or} \tag{60}$$

$$W = 1 + S - \frac{E}{T}, \tag{61}$$

implying

$$W - S \mapsto 0 \ \text{ for } \ T \mapsto \infty, \tag{62}$$

which is a physically sensible result and

$$W - S \quad \mapsto 1 \text{ for } T \mapsto 0, \tag{63}$$

as it should, since $S = 0$ at $T = 0$ (third law of thermodynamics), while W attains there its Lieb's lower bound of unity.

One finds in Equation (60) some degree of resemblance to thermodynamics's first law. To reassure ourselves on this respect, we slightly changed our underlying canonical probabilities μ, multiplying it by a factor $\delta\mathcal{F} =$ random number/100. Specifically, we generated random numbers according to the normal distribution and divided them by 100 to obtain the above factors $\delta\mathcal{F}$. This process leads to new "perturbed" probabilities $P = \mu + \delta\mu$, conveniently normalized. With them we evaluate the concomitant changes dS, dE (we do this 50 times, with different random numbers in each instance). We were then able to numerically verify that the difference $\mathrm{d}S - \beta\mathrm{d}E \sim 0$. The concomitant results are plotted in Fig. 2) Since, as stated, numerically $\mathrm{d}S = (1/T)\,\mathrm{d}E$, this entails, from Equation (60), $\mathrm{d}I_\tau/I_\tau \simeq 0$. The physical connotations are as follows: if the *only* modification effected is that of a change $\delta\mu$ [16] in the canonical distribution μ, this implies that the system undergoes a heat transfer process [16] for which thermodynamics' first law implies $\mathrm{d}U = T\mathrm{d}S$. This is numerically confirmed in the plots of Fig. 2. The null contribution of $\ln I_\tau$ to this process suggests that delocalization (not a thermodynamic effect, but a *dynamic* one) can be regarded as behaving (thermodynamically) like a kind of "work".

Now, since (a) $I_\tau = 1 - e^{-\beta\hbar\omega}$, and (b) the mean energy of excitation is $E = \hbar\omega/(\exp(\beta\hbar\omega) - 1)$, one also finds, for the quantum-semiclassical difference (QsCD) $S - W$ the result

$$W - S = 1 - \frac{I_\tau - 1}{I_\tau}\ln(1 - I_\tau) = F_1(I_\tau) \tag{64}$$

Moreover, since $0 \leq F_1(I_\tau) \leq 1$, we see that, always, $W \geq S$, as expected, since the semiclassical treatment contains less information than the quantal one. Note that the QsCD can be expressed exclusively in Fisher's information terms. This is, the quantum-semiclassical entropic difference $S - W$ may be given in I_τ−terms only. Figure (3) depicts S, βE, and $\ln D$ *vs.* the dimensionless quantity $t = T/\hbar\omega$. Accordingly, entropy is apportioned in such a way that

- part of it originates from excitation energy and
- the remaining is accounted for by phase space delocalization.

A bit of algebra allows one now to express the rate of entropic change per unit temperature increase as

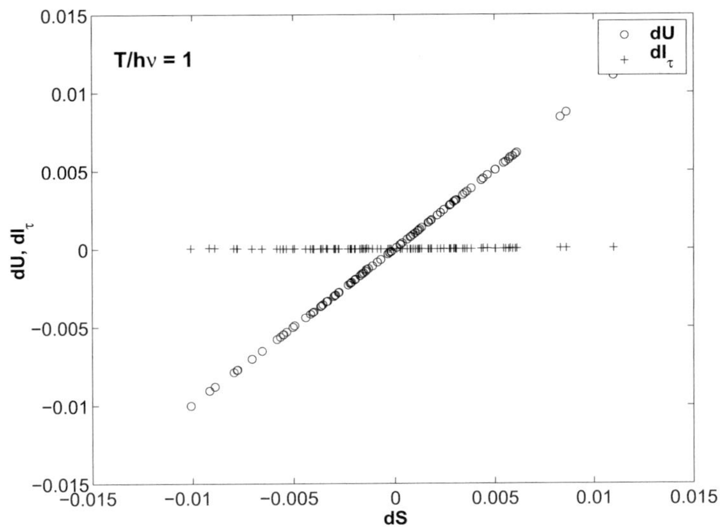

Figure 2: Numerical computation results for the HO: changes $\mathrm{d}I_\tau$ and $\mathrm{d}U$ *vs.* $\mathrm{d}S$ that ensue after randomly generating variations δp_i in the underlying microscopic canonical probabilities p_i.

$$\frac{\mathrm{d}S}{\mathrm{d}T} = \beta \frac{\mathrm{d}E}{\mathrm{d}T} = \beta\, C = \hbar\omega \frac{1}{T}\frac{\mathrm{d}D}{\mathrm{d}T}, \tag{65}$$

entailing

$$C = \hbar\omega \frac{\mathrm{d}D}{\mathrm{d}T}. \tag{66}$$

In the case of the one dimensional HO we see that *the specific heat measures delocalization change per unit temperature increase.* Also, $\mathrm{d}E/\mathrm{d}T \propto \mathrm{d}D/\mathrm{d}T$, providing us with a very simple relationship between mean excitation energy changes and delocalization ones.

$$\frac{\mathrm{d}E}{\mathrm{d}D} = \hbar\omega. \tag{67}$$

7 Heisenberg's thermal uncertainties

Additional considerations are in order with regards to thermal uncertainties, that express the effect of temperature on Heisenberg's celebrated relations (see, for instance [6, 33, 34, 35]). We use now a result obtained in [9] (Equation (3.12)), where the authors cast Wehrl's information measure in terms of the coordinates' variances $\Delta_\mu x$ and $\Delta_\mu p$, obtaining

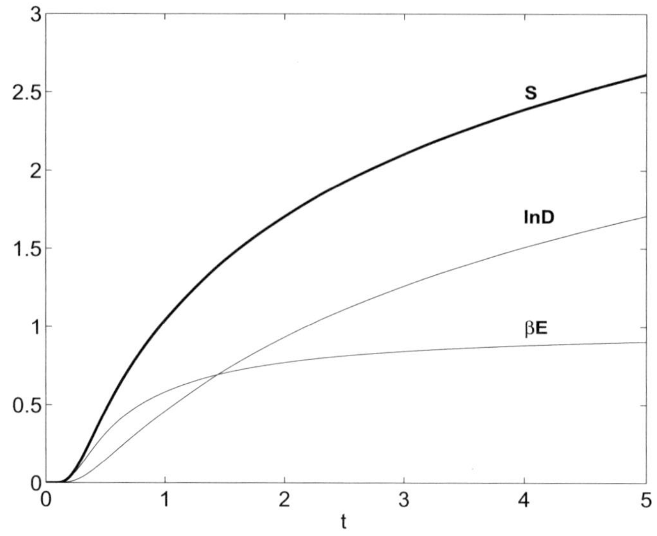

Figure 3: S, βE, and $\ln D$ as a function of $t = T/\hbar\omega$.

$$W = \ln\left\{\frac{e}{\hbar}\,\Delta_\mu x\,\Delta_\mu p\right\}. \tag{68}$$

In the present context, the relation $W = 1 - \ln I_\tau$ allows us to conclude that [17]

$$I_\tau\,\Delta_\mu x\,\Delta_\mu p = \hbar, \tag{69}$$

which can be regarded as a "Fisher uncertainty principle" and adds still another meaning to I_τ: since, necessarily, $\Delta_\mu x\,\Delta_\mu p \geq \hbar/2$, it is clear that $I_\tau/2$ is the "correcting factor" that permits one to reach the uncertainty's lower bound $\hbar/2$, a rather interesting result.

Phase space "localization" is possible, with Husimi distributions, only up to $\hbar$ [14]. This is to be compared to the uncertainties evaluated in a purely quantal fashion, without using Husimi distributions, and in particular with a previous result in [17]. With the virial theorem [16] one can easily ascertain in [17] that

$$\Delta x\,\Delta p = \frac{\hbar}{2}\,\frac{e^{\beta\hbar\omega}+1}{e^{\beta\hbar\omega}-1}, \tag{70}$$

together with (69) yields

$$\Delta_\mu x\,\Delta_\mu p = \frac{2\,\Delta x\,\Delta p}{1+e^{-\beta\hbar\omega}}. \tag{71}$$

Thus We see that, as $\beta \to \infty$, $\Delta_\mu \equiv \Delta_\mu x\, \Delta_\mu p$ is twice the minimum quantum value for $\Delta x \Delta p$, and $\Delta_\mu \to \hbar$, the "minimal" phase-space cell. The quantum and semiclassical results do coincide at very high temperature though. Indeed, one readily verifies [17] that Heisenberg's uncertainty relation, as a function of both frequency and temperature, is governed by a thermal "uncertainty function" F that acquires the aspect

$$F(\beta, \omega) = \Delta x\ \Delta p = \frac{1}{2}\left[\Delta_\mu + \frac{E}{\omega}\right]. \tag{72}$$

Coming back to results derivable within the present context, we realize here that F can be recast as

$$F(\beta, \omega) = \frac{1}{2}\left[\hbar\, D + \frac{E}{\omega}\right], \tag{73}$$

so that, for T varying in $[0, \infty]$, the range of possible $\Delta x\ \Delta p$-values is $[\hbar/2, \infty]$. *Equation* (73) *is a "Heisenberg–Fisher" thermal uncertainty relation* (for a discussion of this concept see, for instance, [6, 33, 34]).

$F(\beta, \omega)$ grows with both E and D. The usual result $\hbar/2$ is attained for minimum D and zero excitation energy. As for $\mathrm{d}F/\mathrm{d}T$, one is able to set $F \equiv F(E, D)$, since $2\mathrm{d}F = \hbar \mathrm{d}D + \omega^{-1}\mathrm{d}E$. Remarkably enough, the two contributions to $\mathrm{d}F/\mathrm{d}T$ are easily seen to be equal and $dF/dT \to (1/\omega)$ for $T \to \infty$. One can also write

$$\left(\frac{\partial F}{\partial D}\right)_E = \frac{\hbar}{2};\ \left(\frac{\partial F}{\partial E}\right)_D = \frac{1}{2\omega}, \tag{74}$$

providing us with a thermodynamic "costume" for the uncertainty function F that sheds some new light onto the meaning of both $\hbar$ and ω. In particular, we see that $\hbar/2$ is the derivative of the uncertainty function F with respect to the delocalization factor D. Increases $\mathrm{d}F$ of the thermal uncertainty function F are of two types

- from the excitation energy, that supplies a C/ω contribution and
- from the delocalization factor D.

8 The purity concept

8.1 Semiclassical specialization

The quantal concept of degree of purity of a general density operator $\hat{\rho}$ is expressed *via* $\mathrm{Tr}\,\hat{\rho}^2$ [36, 37]. Its inverse, the so-called participation ratio

$$R = \frac{1}{\mathrm{Tr}\hat{\rho}^2}, \tag{75}$$

is particularly convenient for calculations [38]. It varies from unity for pure states to N for totally mixed states [38]. It may be interpreted as the effective number of pure states that enter a quantum mixture. Here we will consider the "degree of purity" d_μ of a semiclassical distribution, given by

$$d_\mu = \int \frac{\mathrm{d}^2 z}{\pi}\, \mu^2(z) \leq 1. \tag{76}$$

Clearly, d_μ coincides with the second moment of the Husimi distribution (44) given by Equation (54), *i.e.*,

$$d_\mu = M_2 = \frac{I_\tau}{2}. \tag{77}$$

Using now (52) we relate the semiclassical degree of purity to the delocalization factor and to the Wehrl temperature T_W

$$d_\mu = \frac{1}{2\,D} = \frac{T_W}{2\hbar\omega}, \tag{78}$$

and also to our semiclassical energy-fluctuations (57)

$$d_\mu = \frac{\hbar\omega}{2\Delta_\mu e}. \tag{79}$$

Since $\hbar\omega \leq T_W \leq \infty$, the "best" purity attainable at the semiclassical level equals one-half.

8.2 Quantum purity

For the quantum mixed HO-state $\hat{\rho} = e^{-\beta\mathcal{H}}/Z$, where $\mathcal{H}$ is the Hamiltonian of the harmonic oscillator and the partition function Z is given by Equation (7) [16], we have a degree of purity $d_{\hat{\rho}}$ given by (see the detailed study by Dodonov [35])

$$d_{\hat{\rho}} = \frac{e^{-\beta\hbar\omega}}{Z^2} \sum_{n=0}^{\infty} e^{-2\beta\hbar\omega n}, \tag{80}$$

leading to

$$d_{\hat{\rho}} = \tanh(\beta\hbar\omega/2), \tag{81}$$

where $0 \leq d_{\hat{\rho}} \leq 1$. Thus, Heisenberg' uncertainty relation can be cast in the fashion

$$\Delta_x \, \Delta p = \frac{\hbar}{2} \coth(\beta\hbar\omega/2), \tag{82}$$

where Δx and Δp are the quantum variances for the canonically conjugated observables x and p [35]

$$\Delta x \, \Delta p = \frac{\hbar}{2} \frac{1}{d_{\hat{\rho}}}, \tag{83}$$

which is to be compared to the semiclassical result that was derived above (*cf.* 71).

We relate now the degree of purity of our thermal state with various physical quantities both in its quantal and semiclassical versions. Using Equations (71) and (77) we get

$$d_\mu = \frac{I_\tau}{2} = (1 - d_\mu)\, d_{\hat{\rho}}, \tag{84}$$

which leads to

$$\begin{aligned} d_{\hat{\rho}} &= \frac{d_\mu}{1 - d_\mu} = \frac{I_\tau}{2 - I_\tau}, \\ d_\mu &= \frac{d_{\hat{\rho}}}{1 + d_{\hat{\rho}}} \end{aligned} \tag{85}$$

such as clearly shows that (i) $d_\mu \leq d_{\hat{\rho}}$, and (ii) for a pure state, again, its semiclassical counterpart has a degree of purity equal 1/2.

Additionally, on account of Equation (69), on the one hand, and since the semiclassical degree of purity reads $d_\mu = I_\tau/2$, on the other one, we are led to an uncertainty relation for mixed states in terms of d_μ, namely,

$$\Delta_\mu x \, \Delta_\mu p = \frac{\hbar}{2} \frac{1}{d_\mu}, \tag{86}$$

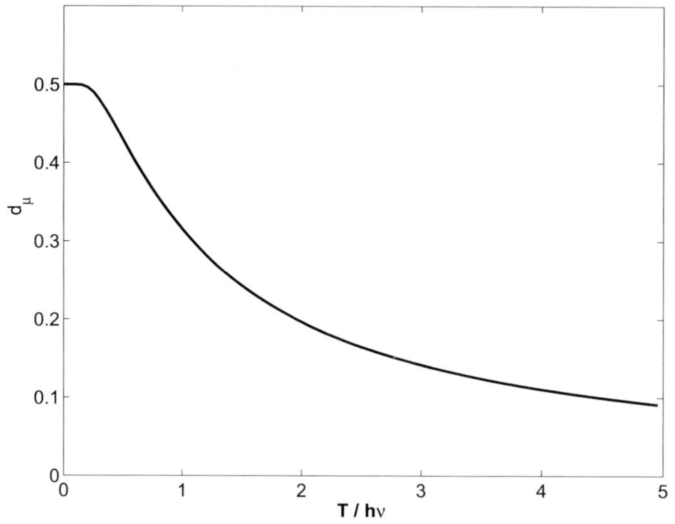

Figure 4: Semiclassical purity d_μ *vs.* $T/h\nu$, a monotonous function.

that tells us just how uncertainty grows as participation ratio $R = 1/d_\mu$ augments. Equation (86) is of semiclassical origin, which makes it a bit different from the one that results form a purely quantal treatment (see [35], Equation (4)). Moreover, notice how information concerning the purely quantal notion of purity $d_{\hat{\rho}}$ is already contained in the semiclassical measure I_τ.

We appreciate the fact that R increases as delocalization grows, a quite sensible result. Fig. 4 depicts $d_\mu(T)$, a monotonously decreasing function, which tells us that degree of purity of a mixed state acts here as a thermometer, and allows then to assign a $T-$value to any of our mixed states.

Also, using once again the result $d_\mu = I_\tau/2$ together with the uncertainties just derived we see that $\beta\hbar\omega = -\ln(1 - I_\tau)$. Thus, we can rewrite Equation (64) in the following form

$$W - S = 1 + \beta\hbar\omega\frac{2d_\mu - 1}{2d_\mu} = 1 + \beta\hbar\omega\frac{d_{\hat{\rho}} - 1}{2d_{\hat{\rho}}}, \tag{87}$$

which casts the difference between the quantal and semiclassical entropies in terms of the degrees of purity. From Equation (87) we can also give the quantal mean excitation energy E in terms of d_μ using (38)

$$E = \frac{\hbar\omega}{2}\frac{1-2d_\mu}{d_\mu} = \frac{\hbar\omega}{2}\frac{1-d_{\hat{\rho}}}{d_{\hat{\rho}}}. \tag{88}$$

9 Summary

We have explored in this review connections between canonical ensemble quantities and two Fisher information measures, associated to the estimation of phase-space location (I_τ) and temperature (I_β). Our most important result is, perhaps, to have shown that there exists a "Fisher-associated" third law of thermodynamics (at least for the HO). From a pure information-theoretic viewpoint, we have, obtained significant results, namely,

1. a connection between Wehrl's entropy and I_τ (*cf.* Equation (25)),
2. an interpretation of I_τ as the HO's ground state occupation probability (*cf.* Equation (26)),
3. an interpretation of I_β proportional to the HO's specific heat (*cf.* Equation (30)),
4. the possibility of expressing the HO's entropy as a sum of two terms, one for each of the above FIM realizations (*cf.* Equation (31)),
5. a new form of Heisenberg's uncertainty relations in Fisher terms (*cf.* Equation (73)),
6. that efficient $|z|$-estimation can be achieved with I_τ at *all* temperatures, as the minimum Cramer–Rao value is always reached (*cf.* Equation (24)).

Our statistical semiclassical treatment yielded, we believe, some new interesting physics that we proceed to summarize. We have, for the HO,

1. established that the semiclassical Fisher measure I_τ contains all relevant statistical quantum information,
2. shown that the Husimi distributions are MaxEnt ones, with the semiclassical excitation energy $\mathcal{E}$ as the only constraint,
3. complemented the Lieb bound on the Wehrl entropy using I_τ,
4. observed in detailed fashion how delocalization becomes the counterpart of energy fluctuations,

5. written down the difference $W - S$ between the semiclassical and quantal entropy also in I_τ−terms,

6. provided a relation between energy excitation and degree of delocalization,

7. shown that the derivative of twice the uncertainty function $F(\beta\omega) = \Delta x \Delta p$ with respect to I_τ^{-1} is the Planck constant $\hbar$,

8. established a semiclassical uncertainty relation in terms of the semiclassical purity d_μ, and

9. expressed both d_μ and the quantal degree of purity in terms of I_τ.

These results are, of course, restricted to the harmonic oscillator. However, this is such an important system that HO insights usually have a wide impact, as the HO constitutes much more than a mere example. Nowadays it is of particular interest for the dynamics of bosonic or fermionic atoms contained in magnetic traps [39, 40, 41] as well as for any system that exhibits an equidistant level spacing in the vicinity of the ground state, like nuclei or Luttinger liquids. The treatment of Hamiltonians including anharmonic terms is the next logical step. We are currently undertaking such a task. To this end analytical considerations do not suffice, and numerical methods are required. The ensuing results will be published elsewhere.

Conflict of interest: None.

Acknowledgements

F. Pennini is grateful for the financial support by FONDECYT, grant 1110827.

References

[1] B.R. Frieden, *Science from Fisher information*, 2nd ed., Cambridge, UK, Cambridge University Press, 2004.

[2] W.K. Wootters, "The acquisition of information from quantum measurements". PhD dissertation, University of Texas at Austin, US, 1980.

[3] B.R. Frieden, B.H. Soffer, "Lagrangians of physics and the game of Fisher information transfer". *Phys. Rev. E* **52**, 2274-2286 (1995).

[4] F. Pennini, A.R. Plastino, A. Plastino, "Rènyi entropies and Fisher informations as measures of nonextensivity in a Tsallis setting". *Physica A* **258**, 446-457 (1998).

[5] B.R. Frieden, A. Plastino, A.R. Plastino, H. Soffer, "Fisher-based thermodynamics: its Legendre transformations and concavity properties.", *Phys. Rev. E* **60**, 48-53 (1999).

[6] F. Pennini, A. Plastino, A.R. Plastino, M. Casas, "How fundamental is the character of thermal uncertainty relations?", *Phys. Lett. A* **302**, 156-162 (2002).

[7] A.R. Plastino, A. Plastino, "Symmetries of the Fokker-Planck equation and the Fisher-Frieden arrow of time.", *Phys. Rev. E* **54**, 4423-4426 (1996).

[8] A. Plastino, A.R. Plastino, H.G. Miller, "On the relationship between the Fisher-Frieden-Soffer arrow of time, and the behaviour of the Boltzmann and Kullback entropies", *Phys. Lett. A* **235**, 129-134 (1997).

[9] A. Anderson, J.J. Halliwell, "Information-theoretic measure of uncertainty due to quantum and thermal fluctuations", *Phys. Rev. D* **481**, 2753-2765 (1993).

[10] M. Dimassi, J. Sjoestrand, *Spectral asymptotics in the semiclassical limit.* Cambridge Univesity Press: Cambridge, UK, 1999.

[11] M. Brack, R.K. Bhaduri, *Semiclassical physics.* Addison-Wesley: Massachuset, US, 1997.

[12] A. Wehrl, "On the relation between classical and quantum entropy", *Rep. Math. Phys.* **16**, 353-358 (1979).

[13] E.H. Lieb, "Proof of an Entropy Conjecture of Wehrl", *Commun. Math. Phys.*62, 35-41 (1978).

[14] J. Schnack, "Thermodynamics of the harmonic oscillator using coherent states", *Europhys. Lett.* **45**, 647-652 (1999).

[15] J.R. Klauder, B.S. Skagerstam, *Coherent states.* World Scientific, Singapore, 1985.

[16] R.K Pathria, *Statistical Mechanics.* Pergamon Press, Exeter, UK, 1993.

[17] F. Pennini, A. Plastino, "Heisenberg–Fisher thermal uncertainty measure", *Phys. Rev. E* **69**, 057101(1:4) (2004).

[18] E.P. Wigner, "On the Quantum Correction For Thermodynamic Equilibrium", *Phys. Rev.* **40**, 749-759 (1932).

[19] J.J. Wlodarz, "Entropy and Wigner Distribution Functions Revisited", *Int. J. Theor. Phys.* **42**, 1075-1084 (2003).

[20] K. Husimi, "Some formal properties of the density matrix", *Proc. Phys. Math. Soc. Japan* **22**, 264-283 (1940).

[21] R.F. O'Connel, E.P. Wigner, "Some properties of a non-negative quantum-mechanical distribution function", *Phys. Lett. A* **85**, 121-126 (1981).

[22] S.S. Mizrahi, "Quantum mechanics in the Gaussian wave-packet phase space representation", *Physica A* **127**, 241-264 (1984).

[23] S.S. Mizrahi, "Quantum mechanics in the Gaussian wave-packet phase space representation II: Dynamics", *Physica A* **135**, 237-250 (1986).

[24] S.S. Mizrahi, "Quantum mechanics in the gaussian wave-packet phase space representation III: From phase space probability functions to wave-functions", *Physica A* **150**, 541-554 (1988).

[25] F. Pennini, A. Plastino, "Escort Husimi distributions, Fisher information and nonextensivity", *Phys. Lett. A* **326**, 20-26 (2004).

[26] H. Cramer, *Mathematical methods of statistics.* Princeton University Press, Princeton, US, 1946.

[27] C.R. Rao, "Information and accuracy attainable in the estimation of statistical parameters", *Bull. Calcutta Math. Soc.* **37**, 81-91 (1945).

[28] E.T. Jaynes, "Information theory and statistical mechanics I", *Phys. Rev.* 106, 620-630 (1957).

[29] E.T. Jaynes, "Information theory and statistical mechanics II" *Phys. Rev.* **108**, 171-190 (1957).

[30] E.T. Jaynes, *Papers on probability, statistics and statistical physics.* Rosenkrantz R. D. Ed.; Dordrecht, Reidel, US, 1987.

[31] A. Katz, *Principles of Statistical Mechanics: The Information Theory Approach.* Freeman and Co., San Francisco, US, 1967.

[32] A. Sugita, H. Aiba, "Second moment of the Husimi distribution as a measure of complexity of quantum states", *Phys. Rev. E* **65**, 36205(1:10) (2002).

[33] B. Mandelbrot, "The Role of Sufficiency and of Estimation in Thermodynamics", *Ann. Math. Stat.* **33**, 1021-1038 (1962).

[34] F. Pennini, A. Plastino, "Power-law distributions and Fisher's information measure", *Physica A* **334**, 132-138 (2004).

[35] V.V. Dodonov, "Purity- and entropy-bounded uncertainty relations for mixed quantum states", *J. Opt. B: Quantum Semiclass. Opt.* **4**, S98-S108 (2002).

[36] W.J. Munro, D.F.V. James, A.G. White, P.G. Kwiatç, "Maximizing the entanglement of two mixed qubits", *Phys. Rev. A* **64**, 30302(1:4) (2003).

[37] U. Fano, "Description of States in Quantum Mechanics by Density Matrix and Operator Techniques", *Rev. Mod. Phys.* **29**, 74-93, 1957.

[38] J. Batle, A.R. Plastino, M. Casas, A. Plastino, "Conditional q-entropies and quantum separability: a numerical exploration", *J. Phys. A: Math. Gen* **35**, 10311-10324 (2002).

[39] M.H. Anderson, J.R. Ensher, M.R. Matthews, C.E. Wieman, E.A. Cornell, "Observation of Bose-Einstein Condensation in a Dilute Atomic Vapor", *Science* **269**, 198-201 (1995).

[40] K.B. Davis *et al.*, "Bose-Einstein Condensation in a Gas of Sodium Atoms", *Phys. Rev. Lett.* **75**, 3969-3973 (1995).

[41] C.C. Bradley, C.A. Sackett, R.G. Hulet, "Bose-Einstein Condensation of Lithium: Observation of Limited Condensate Number", *Phys. Rev. Lett.* **78**, 985-989 (1997).

Send Orders for Reprints to reprints@benthamscience.net
Concepts and Recent Advances in Generalized Information Measures and Statistics, 2013, 283-308

CHAPTER 12

Statistical Complexity of Chaotic Pseudorandom Number Generators

Hilda A. Larrondo [a,e,*], **Luciana De Micco** [a,e], **Claudio M. González** [a], **Angelo Plastino** [b,e], **Osvaldo A. Rosso** [c,d,e]

[a] Facultad de Ingeniería, Universidad Nacional de Mar del Plata, Juan B. Justo 4302, 7600 Mar del Plata, Argentina

[b] Instituto de Física, Facultad de Ciencias Exactas, Universidad Nacional de La Plata C.C. 727, 1900 La Plata, Argentina

[c] Departamento de Física, Instituto de Ciências Exatas, Universidade Federal de Minas Gerais Av. Antônio Carlos, 6627 - Campus Pampulha. 31270-901 Belo Horizonte - MG, Brazil

[d] Chaos & Biology Group, Instituto de Cálculo, Facultad de Ciencias Exactas y Naturales, Universidad de Buenos Aires Pabellón II, Ciudad Universitaria, 1428 Ciudad Autónoma de Buenos Aires, Argentina

[e] Consejo Nacional de Investigaciones Científicas y Técnicas (CONICET), Argentina

Abstract

This chapter deals with the use of the Statistical Complexity Measure, as defined by López Ruiz, Mancini and Calbet [Phys. Lett. A 209 (1995) 321–326] and modified by Rosso and coworkers [P. W. Lamberti, M. T. Martín, A. Plastino, O. A. Rosso; Physica A 334 (2004) 119–131] to characterize pseudo random number generators (*PRNG's*) obtained from chaotic dynamical systems. It is shown that two probability distribution functions are required for a proper characterization: the first one is based on the histogram and is used to characterize the uniformity of the values in the time series; the second one is based on the permutation procedure proposed by Bandt and Pompe [Phys. Rev. Lett. 88 (2002) 174102] and characterize the uniformity of patterns of several consecutive values of the time series.

Keywords: Chaos; Random number generators; Entropy; Statistical Complexity.

* Corresponding author. E-mail: larrondo@fi.mdp.edu.ar

Andres M. Kowalski, Raul D. Rossignoli and Evaldo M. F. Curado (Eds)

1 Introduction

Let us briefly discuss at the very beginning of this chapter five controversial questions one wants to answer.

(1) Is the world deterministic or stochastic?

In his famous lectures Feynman describes a physicist as an observer watching a chess game, making models and experiments to discover the laws behind the players' moves. This vision implies that those rules really exist and our limited knowledge precludes us to know them in a complete way.

Statistical mechanics is founded on the opposite view. The physicist observes the game but they do not care about the existence of detailed rules. Predictions are made as a list of possible events and the probabilities for each one. A lot of theoretical and practical knowledge came from these ideas, and Statistical Mechanics appears today as one of the most fruitful branches of physics.

Two theories put this controversy between determinism and stochasticity on its top: quantum mechanics and deterministic chaos. The last one will be the subject of this chapter. Even if being from a quite different physical origin, time series arising from chaotic systems share with those generated by stochastic processes several properties that make them almost indistinguishable: *a)* a wide-band power spectrum, *b)* a delta-link linear autocorrelation function, *c)* an irregular behavior of the measures signals. In fact this similitude has made it possible to replace stochastic process by chaotic systems in many practical applications.

Wold proved that any stationary time series can be decomposed into two different parts: the first (deterministic) part can be exactly described by a linear combination of its own past, the second (random) is a moving average component of a finite order [1]. Chaotic systems always produce time series with a physical structure. This structure may be discovered and measured by statistical complexity measures [2] (see Chapter 8). In Ref. [3], time series generated by different chaotic and stochastic systems were studied. It was shown that using the Bandt and Pompe procedure to assign a probability distribution function (PDF) to the time series [4], it is possible to distinguish the chaotic systems from stochastic ones (*i.e.*, correlated and uncorrelated noises) when they are localized in the entropy-complexity plane. In point of fact, chaos is representative of deterministic processes in which time-causality constitutes an important fact that must be taken into account for a proper characterization.

(2) What are random number generators (RNG) and pseudo random number generators

(PRNG)?

Randomness is the opposite of deterministic. An ideal random number generator must produce numbers from a set of possible values with two basic properties:

- the values must be equally probable, *i.e.*, the sequence of numbers must have a uniform histogram;
- each number of the sequence must be statistically independent of the others, *i.e.*, it is impossible to predict the next value on the basis of the previous ones.

Mathematical algorithms can only produce pseudo random numbers. Some researches speculate that physical sources can produce truly random numbers. Some remarkable websites to get *physical* random numbers are [5], where random numbers are produced by measuring atmospheric noise, [6] that uses a radioactive decay, and [7] where a charge-coupled device (CCD) inside a dark can. But do physical noise sources produce true random numbers? If we accept a deterministic model of the universe the answer is no. The only difference between physical or algorithmic time series is that the putative underlying model may be unknown in the physical case, but in fact algorithmic time series may also be underlaid by a model that those who analyze them are unaware of.

The relevant issue in our opinion is to analyze the time series and quantify its statistical characteristics. The existence or not of a deterministic model is not relevant at all. This point of view was proposed by Marsaglia [8]. Quoting him, his DIEHARD famous pseudo-random number generator was *produced by combining two or more of the most promising deterministic generators with sources of random noise from three physical devices for those who feel that physical sources of randomness are better than deterministic sources.* But let us stress that the statistical quality of DIEHARD is governed by the deterministic part!

Another important issue is the alphabet of the source of random numbers. In this review we will consider the case of a finite alphabet with N different simbols (numbers), because that is the case in all the applications, as far as a computer has a finite precision and it is also impossible to measure a noise source with infinite precision.

(3) Can a chaotic dynamical system behave like and acceptable PRNG?

Here we are interested in chaotic PRNG. The interesting thing with deterministic chaos is that it faces us with a new situation: in spite the model is deterministic and simple, an unpredictable behavior of the variables may arise due to the sensitivity to initial conditions.

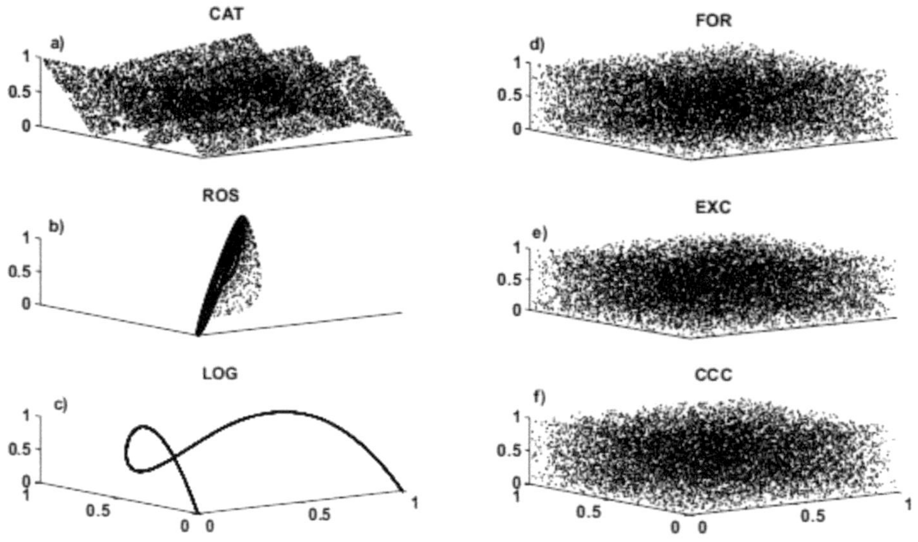

Figure 1: 3D distribution for for three chaotic (see (a) to (c)) and, three non-chaotic ((d) to (f)) PRNG's.

Then statistical mechanics is the best theory to analyze their long time behavior.

In other words: if we could use real numbers with an infinite number of significance figures chaotic systems are strictly deterministic, but real numbers are a mental idealization created by human minds, as the rigid body, the Galileo's inertia law, and many other very fruitful ideas. When evaluations are made, numbers must be represented by a discrete number of significance figures and the sensitivity to initial conditions or the "butterfly effect" Lorenz described in his famous paper [9] produces unpredictable behavior. The sensitivity to initial conditions and the broadband power spectrum of chaotic time series make them good candidates to generate PRNG's with a behavior very similar to physical noise sources. In this review we will focus on discrete chaotic dynamical systems (or maps).

(4) How can we measure the quality of a PRNG?

The first requirement pointed in *(2)*, an uniform histogram, is easy to be tested and the Shannon Entropy is the obvious quantifier. The second requirement, refers to statistical independence. This is hard to theoretically substantiate. Numerically, it is impossible. Instead, several basic properties are tested: long cycle length, uniformity of the power spectrum, mixing constant, *etc.* Also the speed of generation, reproducibility and portability are essential requirements. In engineering applications, for example, it is usual to replace this second requirement for a flat power spectrum. The reason is, a uniform spectrum means that no periodicities exist and the next value may be any of the allowed ones. For discrete sources the power spectrum is replaced by the square of the absolute value of the fast Fourier transform (FFT) of the time series.

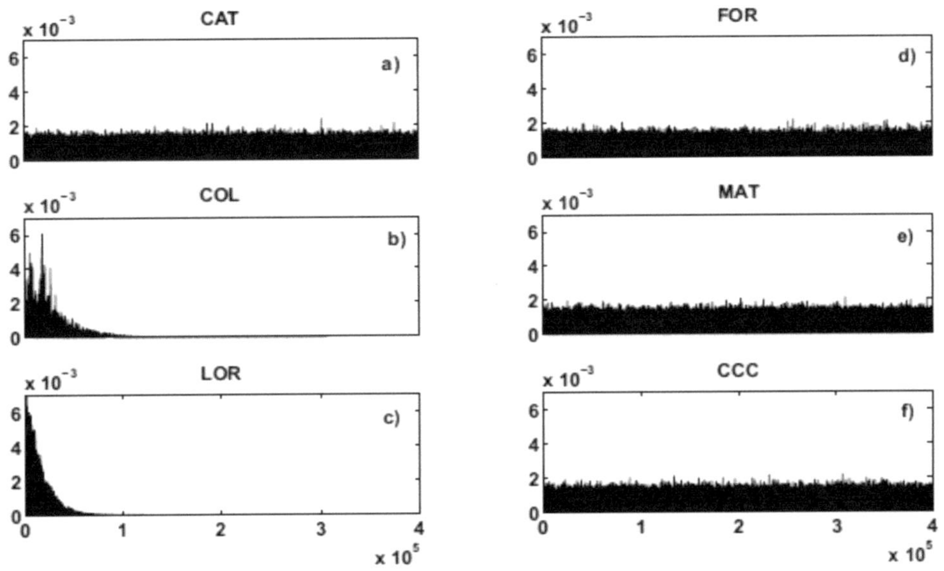

Figure 2: Power spectrum for for three non-chaotic (a) to (c) and three chaotic (d) to (f) PRNG's.

Some general purpose test suites, readily available to researchers in academia and industry who wish to analyze their newly developed PRNG are: tests by George Marsaglia [8], *Crypt-XS* by Helen Gustafson of the Queensland University of Technology [10], the National Institute of Standards and Technology Statistical Test Suite [11] and Rukhin suite of statistical procedures [12]. Additional requirements may be imposed in a specific application.

Of course a statistical test can never prove that a sequence generated by a PRNG is random (*because it is not !*), but it helps to detect certain kinds of weaknesses a generator may have. Furthermore none of these tests can prove that a given generator is reliable in all applications. Vattulainen *et al.* [13], for example, proposed three additional physical tests to detect deficiencies of several PRNG's used in Monte Carlo simulations.

In previous works we studied the ability of different quantifiers to characterize the uniformity of the histogram and the statistical independence in the case of chaotic PRNG's [14]. There we studied quantifiers based on Information Theory, recurrence plots [15] and intrinsic computation [16]. In this chapter we will focus on quantifiers based on Information Theory only.

In [14] we concluded that a very good classification scheme of PRNG's may be done using Information Theory quantifiers but it is mandatory to use two different PDF's: one to measure the uniformity of the outputs and the other one to measure correlation between consecutive samples. In this chapter we will focus on this approach where the Shannon-Jensen Statistical Complexity plays a fundamental role.

(5) What is the strategy to improve the quality of a chaotic PRNG?

As we pointed out several times any PRNG must have a uniform histogram and statistical independence of consecutive values. For a large family of chaotic maps, the histogram is characteristic of the map and it is more precisely defined as the *invariant measure*. The invariant measure is one of the eigenfunctions of the Perron-Frobenius operator of the map. The statistical independence of consecutive values of the time series may be estimated by means of the study of the mixing properties of the map. A mixing constant r_{mix} given by one of the eigenvalues of the Perron-Frobenius operator of the map [17, 18, 19], may be used to compare mixing maps.

When a chaotic time series has not a constant invariant measure, or a null mixing constant, or both, randomizing techniques may be designed to improve it. In this chapter we only consider one of this techniques called skipping, that improves the mixing but does not change the histogram. It works fine when the chaotic time series has a uniform histogram as is the case in piecewise linear maps. More details about more general randomization processes can be seen in [20].

This chapter is organized as follows: in Section 2 we briefly describe fourteen algorithmic PRNG. Half of them are non chaotic and half of them are chaotic. Results for these 14 PRNG's are used to exemplify the analysis tools. In Section 3 we briefly describe how the statistical description of a chaotic map may be done by studying the Perron-Frobenius operator associated with the map. Section 4 deals with classical tests to characterize chaotic PRNG's and the quantifiers based on Information Theory are introduced. In section 5 we describe one randomization technique called skipping, used to improve the mixing and consequently the statistical independence between consecutive values, for chaotic maps.

The characterization of the PRNG considered in the entropy-complexity plane is given in Section 6. Finally conclusions are presented in Section 7.

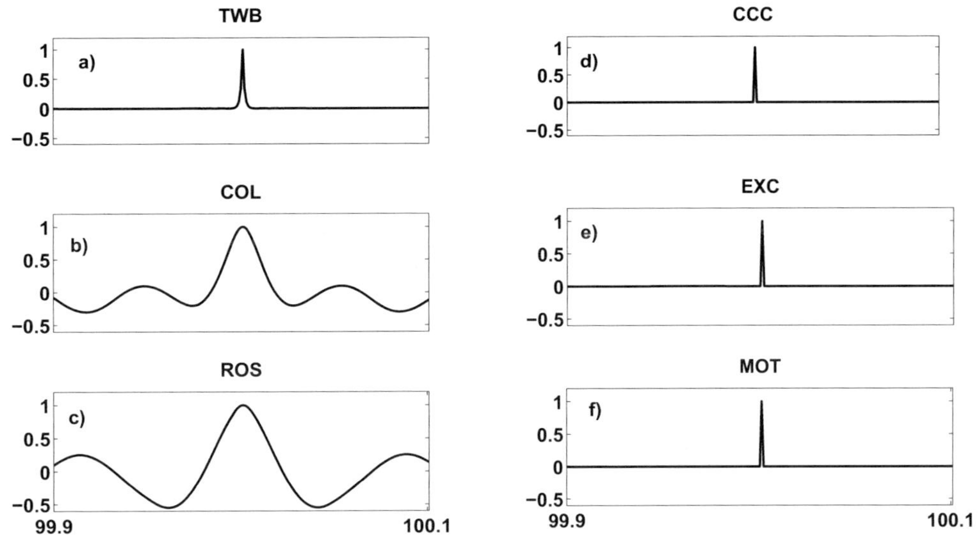

Figure 3: Linear autocorrelation for three chaotic (see (a) to (c)) and three non-chaotic ((d) to (f)) PRNG's.

2 Algorithmic Pseudo Random Number Generators

Most non chaotic algorithmic PRNGs are based on three methods [8]:

- *Congruential:* These generators use a linear transformation on the ring of reduced residues of some modulus m, to produce a sequence of integers, $x_1,\ x_2,\ x_3,\ \cdots$, with $x_n \equiv a\ x_{n-1} + b \mod m$.

- *Shift-register:* These generators iterates a binary $1 \times n$ vector β by means of a linear transformation with a matrix T. β is the seed and T must be a non singular $n \times n$ matrix having order $2^n - 1$ to produce a sequence $\beta,\ \beta T,\ \beta T^2,\ \cdots$, with period $2^n - 1$.

- *Lagged-Fibonacci:* These generators use an initial set of elements $x_1,\ x_2,\ \cdots,\ x_r$, and two lags r and s with $r > s$. Successive elements are generated by the recursion

$x_n = x_{n-r} \bullet x_{n-s}$, where $\bullet$ is some binary operation, $n > r$, the initial seed elements are computer words and the binary operation might be $+$, $-$, $*$ or $\oplus$ (exclusive or). For operations $+$ or $-$ the x's might be integers mod 2^k or single, or double precision reals mod 1. For the operations $*$ the x's might be odd integers mod 2^k. Each lagged-Fibonacci generator depends on details of the particular binary operation and the set of elements it operates on.

Best results are obtained by combining two or more simple generators through a convenient computer operation such as $+$, $-$, $*$ or $\oplus$.

In this review we will consider seven non chaotic algorithmic generators widely used:

- Excel© RNG (EXC).

- RNG available in Intel© fortran compiler (FOR).

- RNG available in Borland © C++ compiler (CCC).

- Matlab© RAND function (MAT).

- *Mother* RNG, available in Marsaglia website [8] (MOT), given by

$$\begin{cases} u_{n+1} = 2111111111\ x_{n-3} + 1492\ x_{n-2} + 1776\ x_{n-1} + 5115\ x_n + c_n\ , \\ x_{n+1} = u_{n+1} \bmod 2^{32}\ , \\ c_{n+1} = floor(u_{n+1}/2^{32})\ . \end{cases} \tag{1}$$

- *Multiple with carry* RNG (MWC) [8], given by

$$\begin{cases} u_{n+1} = 2131995753\ x_n + c_n\ , \\ x_{n+1} = u_{n+1} \bmod 2^{32}\ , \\ c_{n+1} = floor(u_{n+1}/2^{32})\ . \end{cases} \tag{2}$$

- *Lehmer* RNG (LEH), given by

$$\begin{cases} u_{n+1} = 16807\ x_n\ , \\ x_{n+1} = u_{n+1} \mod 2^{31} - 1\ . \end{cases} \tag{3}$$

All these algorithmic non-chaotic PRNGs are widely used and they pass most tests by Marsaglia [8].

We will also consider seven chaotic generators; the first three are continuous dynamical systems discretized by the Euler procedure and the last four are chaotic maps. They are:

- *Lorenz discrete dynamical system* (LOR):

$$\begin{cases} x_{n+1} = x_n + k\ [-\delta\ (x_n - y_n)]\ , \\ y_{n+1} = y_n + k\ [-x_n\ z_n + r\ x_n - y_n]\ , \\ z_{n+1} = z_n + k\ [x_n\ y_n - b\ z_n]\ . \end{cases} \tag{4}$$

with $k = 1/64$, $\delta = 8$, $r = 24$, $b = 2$.

- *Collpits discrete dynamical system* (COL):

$$\begin{cases} x_{n+1} = x_n + k\ [a\ z_n - b\ u_n]\ , \\ y_{n+1} = y_n + k\ [a\ z_n + c\ u_n - y_n - d]\ , \\ z_{n+1} = z_n + k\ [e\ d - z_n - x_n - y_n]\ , \\ u_{n+1} = \begin{cases} 0 & \text{for } y_n \geq -1\ , \\ -y_n - 1 & \text{for any other case.} \end{cases} \end{cases} \tag{5}$$

with $k = 1/128$, $a = 12$, $b = 800$, $c = 4$, $d = 8$, $e = 8$.

- *Rössler discrete dynamical system* (ROS):

$$\begin{cases} x_{n+1} = x_n + k\ [-y_n - z_n]\ , \\ y_{n+1} = y_n + k\ [x_n + a\ y_n]\ , \\ z_{n+1} = z_n + k\ [b + x_n\ z_n - \mu\ z_n]\ . \end{cases} \tag{6}$$

with $k = 1/16$, $a = 1/4$, $b = 1/2$, $\mu = 8$.

- *Cat map* (CAT):

$$\begin{cases} x_{n+1} = x_n + y_n - floor(x_n + y_n) \ , \\ y_{n+1} = x_n + 2\ y_n - floor(x_n + 2\ y_n) \ . \end{cases} \tag{7}$$

- *Tent map* (TEN):

$$x_{n+1} = \begin{cases} (2 * x_n + 1 - a)/(1 + a) & \text{if } -1 \leq x_n \leq a \ , \\ (-2 * x_n + 1 + a)/(1 - a) & \text{if } a < x_n \leq 1 \ . \end{cases} \tag{8}$$

with $a = 0.26$.

- *Logistic map* (LOG):

$$x_{n+1} = r\ x_n\ (1 - x_n) \ . \tag{9}$$

with $r = 4$.

- *Three way Bernoulli map* (TWB):

$$x_{n+1} = \begin{cases} 3x_n & \text{if } 0 \leq x_n \leq 1/3 \ , \\ 3x_n - 1 & \text{if } 1/3 < x_n \leq 2/3 \ , \\ 3x_n - 2 & \text{if } 2/3 < x_n \leq 1 \ . \end{cases} \tag{10}$$

Some of these chaotic PRNG's do not pass the tests by Marsaglia [8] but they may be randomized as will be explained in Section 5.

3 Perron-Frobenius approach for chaotic maps

The statistical properties of chaotic maps have been characterized by means of the Perron-Frobenius operator associated with the map [21, 22, 23]. This issue will not be detailed in this chapter because there exists an extensive bibliography about it (see for example [17, 18, 19]).

We only give here some general aspects. Let f be a chaotic map on the interval $[0, 1]$. Suppose the map has an invariant measure $\mu(x)$. Then the map is *ergodic* if for any

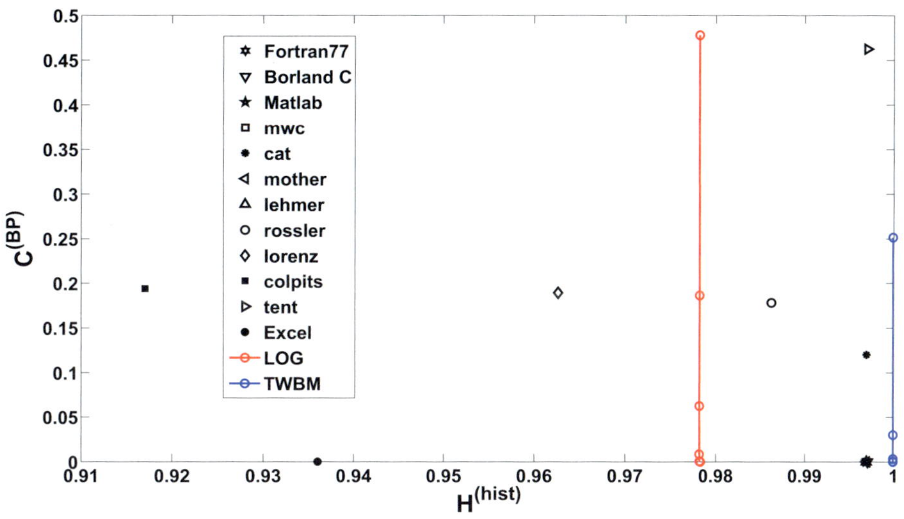

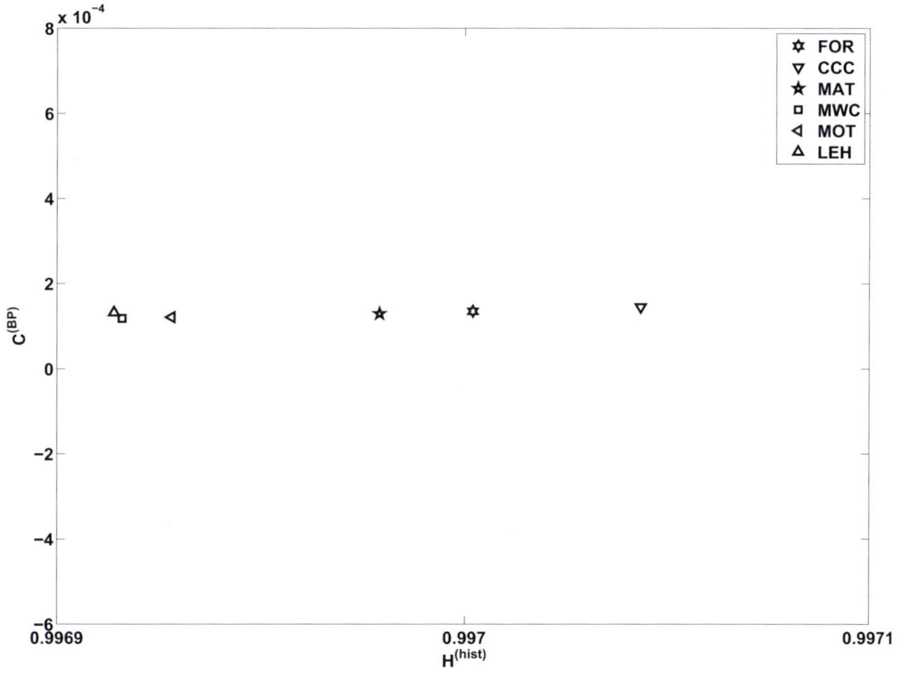

Figure 4: (a) Entropy complexity plane $H^{(hist)} \times C^{(BP)}$ for all the PRNGs of Section 2. (b) Zoom of Fig. (a) around the point (0.997,0), where several PRNGs are very close to each other.

integrable test function $Q(x)$, and for an arbitrary initial condition x_0 (up to a set of zero

μ-measure), the time average is equal to the ensemble average:

$$\overline{Q} = \langle Q \rangle \ . \tag{11}$$

Equation (11) is a consequence of the famous Birkhoff ergodic theorem [24]. *Mixing* is an even stronger requirement than ergodicity. A map is called "*mixing*" if any smooth initial probability density $\rho(x)$ converges to the invariant measure $\mu(x)$ after enough successive iterations. Mixing implies ergodicity. The reverse, however, is not true [25].

There exists an equivalent definition of mixing *via correlation functions.* Let $\phi_1(x)$ and $\phi_2(x)$ be two integrable test-functions and define the generalized correlation function of the map f by

$$C(\phi_1, \phi_2, n) = \lim_{J\to\infty} \frac{1}{J} \sum_{j=0}^{J-1} \phi_1(x_{j+n})\phi_2(x_j) - \langle\phi_1\rangle\langle\phi_2\rangle \ , \tag{12}$$

where

$$\langle\phi_i\rangle = \lim_{J\to\infty} \frac{1}{J} \sum_{j=0}^{J-1} \phi_i(x_j) \ . \tag{13}$$

The map is mixing if, for arbitrary ϕ_1 and ϕ_2,

$$\lim_{n\to\infty} C(\phi_1, \phi_2, n) = 0 \ . \tag{14}$$

Let us stress that it is not easy to prove that f is a *mixing map* because the mixing condition given in equation (14) must be fulfilled for *arbitrary* test functions. Formally, every mixing map f has an associated determines the time evolution of any initial density $\rho_0(x)$ towards the invariant measure $\mu(x)$:

$$\rho_{n+1} = \mathcal{L}[\rho_n] \ . \tag{15}$$

The explicit formal expression for the Perron-Frobenius operator for a one-dimensional map f is given in [25].

$$\mathcal{L}[\rho_y] = \sum_{x\epsilon f^{-1}(y)} \frac{[\rho_0(x)]}{|f'(x)|} \ . \tag{16}$$

This operator $\mathcal{L}$ has a set of eigenfunctions $\psi_\alpha(x)$ and eigenvalues η_α. The invariant measure $\mu(x)$ is the eigenfunction corresponding to the largest eigenvalue $\eta_0 = 1$. The full set of eigenfunctions and eigenvalues may be used as a basis to express any density:

$$\begin{aligned} \rho_0(x) &= \sum_\alpha c_\alpha \psi_\alpha(x) \ , \\ \rho_n(x) &= \mathcal{L}^n \rho_0(x) = \sum_\alpha \eta_\alpha^n c_\alpha = c_0\psi_0(x) + R_n \ . \end{aligned} \tag{17}$$

The eigenvalue with the second largest absolute value, η_1, has a "distinguished" physical meaning: it is related with the *mixing constant* r_{mix} that governs the relaxation of *exponentially mixing* maps:

$$|R_n| \sim |\eta_1|^n \sim \exp(-n/r_{mix}) \; . \tag{18}$$

In the case of mixing maps the natural invariant measure $\mu(x)$ is identical to the normalized histogram of the time series. Consequently the first main requirement of a RPNG is $\mu(x) = const.$ The mixing constant r_{mix} gives the characteristic transient time [18, 26, 27, 28, 29, 30, 31]. A mixing constant r_{mix} pretends to be an indirect measure of statistical independence: if $r_{mix} = 0$ the maps mixes the consecutive values making them *look as* statistically independent random variables. For a PRNG the ideal value is $r_{mix} = 0$.

Usually, the analytical expression of the invariant measure $\mu(x)$ of a given map f is not known. Exceptions are the logistic map in full chaos, and piecewise-linear maps. The mixing constant r_{mix} has been analytically obtained only for piecewise-linear maps. For other maps it must be numerically obtained by means of a piecewise linear approximation of the map.

4 Classical Tests and quantifiers based on Information Theory

4.1 Classical Tests

Let us start describing four classical tests usually applied to PRNGs, that are also useful for the study of chaotic PRNGs:

- *$2D$ and $3D$ distribution:*

  These are excellent graphic or visual tools to discover hidden structures produced by nonlinear correlations in the time series. They are obtained by plotting vectors $\vec{x}_i$ with components equal to successive values of the time series, *i.e.*, $\vec{x}_i = (x_i, x_{i+1})$ in the case of $2D$ and $\vec{x}_i = (x_i, x_{i+1}, x_{i+2})$ in the case of $3D$. The sequence $\{\vec{x}_i\}$ must fill the complete $2D$ or $3D$ space in an uniform way. It only means that any combination of 2 or 3 successive values appear the same number of times. Note that to assure randomness based on this test it is necessary to extend the procedure to infinite dimensions.

In Figs. 1 it is shown the $3D$ distribution for some of the non-chaotic and chaotic systems reported in this chapter. Several chaotic systems do not fill the complete $2D$ or $3D$ space in an uniform way. This is an evidence that the system does not constitute a good PRNG and must be randomized.

- *Power Spectrum:*

  The Fast Fourier Transform (FFT) $\{X_k\}$ of the time series $\{x_n\}$ is performed and the corresponding discrete power spectrum $P_s = X_k \; X_k^*$ is computed. A flat power spectrum, with equal frequency contribution for all frequencies is indicative of statistical independence between successive values. Consequently a good PRNG must have a uniform power spectrum.

In Figs. 2 it is shown the power spectrum for several non-chaotic and chaotic systems reported in this chapter.

- *Autocorrelation Function:*

  The discrete linear autocorrelation function given by

$$Corr(x,x)_j = \sum_k x_{nk} \; x_{nk+j} \;, \tag{19}$$

where j is the lag. An almost constant value of $Corr(x,x)_j$, independent of the lag j, will be indicative of an uncorrelated random series. But this test only considers linear correlations and it is not a good test for chaotic PRNGs.
In Fig. 3 we show the autocorrelation function for several PRNG's studied in this chapter. Note that for example LOG has a delta like autocorrelation in spite consecutive values are not statistically independent and are strongly correlated by the quadratic map.

- *Marsaglias's tests:*

  Marsaglia proposed a set of tests he called *stringent tests.* They are based on the fact that a random number generator is supposed to produce a sequence of statistical independent and identically distributed random variables $x_1, x_2, x_3, \cdots$. Any function of elements of that sequence may serve as a test. Marsaglia's tests provide a matrix of 127 values to characterize the PRNG. The idea is that a good PRNG must pass all the tests but it is not easy to compare widely different PRNG's by the values obtained. Quoting Marsaglia: "If the RNG is to be used for a particular problem one should try to create a test based on a similar problem for which the underlying distributions are known or lacking that at least compared with results produced by widely different RNG's".

For those interested in test suites we refer to [8, 10, 11, 32].

4.2 Information Theory quantifiers for chaotic maps

As we need to characterize respectively the uniformity of the histogram and the statistical independence, our approach is to use two quantifiers and represent the PRNG's on a plane with one axis for each quantifier. The goal of our approach is to get an easy way to compare PRNG's of different origin. The uniformity of the histogram is obviously measured by the Shannon Entropy. To quantify the statistical independence between consecutive values we studied several quantifiers for chaotic systems in previous works the FFT, the mixing constant, recurrence plots quantifiers, intrinsic computation quantifiers, *etc.* We will focus in this chapter to the use of quantifiers based on Information Theory [14, 20, 33, 34, 35] (see also Chapter 2 and 8).

Information Theory quantifiers are appropriate functionals of the probability distribution function (PDF). Let $\{x_i\}$ be the time series under analysis, with length M. There are infinite possibilities to assign a PDF to a given time series, a subject that will be given due consideration below. In the meantime, suppose that the PDF is discrete and is given by $P = \{p_i; i = \cdots, N\}$. One defines then various quantities, namely,

- *Normalized Shannon Entropy $H[P]$:*
  Let $S[P]$ be the Shannon Entropy

$$S[P] = -\sum_{i=1}^{N} p_i \ \ln(p_i) \ . \tag{20}$$

Is is well known that the maximum $S_{max} = \ln(N)$ is obtained for $P_e = \{1/N, \cdots, 1/N\}$, that is, the uniform PDF. A "normalized" entropy $H[P]$ can also be defined in the fashion

$$H[P] = S[P]/S_{max} \ . \tag{21}$$

- *Shannon-Jensen Statistical Complexity Measure* $C[P]$*:*

For the statistical complexity measure, in this chapter we adopt the functional form definition given by López Ruiz-Mancini-Calbet [36] (see Chapter 7) with the modifications advanced by Lamberti and co-workers in [2] (see Chapter 8), so as to ensure that the concomitant SCM-version becomes *(i)* able to grasp essential details of the dynamics, *(ii)* an intensive quantity (in the thermodynamical sense and, *(iii)* capable of discerning both among different degrees of periodicity and chaos [3]. The ensuing measure, to be referred to as the Shannon-Jensen statistical complexity (see Chapter 8), is a functional $C[P]$ that reads

$$C[P] = Q_J[P, P_e] \cdot H[P] \ , \tag{22}$$

where Q_J is the "disequilibrium", defined in terms of the so-called extensive Jensen-Shannon divergence (which induces a squared metric) [2]. One has

$$Q_J[P, P_e] = Q_0 \cdot \{S[(P + P_e)/2] - S[P]/2 - S[P_e]/2\} \ , \tag{23}$$

with Q_0 a normalization constant ($0 \leq Q_J \leq 1$) that reads

$$Q_0 = -2\left\{\left(\frac{N+1}{N}\right)\ln(N+1) - 2\ln(2N) + \ln N\right\}^{-1} \ . \tag{24}$$

We see that the disequilibrium Q_J is an intensive quantity that reflects on the systems's "architecture", being different from zero only if there exist "privileged", or "more likely" states among the accessible ones. $C[P]$ quantifies the presence of correlational structures as well [2, 37]

The opposite extremes of perfect order and maximal randomness possess no structure to speak of and, as a consequence, $C[P] = 0$. In between these two special instances a wide range of possible degrees of physical structure exist, degrees that should be reflected in the features of the underlying probability distribution. In the case of a PRNG the "ideal" values are $H[P] = 1$ and $C[P] = 0$.

As pointed out above, P itself is not a uniquely defined object and several approaches have been employed in the literature so as to "extract" P from the given time series. Just to mention some frequently used extraction procedures: *a)* frequency count [38] *b)* procedures based on amplitude statistics (histograms) [20], *c)* binary symbolic-dynamics [39], *d)* Fourier analysis [40], *e)* wavelet transform [41, 42], *f)* partition entropies [43], *g)* permutation entropy [4, 44], *h)* discrete entropies [45], *etc.* There is ample liberty to choose among them. De Micco *et al.* [20] proposed two probability distributions as relevant for testing respectively the uniformity of $\mu(x)$ and the mixing constant: *(a)* a P based on time series' histograms and *(b)* a P based on ordinal patterns permutation ordering that derives from using the Bandt-Pompe method [4].

For extracting P *via* the histogram divide the interval $[0, 1]$ into a finite number N_{bin} of non overlapping subintervals A_i: $[0, 1] = \bigcup_{i=1}^{N_{bin}} A_i$ and $A_i \bigcap A_j = \emptyset \ \forall i \neq j$. Note that N in equation (20) is equal to N_{bin}. Of course, in this approach the temporal order of the time-series plays no role at all. The quantifiers obtained *via* the ensuing PDF are called in this paper $H^{(hist)}$ and $C^{(hist)}$. Let us stress that for time series within a finite alphabet it is relevant to consider an optimal value of N_{bin} (see *i.e.*, [20]).

In extracting P by recourse to the Bandt-Pompe method the resulting probability distribution P is based on the details of the attractor-reconstruction procedure. *Causal information* is, consequently, duly incorporated into the construction-process that yields P. The quantifiers obtained *via* the ensuing PDF are called in this paper $H^{(BP)}$ and $C^{(BP)}$. A notable Bandt-Pompe result consists in getting a clear improvement in the quality of Information Theory-based quantifiers [3, 34, 35, 46, 47, 48, 49, 50].

The extracting procedure is as follows. For the time-series $\{x_t : t = 1, \cdots, M\}$ and an embedding dimension $D > 1$, one looks for "ordinal patterns" of order D [4, 44, 51] generated by

$$(s) \mapsto \left(x_{s-(D-1)}, x_{s-(D-2)}, \cdots, x_{s-1}, x_s\right) \ , \tag{25}$$

which assign to each "time s" a D-dimensional vector of values pertaining to the times $s, s-1, \cdots, s-(D-1)$. Clearly, the greater the $D-$value, the more information on "the past" is incorporated into these vectors. By the "ordinal pattern" related to the time (s) we mean the permutation $\pi = (r_0, r_1, \cdots, r_{D-1})$ of $(0, 1, \cdots, D-1)$ defined by

$$x_{s-r_{D-1}} \leq x_{s-r_{D-2}} \leq \cdots \leq x_{s-r_1} \leq x_{s-r_0}. \tag{26}$$

In order to get a unique result we consider that $r_i < r_{i-1}$ if $x_{s-r_i} = x_{s-r_{i-1}}$. Thus, for all the $D!$ possible permutations π of order D, the probability distribution $P = \{p(\pi)\}$ is defined by

$$p(\pi) = \frac{\sharp\{s|s \leq M - D + 1; (s) \quad \texttt{has type } \pi\}}{M - D + 1}. \tag{27}$$

In the last expression the symbol $\sharp$ stands for "number".

The advantages of the Bandt-Pompe method reside in *a)* its simplicity, *b)* the associated extremely fast calculation-process, *c)* its robustness in presence of observational and dynamical noise, and *d)* its invariance with respect to nonlinear monotonous transformations. The Bandt-Pompes methodology is not restricted to time series representative of low dimensional dynamical systems but can be applied to any type of time series (regular, chaotic, noisy, or reality based), with a very weak stationary assumption (for $k = D$, the probability for $x_t < x_{t+k}$ should not depend on t [4]. One also assumes that enough data are available for a correct phase space reconstruction. Of course, the embedding dimension D plays an important role in the evaluation of the appropriate probability distribution because D determines the number of accessible states $D!$. Also, it conditions the minimum acceptable length $M \gg D!$ of the time series that one needs in order to work with a reliable statistics. In relation to this last point Bandt and Pompe suggest, for practical purposes, to work with $3 \leq D \leq 7$ with a time lag $\tau = 1$. This is what we do here (in the present work $D = 6$ is used).

d	LOG	TWBM
1	0.56789	0.333333333
2	0.31848	0.111111111
3	0.13290	0.037037037
4	0.05788	0.012345679
5	0.03646	0.004115226
6	0.01791	0.001371742
7	0.01152	0.000457247
8	0.00515	0.000152416

Table 1: r_{mix} as a function of the iteration-order d for the TWBM and LOG chaotic maps, respectively.

$\mathcal{S}_{IN}$	$\mathcal{S}_{OUT} \equiv f^d$				
($\mathbb{R}$)	$d = 2$	$d = 3$	$d = 4$	$d = 5$	$d = 6$
0.010559404					
0.041791613	0.041791613				
0.160180296		0.160180296			
0.538090276	0.538090276		0.538090276		
0.994196523				0.994196523	
0.023079185	0.023079185	0.023079185			0.023079185
0.090186145					
0.328210418	0.328210418		0.328210418		
0.881953358		0.881953358			
0.416446528	0.416446528			0.416446528	
0.972075269					
0.10857976	0.10857976	0.10857976	0.10857976		0.10857976
0.387160782					
0.949069244	0.949069244				
0.193347257		0.193347257		0.193347257	
0.62385638	0.62385638		0.62385638		

Table 2: Illustrating the **Skipping** procedure

5 Skipping randomization procedure

It is possible to show that the invariant measure of the iterated map f^d is identical to the invariant measure of the original map f. Also, the mixing constant r_{mix} for f^d is lower that the mixing constant for f. As an example, in Table 1 the value or r_{mix} of f and f^d is shown for two of the chaotic maps studied in this chapter.

The iteration of a map is then the simplest randomization procedure proposed in the literature, used to diminish r_{mix} or equivalently to increase the statistical independence of consecutive values. This procedure is also known as **Skipping** because iterating a map is tantamount to skipping values in the original time series.

The procedure is as follows: let f be a chaotic map. Starting from a randomly chosen initial condition, the map is iterated generating the chaotic time-series (CHTS) $\mathcal{S}_{IN} =$

$\{x_0, x_1, \cdots\}$. This CHTS is to be regarded as the input for the randomizing process (see Table 2). Let us assume that x_i is a floating-point number (in the IEEE normalized representation). Without loss of generality, we consider values restricted to the interval $[0, 1]$. As previously stated, the central idea here is that of employing the well-known symbolic dynamics-randomization process called Skipping to obtain the new time-series (STS) $\mathcal{S}_{OUT}$, obtained by discarding $d-1$ values of $\mathcal{S}_{IN}$, are "skipped" to get the STS $\mathcal{S}_{OUT}$ which originates the name **Skipping** for this technique. In other words, one employs, instead of the original map f, its d-times iterated one f^d. This randomization technique is routinely (and successfully) used with piecewise linear maps in many applications [18]. In Table 2, an example with different values of d is displayed.

6 Characterization of PRNG's by means of the Entropy-Complexity plane

In this section we present the main result reported in this chapter. That is a representation of all the PRNG's on a plane with one axis used to characterize the uniformity of the histogram and the other axis used to characterize the statistical independence.

The uniformity of the histogram is measured by $H^{(hist)}$, the normalized Shannon entropy of the histogram. The statistical independence is measured by $C^{(BP)}$, the Shannon-Jensen statistical complexity measure using the PDF obtained by the Bandt-Pompe procedure. Figs. 4 shows our results. We represent on the plane $H^{(hist)} \times C^{(BP)}$ all the PRNG's presented in Section 2. The ideal PRNG is one with $H^{(hist)} = 1$ and $C^{(BP)} = 0$ and Fig. 4 shows that the best PRNG is TWB, after the skipping procedure is applied. Near the ideal there are several PRNGs: MOT, FOR, CCC and MAT. LOG does not produce a good result even after the skipping procedure. The reason is the skipping procedure does not change the histogram of the map and consequently it is impossible to improve the value of $H^{(hist)}$ by this randomization procedure. For other strategies that can be used for maps with nonuniform histogram like LOG see [20].

It is interesting to see that EXC and also LOR, ROS and COL are not good choices. All these generators have low entropy and high complexity. For LOR, ROS and COL it is possible to use a bigger step in the Euler procedure to get better results [52].

7 Conclusions

In summary two main characteristics define the quality of a PRNG: equiprobability of all the values and statistical independence between consecutive outputs. Both characteristics may be quantified by means of an Information Theory approach but two different PDF's are required. Other quantifiers have been also studied in previous works: $3D$ distributions, power spectrum uniformity, r_{mix} constant, recurrence plots, intrinsic computation quantifiers [14]. The advantage of a procedure based only on Information Theory quantifiers is to use very well understood concepts as Entropy and Statistical Complexity. In fact the first above mentioned requirement - the equiprobability of all the values - is equivalent to a uniform histogram and implies a normalized Shannon entropy $H^{(hist)} = 1$. Consequently one PRNG will be better than other if it has $H^{(hist)}$ closer to the ideal value 1.

To quantify the statistical independence it is required to use a causal PDF, it means a PDF for trajectories and not for individual values, in order to capture correlations between consecutive outputs. Our approach is to define a probability distribution function by means of the Bandt-Pompe procedure and evaluate the Shannon-Jensen statistical complexity $C^{(BP)}$ of that distribution. The ideal PRNG will have $C^{(BP)} = 0$. The representation of different PRNG's on a plane with axis $H^{(hist)}$ and $C^{(BP)}$ allows one to easily compare them with each other. The ideal PRNG must be represented by the point $(1, 0)$ in this plane.

In the present work we used the skipping randomization procedure in order to improve the PRNG performance. Its viability was checked using the entropy-complexity plane $H^{(hist)} \times C^{(BP)}$. We have shown that the TWB, MOT, FOR, CCC and MAT are very near the ideal point $(1, 0)$ after skipping procedure. However, the proposed procedure for improvement the PRNG performance do not work for the case of LOG due to the characteristics that its PDF-histogram present. Something similar happen with EXC, LOR, ROS and COL due to they present low entropy and high complexity values.

It is interesting to note that the representation points of piecewise chaotic maps can almost reach this ideal point by means of the skipping procedure. The reason is skipping diminishes correlation between consecutive values. For other chaotic maps more involved randomization techniques are required [20] Let us finally stress that statistical quality is the main requirement for a PRNG used in many applications such as Monte Carlo simulations, electromagnetic compatibility improvement [53], filtering by random sampling [29], *etc.* But it is not enough to guarantee good cryptographic properties [54]. Cryptographic properties are not the subject of this chapter.

Conflict of interest: None.

Acknowledgments

This work was partially supported by Universidad Nacional de Mar del Plata, ANPCyT (PICT 2010-2335) and CONICET (PIP 112-200801-1420 CCTMDP). OAR acknowledges partial support from CONICET, Argentina, and CAPES, PVE fellowship, Brazil.

References

[1] Wold H. *A Study in the Analysis of Stationary Time Series.* Almqvist and Wiksell Book Co., Upsala, Sweden, 1938.

[2] Lamberti, P.W.; Martín, M.T.; Plastino, A.; Rosso, O.A. Intensive entropic non-triviality measure. *Physica A* **334** 119–131 (2004).

[3] Rosso, O.A.; Larrondo, H.A.; Martín, M.T.; Plastino, A.; Fuentes, M.A. Distinguishing noise from chaos. *Phys. Rev. Lett.* **99** 154102 (2007).

[4] Bandt, C.; Pompe, B. Permutation entropy: a natural complexity measure for time series. *Phys. Rev. Lett.* **88** 174102 (2002).

[5] http://www.random.org

[6] http://www.fourmilab.ch/hotbits/

[7] http://www.lavarnd.org

[8] Marsaglia, G. *The marsaglia random number cdrom including the diehard battery of tests of randomness.* http://www.stat.fsu.edu/pub/diehard/, 1995.

[9] Lorenz, E.N. Deterministic non periodic flow. *Journal of the Atmospheric Sciences* **20** 130–141 (1963).

[10] Gustafson, II.M.; Dawson, E.P.; Nielsen, L.; Caelli, W.J. A computer package for measuring the strength of encryption algorithms. *J. Comput. Security* **13** 687–697 (1994).

[11] Soto, J. *Statistical testing of random number generators, available online.* http://www.itl.nist.gov/div893/staff/soto/jshome.html.

[12] Rukhin, A.L. Testing randomness: a suite of statistical procedures. *Theory Probab. Appl.* **45** 111–132 (2000).

[13] Vattulainen, I.; Ala-Nissila, T.; Kankaala, K. Physical tests for random numbers in simulations. *Phys. Rev. Lett.* **73** 2513–2516 (1994).

[14] De Micco, L.; Larrondo, H.A.; Plastino, A.; Rosso, O.A. Quantifiers for randomness of chaotic pseudo random number generators. *Philosophical Transactions of the Royal Society A* **367** 3281–3296 (2009).

[15] Marwan, N.; Romano, M.C.; Thiel; M., Kurths, J. Recurrence plots for the analysis of complex systems. *Physics Reports* **438** 237–329 (2007).

[16] Feldman, D.P.; McTague, C.S.; Crutchfield, P. The organization of intrinsic omputation: complexity-entropy diagrams and the diversity of natural information processing. *Chaos* **18** 043106 (2008).

[17] Beck, C.; Schlögl, F. *Thermodynamics of chaotic systems: an introduction.* Cambridge University Press, 1997.

[18] Setti, G.; Mazzini, G.; Rovatti, R.; Callegari, S. Statistical modeling of discrete-time chaotic processes: Basic finite-dimensional tools and applications. *Proceedings of the IEEE* **90** 662–690 (2002).

[19] Lasota, A.; Mackey, M.C. Chaos, Fractals, and Noise: Stochastic Aspects of Dynamics. *Applied Mathematical Sciences* **97**. Springer Verlag, 2nd. edition, 1994.

[20] De Micco, L.; González, C.M.; Larrondo, H.A.; Martín, M.T.; Plastino, A.; Rosso, O.A. Randomizing nonlinear maps via symbolic dynamics. *Physica A* **387** 3373–3383 (2008).

[21] Dellnitz, M.; Froyland, G.; Sertl, S. On the isolated spectrum of the Perron-Frobenius operator. *Nonlinearity* **13** 1171–1188 (2000) .

[22] Ding, J.; Zhou, A. Finite approximations of Frobenius-Perron operators. A solution to Ulam's conjecture to multi-dimensional transformations. *Physica D* **92** 61–68 (1996).

[23] Pingel, D.; Schmelcher, P. Theory and examples of the inverse Frobenius-Perron problem for complete chaotic maps. *Chaos* **9** 357–66 (1999).

[24] Cornfeld, P.; Fomin, S.V.; Sinai, Ya. G. *Ergodic Theory.* Springer, New York, 1982.

[25] Beck, C. Ergodic properties of a kicked damped particle. *Commun. Math. Phys.* **130** 51–60 (1990).

[26] De Micco, L.; Petrocelli, R.A.; Larrondo, H.A. Constant envelope wideband signals using arbitrary chaotic maps. *Proceedings of XII RPIC*, 2007.

[27] De Micco, L.; Arizmendi, C.M.; Larrondo, H.A. Zipping characterization of chaotic sequences used in spread spectrum communication systems. *Institute of Physics Conference Proceedings* **913** 139–144 (2007).

[28] De Micco, L.; Petrocelli, R.A.; Carrica, D.O.; Larrondo, H.A. Muestreo caótico para la adquisición de señales de baja frecuencia con ruido de alta frecuencia. *Proceedings of XII RPIC*, 2007.

[29] Petrocelli, R.A.; De Micco, L.; Carrica, D.O.; Larrondo, H.A. Acquisition of low frequency signals immersed in noise by chaotic sampling and fir filters. *IEEE International Symposium on Intelligent Signal Processing 2007. WISP2007* 1–6 (2007).

[30] Rovatti, R.; Mazzini, G.; Setti, G. On the ultimate limits of chaos-based asynchronous DS-CDMA-I: Basic definitions and results. *IEEE Transactions on Circuits and Systems I: Fundamental Theory and Applications* **51** 1336–1347 (2004).

[31] Rovatti, R.; Mazzini, G.; Setti, G. On the ultimate limits of chao-sbased asynchronous DS-CDMA-II: Analytical results and asymptotics. *IEEE Transactions on Circuits and Systems I: Fundamental Theory and Applications* **51** 1348–1364 (2004).

[32] Rukhin, A.L. *Random number generation.* Available from http://www.iro.umontreal.ca/lecuyer/myftp/papers/handstat.pdf, 2004.

[33] González, C.M.; Larrondo, H.A.; Rosso, O.A. Statistical complexity measure of pseudorandom bit generators. *Physica A* **354** 281–300 (2005).

[34] Larrondo, H.A.; González, C.M.; Martín, M.T.; Plastino, A.; Rosso, O.A. Intensive statistical complexity measure of pseudorandom number generators. *Physica A* **356** 133–138 (2005).

[35] Larrondo, H.A.; Martín, M.T.; González, C.M.; Plastino, A.; Rosso, O.A. Random number generators and causality. *Phys. Lett. A* **352** 421–425 (2006).

[36] López-Ruiz, R.; Mancini, H.L.; Calbet, X. A statistical measure of complexity. *Phys. Lett. A* **209** 321–326 (1995).

[37] Martín, M.T.; Plastino, A.; Rosso, O.A. Statistical complexity and disequilibrium. *Phys. Lett. A* **311** 126–132 (2003).

[38] Rosso, O.A.; Craig, H.; Moscato, P. Shakespeare and other english renaissance authors as characterized by Information Theory complexity quantifiers. *Physica A* **388** 916–926 (2009).

[39] Mischaikow, K.; Mrozek, M.; Reiss, J.; Szymczak, A. Construction of symbolic dynamics from experimental time series. *Phys. Rev. Lett.* **82** 1114–1147 (1999).

[40] Powell, G.E.; Percival, I.C. A spectral entropy method for distinguishing regular and irregular motion of hamiltonian systems. *J. Phys. A: Math. Gen.* **12** 2053–2071 (1979).

[41] Blanco, S.; Figliola, A.; Quian Quiroga, R.; Rosso, O.A.; Serrano, E. Time-frequency analysis of electroencephalogram series (III): Wavelet packets and Information Cost Function. *Phys. Rev. E* **57** 932–940 (1998).

[42] Rosso, O.A.; Blanco, S.; Jordanova, J.; Kolev, V.; Figliola, A.; Schürmann, M.; Başar, E. Wavelet entropy: a new tool for analysis of short duration brain electrical signals. *Journal of Neuroscience Methods* **105** 65–75 (2001).

[43] Ebeling, E.; Steuer, R. Partition-based entropies of deterministic and stochastic maps. *Stochastics and Dynamics* **1** 1–17 (2001).

[44] Keller, K.; Sinn, M. Ordinal analysis of time series. *Physica A* **356** 114–120 (2005).

[45] Amigó, J.M.; Kocarev, L.; Tomovski, I. Discrete entropy. *Physica D* **228** 77–85 (2007).

[46] Kowalski, A.M.; Martín, M.T.; Plastino, A.; Rosso, O.A. Bandt-Pompe approach to the classical-quantum transition. *Physica D* **233** 21–31 (2007).

[47] Rosso, O.A.; Zunino, L.; Pérez, D.G.; Figliola, A.; Larrondo, H.A.; Garavaglia, M.; Martín, M.T.; Plastino, A. Extracting features of gaussian selfsimilar stochastic processes via the Bandt and Pompe approach. *Phys. Rev. E* **76** 061114 (2007).

[48] Rosso, O.A.; Vicente, R.; Mirasso, C.R. Encryption test of pseudo-aleatory messages embedded on chaotic laser signals: an Information Theory approach. *Phys. Lett. A* **372** 1018–1023 (2008).

[49] Zunino, L.; Pérez, D.G.; Martín, M.T.; Plastino, A.; Garavaglia, M.; Rosso, O.A. Characterization of gaussian self-similar stochastic processes using wavelet-based informational tools. *Phys. Rev. E* **75** 021115 (2007).

[50] Zunino, L.; Pérez, D. G.; Martín, M.T.; Garavaglia, M.; Plastino, A.; Rosso, O.A. Permutation entropy of fractional Brownian motion and fractional Gaussian noise. *Physics Letters A* **372** 4768–4774 (2008).

[51] Keller, K.; Lauffer, H. Symbolic analysis of high-dimensional time series. *Int. J. Bifurcation and Chaos* **13** 2657–2668 (2003).

[52] De Micco, L.; Zabaleta, O.G.; González, C.M.; Arizmendi, C.M.; Larrondo, H.A. Estocasticidad de un atractor caótico determinista implementado en FPGA. *Proceedings IBERCHIP 2010.*

[53] De Micco, L.; Petrocelli, R.A.; Rosso, O.A.; Plastino, A.; Larrondo, H.A. Mixing chaotic maps and electromagnetic interference reduction. *Int. J. App. Math. Stat*, **26**, 106–120 (2012).

[54] Liberatori, M.C.; Castieira Moreira, J.; Petruzzi, D.M.; Honary, B. Trellis-hopping turbo coding. *IEE Proc.-Commun.* **153** 966–975 (2006).

CHAPTER 13

Analysis of an EL Niño-Southern Oscillation proxy record using Information Theory quantifiers

Laura C. Carpi [a,e], Patricia M. Saco [a,b], Alejandra Figliola [c,g], Eduardo Serrano [d], Osvaldo A. Rosso [e,f,g,*]

[a] Civil, Surveying and Environmental Engineering, The University of Newcastle, University Drive, Callaghan NSW 2308, Australia

[b] Departamento de Hidráulica, Facultad de Ciencias Exactas, Ingeniería y Agrimensura, Universidad Nacional de Rosario, Avenida Pellegrini 250, 2000 Rosario, Argentina

[c] Instituto del Desarrollo Humano, Universidad Nacional de General Sarmiento Juan María Gutiérrez 1150, Los Polvorines, Buenos Aires, Argentina

[d] Centro de Matemática Aplicada, ECyT, Universidad Nacional de San Martín, Irigoyen 3100, San Martín, Buenos Aires, Argentina

[e] Departamento de Física, Instituto de Ciências Exatas, Universidade Federal de Minas Gerais, Av. Antônio Carlos, 6627 - Campus Pampulha, 31270-901 Belo Horizonte - MG, Brazil

[f] Chaos & Biology Group, Instituto de Cálculo, Facultad de Ciencias Exactas y Naturales, Universidad de Buenos Aires, Pabellón II, Ciudad Universitaria, 1428 Ciudad Autónoma de Buenos Aires, Argentina

[g] Consejo Nacional de Investigaciones Científicas y Técnicas (CONICET), Argentina

Abstract

Quantifiers based on Information Theory, the normalized Shannon entropy, the Fisher information measure and the Shannon-Jensen statistical complexity are used to characterize changes in the dynamical behavior of EL Niño Southern-Oscillation (ENSO) during the Holocene. For this purpose we analyze the ENSO proxy record corresponding to Pallcacocha Lake sedimentary data for the Holoceno period (11,000 yr BP to present). By recourse of these quantifiers we found evidence of a shift in dynamics and cyclic behavior of the ENSO proxy, which is consistent with the results of Moy *et al.*

* Corresponding author. E-mail: oarosso@gmail.com

Andres M. Kowalski, Raul D. Rossignoli and Evaldo M. F. Curado (Eds)

[Nature **420** (2002) 162-165]. In addition, we have been also able to localize these cycles in time and to analyze connections to epochs of rapid climate change (RCC) during the Holocene.

Keywords: Entropy; Statistical Complexity; Time series analysis; Probability distribution; ENSO dynamics.

1 Introduction

The understanding of weather and climate has been pursued since early human history. Ancient civilizations along great rivers like the Euphrates, Tigris, Nile and Indus strongly relied on annual rainfall and floods for their growth and prosperity [1]. The Egyptian civilization for example, located in the northeastern Sahara since Predynastic time, depended on the Nile as the primary and vital source of water [2].

For this reason, periods of rapid climate change were catastrophic for some empires. For example, it has been discussed in the literature [3], that the decline of the Mycenaean civilization during the late thirteenth century (beginning of the Greek Dark Ages) was caused by a prolonged drought and not because of the invasion of less civilized Dorian tribes. The end of the Old Kingdom in Egypt was simultaneous with changes in the annual flood of the Nile, and that this event may in part have led its collapse [4]. It has also been argued [5] that changes in climate were the most important cause for the fall of the Roman Empire, and that the decline of the Mayan civilization in the ninth century was caused by long lasting periods of drought [6].

Climate change has also had dramatic effects on animal species. One example is the woolly mammoth that inhabited Eurasia and North America from late Middle Pleistocene (300,000 yr BP [1]) and vanished in the Holocene (3,600 BP). They had a drastic decline in size population between the late Pleistocene and the Holocene caused by a collapse of the climatically suitable area. This fact brought a significant drop in mammoth size population and made the animals more vulnerable to human hunting [7].

The study of climate change throughout the history of Earth has been possible using paleoclimatic data, with records taken from fossil corals, ice sheets, tree rings, sediments, rocks, *etc.* Climate has been in constant flux for the past few billion years with events ranging from ice ages to long periods of warmth and changes occurring either rapidly (such

[1] BP: before present

as ice-sheet disintegration in the poles) or gradually over a long period of time [8]. Studies using paleoclimatic data show that the El Niño-Southern Oscillation (ENSO) has been present in the Earth's climate for at least the past 130,000 years [9].

The ENSO system can be considered as a self-sustained oscillator comprising the coupled tropical ocean atmosphere system [1, 10]. A sea surface temperature (SST) gradient between the western Pacific warm pool east of Indonesia and the eastern Pacific cold tongue of the Peruvian coast induces the easterly winds of the Walker circulation. These winds in turn stabilize the temperature gradient by driving surface water masses westwards and upwelling of cold deep water close to South America. Perturbations induced by the annual cycle and noise, excite the so-called Kelvin and Rossby waves. Their propagation produce variability on interannual timescales. The El Niño phenomenon sets on in early summer and usually peaks during winter time. It was given the name El Niño, that means in Spanish "the boy"and refers to the Christ child as this periodic warming is seen in the Pacific near South America around Christmas. El Niño event occurs every two to seven years and usually causes important socio economic impacts. El Niño events are usually followed by a stronger than normal Walker circulation and temperature gradient, called La Niña, the Spanish for "the girl". El Niño and La Niña events are opposite states of the ENSO system: El Niño occurs when the equatorial Pacific is warmer than average and La Niña occurs when it is cooler than average. Once an El Niño or La Niña event develops, it tends to continue for about a year.

Although there is an increasing number of studies evaluating the behavior of ENSO over millennial time scales, the understanding of its complex dynamics is still limited. Here, we study the dynamics of the Holocene proxy ENSO record corresponding to the Pallcacocha Lake sedimentary data [11, 12]. Specifically, we analyze temporal changes in dynamics over the last 11,000 years, using different Information Theory quantifiers. Using wavelet analysis on this ENSO proxy record, Moy *et al.* [11, 12] found a millennial-scale oscillation coherent throughout the Holocene but displaying less significant variance in the early Holocene. They identified a shift around 5,000 BP from a 1,500-yr period in the middle/early Holocene to a 2,000-yr period in the late Holocene. Tsonis [14] analyzed this record using several non-linear dynamic techniques and concluded that the shift in behavior around 5,000 BP could be explained as a bifurcation in the dynamics given by a transition from chaotic to hyper-chaotic dynamics. In the present work, using Information Theory quantifiers we found evidence of a shift in dynamics and cyclic behavior which is consistent with the results of Moy *et al.* [11]. In addition, we have been also able to localize these cycles in time and to further analyze possible connections to epochs of rapid climate change (RCC) during the Holocene [15, 16].

This paper is organized as follows. In Section 2 we describe The Pallcacocha Lake data sedimentary data. The Information Theory quantifiers and the methodologies for the computation of the necessary probability distributions are presented in Sections 3 and 4 respectively. Section 5 presents the results of the The Pallcacocha Lake data analysis and the connections to epochs of rapid climate change (RCC). Conclusions are given in Section 6.

2 The El Niño-Southern Oscillation data

Here we study the dynamics of the Holocene proxy ENSO record corresponding to Pallcacocha Lake sedimentary data [11, 12]. The proxy record was obtained from the analysis of clastic laminae deposition in sediment two 8-m cores retrieved from the Pallcacocha Lake in Ecuador. Moy *et al.* [11] explain that during warm ENSO events, convective precipitation triggers erosion and debris-flow activity increasing the sediment load that is contributed to the lake. It was assumed that the light-colored, inorganic clastic sediments laminae in the sediment core were deposited during the ENSO-driven episodes. This hypothesis was based on the observation that light-colored laminae deposited in the more recent 200 years generally correlated with known moderate to severe El Niño events from instrumental and historical records. To quantify the distribution of the light-colored laminae in the lake sediments, the surface of the cores sections was digitally scanned and the red color intensity was used to generate the record of ENSO variability. Then an age model based on radiocarbon chronology was used to create the time series from 11,000 calendar years BP to today of red color intensity (Holocene period).

As is explained by Moy *et al.* [11, 12], the age-model for the new record is based on the same radiocarbon chronology used by Rodbell *et al.* [17]. The laminae dated by AMS ^{14}C of terrestrial microfossils in the record [17] are distinctive and could be confidently identified in the new cores. By creating a composite section from overlapping drives, Moy *et al.* were able to improve on the original age model. They adopted the constant carbon accu-

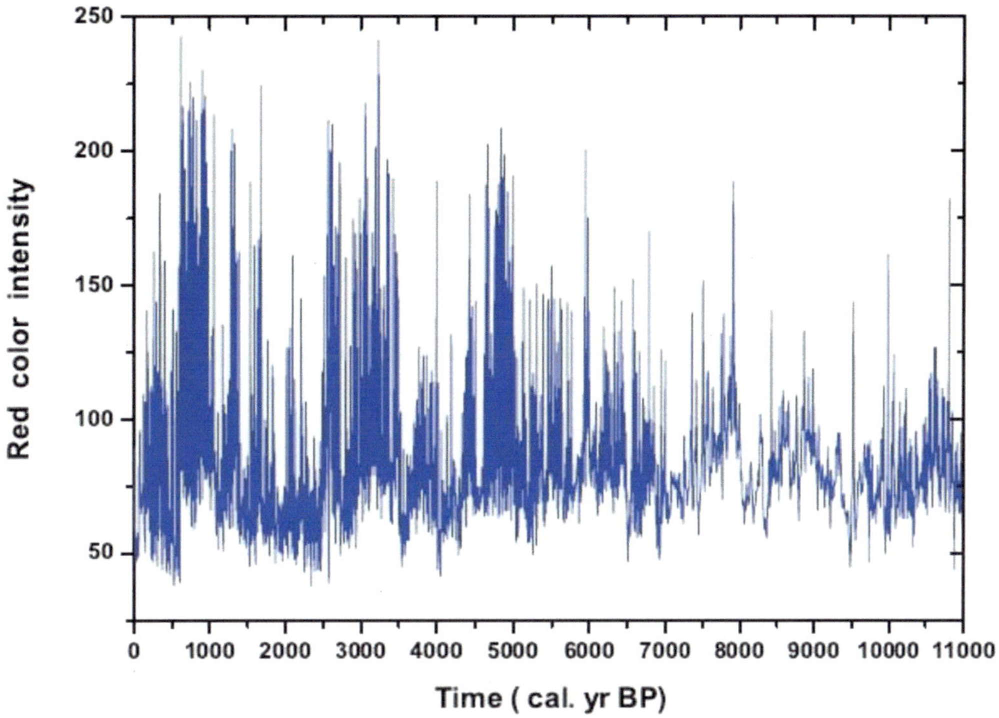

Figure 1: ENSO proxy data from the Pallcacocha Lake in Ecuador [11, 12]. The original data [12] were interpolated to a sample time of one year using a cubic Hermite polynomial.

mulation model and an event model to allocate time between dated intervals [17, 18]. The constant carbon accumulation model assumes that the rate of organic carbon deposition has remained nearly constant between radiocarbon-dated intervals through the Holocene, and this continuous sedimentation was punctuated by nearly instantaneous clastic depositional events.

The original Pallcacocha Lake data [12] has been interpolated to a sample time of one year using a cubic Hermite polynomial. The corresponding time series of $M = 11000$ *data* has been considered in the analysis and is displayed in Fig. 1 ($T_{sample} = 1\ year$). Visual inspection of this ENSO proxy data from the Pallcacocha Lake indicates that during the last 5,000 years there is a switch to more frequent and strong El Niño events (high red colour intensity).

3 Information Theory quantifiers

3.1 Shannon entropy measure

The Statistical Mechanics formulation of entropy is essentially equivalent to that used in Information Theory (see Chapter 2). Moreover, the Information Theory approach enlights many aspects of Statistical Mechanics. The information content of a system is typically evaluated from probability distribution function (PDF). Let $f(x)$ a given continuous probability distribution associated to some measurable or observable property of the system under study. The *Shannon Entropy* S is given by [19]:

$$S[f] \;=\; -\int f\;\ln(f)\;dx\;. \qquad (1)$$

Consider now a given time series $X = \{x_t, t = 1, \cdots, M\}$ of length M and the corresponding discrete probability distribution set $P = \{p_i;\; i = 1, \cdots, N\}$ describing the empirical distribution of X, in which N is the number of possible states. The corresponding Shannon entropy is evaluated as:

$$S[P] \;=\; -\sum_{i=1}^{N} p_i\;\ln(p_i)\;. \qquad (2)$$

The Shannon entropy S is regarded as the measure of the uncertainty associated to the physical processes described by the probability distribution P. If $S[P] = 0$ we are in a position to predict with certainty *which* of the possible outcomes i whose probabilities are given by the p_i will actually take place. Our knowledge of the underlying process described by the probability distribution is in this instance maximal. On the other hand, our ignorance is maximal for a uniform distribution, $P_e = \{1/N, \cdots, 1/N\}$, and in this case $S[P_e] = S_{max} = \ln(N)$. The "Normalized Shannon Entropy" is defined as

$$\mathcal{H}[P] = S[P]/S_{max}\;. \qquad (3)$$

3.2 Fisher information measure

Shannon entropy is a measure of the "global character" (smoothness) in the PDF, in the sense that it is very little sensitive to strong changes on the distribution over a small-sized region. Such is not the case for the *Fisher Information Measure* $\mathcal{F}$ [20], which constitutes a quantifier of the gradient content of the distribution f, and is therefore much more sensitive to strongly localized perturbations. It is defined as:

$$\mathcal{F}[f] = \int \frac{|\vec{\nabla} f|^2}{f} \, dx \; . \tag{4}$$

Fisher information can be variously interpreted as a measure of the ability to estimate a parameter, as the amount of information that can be extracted from a set of measurements, and also as a measure of the state of disorder of a system or phenomenon [21, 22]. Its most important property is the so-called Cramer-Rao bound, that for the sake of simplicity, we recapitulate in only one-dimension. The classical Fisher information associated with translations of a one-dimensional observable x with corresponding probability density $f(x)$ is [23]

$$I_x = \int \mathrm{d}x \, f(x) \left(\frac{\partial \ln f(x)}{\partial x} \right)^2 , \tag{5}$$

which obeys the above referred to Cramer-Rao inequality

$$(\Delta x)^2 \geq I_x^{-1} \tag{6}$$

involving the variance of the stochastic variable x [23],

$$(\Delta x)^2 = \langle x^2 \rangle - \langle x \rangle^2 = \int \mathrm{d}x \, f(x) \, x^2 - \left(\int \mathrm{d}x \, f(x) \, x \right)^2 . \tag{7}$$

The gradient operator significantly determines the contribution of the aforementioned variations of the distribution over its domain to the value of the Fisher information. This is the reason of such a measure $\mathcal{F}$ being usually referred as a "local" descriptor, according to its sensitivity to local characteristics of f. In fact, note that the Shannon entropy decreases for a skewed distribution, instead the Fisher information increases for a skewed distribution. This sensitivity can be useful for situations in which there is a notion of ordering. Such a situation arises in dynamical systems, for which time is a natural ordering variable.

Let $P = \{p_i; \; i = 1, \cdots, N\}$ be a discrete probability distribution set, with N the number of possible states of the system under study. For the evaluation of the Fisher information measure in the case of a discrete PDF, we follow the proposal of Ferri and coworkers [24]

$$\mathcal{F}[P] = \frac{1}{4} \sum_{i=1}^{N-1} 2 \frac{(p_{i+1} - p_i)^2}{(p_{i+1} + p_i)} \; . \tag{8}$$

The associated problem of loss of information due to the discretization has been studied for various measures of information (see for instance [25, 26, 27] and references therein). This entails the loss of the invariance of the Fisher information measure under displacement, which is of no importance for the purpose of our current study.

In the case of a system in a very ordered state, represented by a very narrow PDF, we have Shannon entropy $S \sim 0$ and Fisher information measure $\mathcal{F} \sim \mathcal{F}_{max}$. In the other extreme, when the system under study is in a very disordered state, represented by an almost flat PDF, we have $S \sim S_{max}$ and $\mathcal{F} \sim 0$. We represent by S_{max} and $\mathcal{F}_{max}$ the maximum values for the Shannon entropy and Fisher information measure. In summary, we can say that the general behavior of the Fisher Information is opposite to the Shannon entropy measure.

3.3 Shannon-Jensen Statistical Complexity

Complexity denotes a state of affairs that one can easily appreciate when confronted with it; however, is rather difficult to define it quantitatively, probably due to the fact that there is no universal definition of complexity. In between the two special instances of perfect order and complete randomness, a wide range of possible degrees of physical structure exists that should be reflected in the features of the underlying probability distribution P. One would like to assume that the degree of correlational structures would be adequately captured by some functional $\mathcal{C}[P]$ in the same way that Shannon's entropy [19] "captures" randomness.

In trying to ascertain what is meant by statistical complexity, different measures have been proposed in the literature [28, 29, 30, 31, 32, 33, 34] (see also Chapters 7 and 8). One can begin by excluding processes that are certainly not complex, such as those exhibiting periodic motion. Additionally, a purely random process cannot be considered complex either, notwithstanding its irregular and unpredictable character, because it does not contain any non-trivial, patterned structure. Statistical complexity is often characterized by the paradoxical situation of a complicated dynamics generated from relatively simple systems. Obviously, if the system itself is already involved enough and is constituted by many different parts, it clearly may support a rather intricate dynamics, but perhaps without the emergence of typical characteristic patterns [35]. Therefore, a complex system does not necessarily generate a complex output. Statistical complexity is therefore related to patterned structures hidden in the dynamics, emerging from a system which itself can be much simpler than the dynamics it generates [35].

Assessing the degree of unpredictability and randomness of a system is not automatically tantamount to adequately grasp the correlational structures that may be present, *i.e.*, to be in a position to capture the relationship between the components of a physical system [31, 33]. These correlational structures influence, of course, the functional form of the probability distribution that is able to describe the physics one is interested in. In other words, randomness, on the one hand, and structural correlations on the other one, are not independent aspects of the dynamics. A suitable candidate to "capture" complexity is an information measure that has come to be called the statistical complexity. This complexity, $\mathcal{C}[P]$, vanishes in the two special extreme instances mentioned above (perfect order and complete randomness).

For a given probability distribution, P, we define the amount of "disorder" as the normalized Shannon entropy $\mathcal{H}[P]$ (see eq.(3)). It follows that a reasonable definition of the statistical complexity measure must adopt some kind of distance $\mathcal{D}$ to the uniform distribution P_e [28, 36, 37]. For such a purpose the "disequilibrium", $\mathcal{Q}$, is defined as

$$\mathcal{Q}[P, P_e] = Q_0 \cdot \mathcal{D}[P, P_e] \ , \tag{9}$$

where Q_0 is a normalization constant such that $0 \leq \mathcal{Q} \leq 1$. The disequilibrium $\mathcal{Q}$ reflects on the system's "architecture", being different from zero if there are "privileged", or more likely states among the accessible ones. Consequently, we will adopt the following functional form, introduced originally by López-Ruiz, Mancini and Calbet (LMC) [28] (see Chapter 7), for the statistical complexity measure:

$$\mathcal{C}[P] = \mathcal{H}_S[P] \cdot \mathcal{Q}[P, P_e] \ . \tag{10}$$

This quantity reflects the interplay between the amount of information stored in the system and its disequilibrium.

In the LMC definition of complexity, the disequilibrium, $\mathcal{Q}[P, P_e]$, is computed using the Euclidean distance from P to the uniform distribution P_e. Martín, Plastino, and Rosso [36] proposed a modification of the LMC-complexity consisting of the use of Wootters statistical distance [38] instead of the Euclidean distance. An alternative definition of the disequilibrium was proposed by Lamberti *et al.* [37] using the Jensen-Shannon divergence, which is the symmetric form of the Kullback-Leiber relative entropy, $K[P_1|P_2]$. (*i.e.*, $\mathcal{J}_S[P_1, P_2] = (K[P_1|P_2] + K[P_2|P_1])/2$). For discussion of other simple alternatives of the disequilibrium Q and the corresponding advantages and disadvantages of each ones see [39] (see also Chapter 8).

The statistical complexity that computes the disequilibrium using the Jensen-Shannon divergence, in the following, referred to as Shannon-Jensen statistical complexity (see also Chapter 8), is the one used in this work and has recently been shown [40] to be suitable for distinguishing time series generated by stochastic and by chaotic systems, and able to classify different degrees of "stochasticity". Moreover, the Shannon-Jensen statistical complexity was shown to be suitable for distinguishing Gaussian from non-Gaussian process, and for distinguishing among different degrees of correlations (colored noises).

Summarizing the above discussion, the Shannon-Jensen statistical complexity for a probability distribution $P = \{p_i; i = 1, \cdots, N\}$, with N being the number of possible states of the system, can be cast as

$$\mathcal{C}[P] = \mathcal{H}[P] \cdot \mathcal{Q}_J[P, P_e], \tag{11}$$

in which the disorder is measured by the normalized Shannon entropy and the disequilibrium $\mathcal{Q}_J$ is defined in terms of the Jensen–Shannon divergence $\mathcal{J}_S$ [37]:

$$\mathcal{Q}_J[P, P_e] = Q_0 \cdot \mathcal{J}_S[P, P_e] \ , \tag{12}$$

where

$$\mathcal{J}_S[P_1, P_2] = \{S[(P_1 + P_2)/2] - S[P_1]/2 - S[P_2]/2\} \tag{13}$$

and Q_0 is a normalization constant, equal to the inverse of maximum possible value of $\mathcal{J}_S[P, P_e]$,

$$Q_0 = -2\left\{\left(\frac{N+1}{N}\right)\ln(N+1) - 2\ln(2N) + \ln N\right\}^{-1} . \tag{14}$$

The complexity measure constructed in this way has the intensive property found in many thermodynamic quantities. We stress the fact that the above defined statistical complexity is the product of two normalized entropies (the Shannon entropy and Jensen-Shannon divergence), but is a nontrivial function of the entropy, because it depends on two different probabilities distributions, the one corresponding to the state of the system, P, and the uniform distribution, P_e. Furthermore, it was shown that for a given value of $\mathcal{H}$, there exists a range of possible statistical complexity values between a minimum $\mathcal{C}_{min}$ and a maximum $\mathcal{C}_{max}$ [41]. Thus, evaluating the complexity provides additional insight in the details of a system's probability distribution, which are not discriminated by randomness measures like entropy [40, 41, 42]. In order to study the time evolution of the statistical complexity, a diagram of $\mathcal{C}$ *versus* $\mathcal{H}$ can be used, the $\mathcal{H} \times \mathcal{C}$ plane (in this case, $\mathcal{H}$ can be regarded as an arrow of time [43]). Also, this kind of diagram has been used to study changes in the dynamics of a system originated by modifications of some characteristic parameters [28, 39, 40, 44, 45, 46, 47, 48, 49, 50, 51, 52, 53, 54, 55, 56].

4 Methodologies for the estimation of the PDFs

An important point for the evaluation of the previous information measurements is the proper determination of the underlying probability distribution function (PDF) P, associated to a given dynamical system or time series. This is an often neglected issue that indeed deserves detailed consideration. Why? Because the probability distribution P and the sample space Ω are inextricably linked. Many schemes have been proposed for a proper selection of the probability space (Ω, P). We can mention, among others: *(a)* procedures based on amplitude statistics (histograms) [57], *(b)* binary symbolic dynamics [58], *(c)* Fourier analysis [59] and, *(d)* wavelet transforms [60, 61]. Their applicability depends on particular characteristics of the data such as stationarity, length of the time series, variation of the parameters, level of noise contamination, *etc.* In all these cases the global aspects of the dynamics can be somehow captured, but the different approaches are not equivalent in their ability to discern all the relevant physical details. One must also acknowledge the fact that the above techniques are introduced in a *rather ad hoc fashion and are not directly derived from the dynamical properties themselves of the system under study,* as can be adequately achieved, for instance, by recourse to the Bandt-Pompe methodology [62].

Bandt and Pompe introduced a successful methodology for the evaluation of the PDF associated to scalar time series data using a symbolization technique [62, 63]. The symbolic data is created by ranking the values of the series and defined by reordering the embedded data in ascending order, which is reconstructed with embedding dimension D (see definition and methodological details below). *Causal information* is, consequently, properly incorporated into the "building-up" process that yields (Ω, P). In this way it is possible to quantify the diversity of the ordering symbols (patterns) derived from a scalar time series, evaluating the so called permutation entropy (the Shannon entropy corresponding to the Bandt and Pompe PDF). The Bandt and Pompe technique is computationally fast and does not require the reconstruction of an attractor in phase space. It is the only technique among those most currently used, that takes into account the temporal structure of the time series generated by the physical process under study. An additional advantage of using this technique is that it is based on a very weak stationarity assumption. This characteristic allows to uncover important details concerning the ordinal structure of the time series [40, 50, 55, 56, 57, 64, 65, 66, 67, 68, 69, 70, 71, 72], and can also uncover information about the temporal correlation [52, 53].

In the current work we use two different approaches to capture different aspects of the ENSO proxy dynamics. The first of these is broadly used and computes P using a histogram of amplitudes. The second one, is based on Bandt-Pompe symbolization dynamics.

4.1 PDF based on histograms

In order to extract a PDF *via* amplitude-statistics, the interval $[a, b]$ (with a and b the minumum and maximum of the time series) is first divided into a finite number $nbin$ of non overlapping subintervals A_i: $[a, b] = \bigcup_{i=1}^{nbin} A_i$ and $A_i \bigcap A_j = \emptyset \; \forall i \neq j$. One then employs the usual histogram-method, based on counting the relative frequencies of the time series values within each subinterval. It should be clear that the resulting PDF lacks any information regarding temporal evolution. The only pieces of information available in this case are the x_t-values that allow one to assign inclusion within a given bin, ignoring their temporal location (this is, the subindex i).

Note that N in equations (2) and (8) is equal to N_{bin} and the division procedure of the interval $[a, b]$ provides the natural order sequence for the evaluation of the PDF gradient involved in the Fisher information measure. Let us also point out that it is relevant to consider a judiciously chosen optimal value for N_{bin} (see *i.e.*, De Micco *et al.* [57]). The normalized Shannon entropy obtained *via* the ensuing histogram PDF is called in this paper $\mathcal{H}^{(Hist)}$.

4.2 PDF based on Bandt and Pompe methodology

To use the Bandt and Pompe [62] methodology for the evaluation of the probability distribution P associated to the time series (dynamical system) under study, one starts by considering partitions of the pertinent D-dimensional space that will hopefully "reveal" relevant details of the ordinal-structure of a given one-dimensional time series $X = \{x_t : t = 1, \cdots, M\}$ with embedding dimension $D > 1$. We are interested in "ordinal patterns" of order D

[62, 63] generated by

$$(s) \mapsto \left(x_{s-(D-1)}, x_{s-(D-2)}, \cdots, x_{s-1}, x_s \right) , \qquad (15)$$

which assigns to each time s the D-dimensional vector of values at times $s, s-1, \cdots, s-(D-1)$. Clearly, the greater the $D-$value, the more information on the past is incorporated into our vectors. By "ordinal pattern" related to the time (s) we mean the permutation $\pi = (r_0, r_1, \cdots, r_{D-1})$ of $[0, 1, \cdots, D-1]$ defined by

$$x_{s-r_{D-1}} \le x_{s-r_{D-2}} \le \cdots \le x_{s-r_1} \le x_{s-r_0} . \qquad (16)$$

In order to get a unique result we set $r_i < r_{i-1}$ if $x_{s-r_i} = x_{s-r_{i-1}}$. This is justified if the values of x_t have a continuous distribution so that equal values are very unusual. Otherwise, it is possible to break these equalities by adding small random perturbations. Thus, for all the $D!$ possible permutations π of order D, their associated relative frequencies can be naturally computed by the number of times this particular order sequence is found in the time series divided by the total number of sequences. The probability distribution $P = \{p(\pi)\}$ is defined by

$$p(\pi) = \frac{\sharp\{s | s \le M - D + 1; \ (s), \texttt{ has type } \pi\}}{M - D + 1} . \qquad (17)$$

In this expression, the symbol $\sharp$ stands for "number".

The procedure can be better illustrated with a simple example; let us assume that we start with the time series $\{1, 3, 5, 4, 2, 5, \ldots\}$, and we set the embedding dimension $D = 4$. In this case the state space is divided into 4! partitions and 24 mutually exclusive permutation symbols are considered. The first 4-dimensional vector is $(1, 3, 5, 4)$. According to Eq. (15) this vector corresponds with $(x_{s-3}, x_{s-2}, x_{s-1}, x_s)$. Following Eq. (16) we find that $x_{s-3} \le x_{s-2} \le x_s \le x_{s-1}$. Then, the ordinal pattern which allows us to fulfill Eq. (16) will be $[0, 1, 3, 2]$. The second 4-dimensional vector is $(3, 5, 4, 2)$, and $[3, 0, 2, 1]$ will be its associated permutation, and so on.

For the computation of the Bandt and Pompe PDF we follow the very fast algorithm described by Keller and Sinn in [63], in which the different ordinal patterns are generated in lexicographic ordering and this is the order sequence used in the evaluation of the PDF gradient involved in the Fisher information measure. The information measures, normalized Shannon entropy, Fisher information measure and Shannon-Jensen statistical complexity, obtained *via* the ensuing Bandt and Pompe PDF are called in this paper $\mathcal{H}^{(BP)}$, $\mathcal{F}^{(BP)}$ and $\mathcal{C}^{(BP)}$ respectively.

The Bandt-Pompes methodology is not restricted to time series representative of low dimensional dynamical systems but can be applied to any type of time series (regular, chaotic, noisy, or reality based), with a very weak stationary assumption (for $k = D$, the probability for $x_t < x_{t+k}$ should not depend on t [62]). It also assumes that enough data are available for a correct embedding procedure. Of course, the embedding dimension D plays an important role in the evaluation of the appropriate probability distribution because D determines the number of accessible states $D!$. Also, it conditions the minimum acceptable length $M \gg D!$ of the time series that one needs in order to work with a reliable statistics.

As shown recently by Amigó *et al.* [73] in the case of deterministic one-dimensional maps not all the possible ordinal patterns (as defined using the Bandt and Pompe methodology) can be effectively materialized into orbits, which in a sense makes these patterns "forbidden". Indeed, the existence of these forbidden ordinal patterns becomes a persistent dynamical property. That is, for a fixed pattern length (embedding dimension) D the number of forbidden patterns present in the time series (unobserved patterns) is independent of the series length M. It is interesting to remark that, this persistence is in opposition to other properties such as proximity and correlation, which die out with time in a chaotic (deterministic) dynamics [73]. For example, in the time series generated by the logistic map with $r = 4$ if we consider patterns of length $D = 3$, the pattern $[2, 1, 0]$ is forbidden. That is, the pattern $x_{k+2} < x_{k+1} < x_k$ never appears [73].

In the case of time series generated by an unconstrained stochastic process (uncorrelated process) every ordinal pattern has the same probability of appearance. That is, if the data set is long enough, all the ordinal patterns will appear. In this case, when the number of observations in the time series $M \to \infty$, the associated probability distribution function should be a uniform distribution, and the number of observed patterns will depend only, on the length M of the time series under study. For correlated stochastic processes it is reasonable to assume that the probability of pattern observation will depend not only on the time series length M but also on the correlation structure [72]. The existence of a non-observed ordinal pattern does not qualify it as "forbidden", only as "missing" due to the time series finite length.

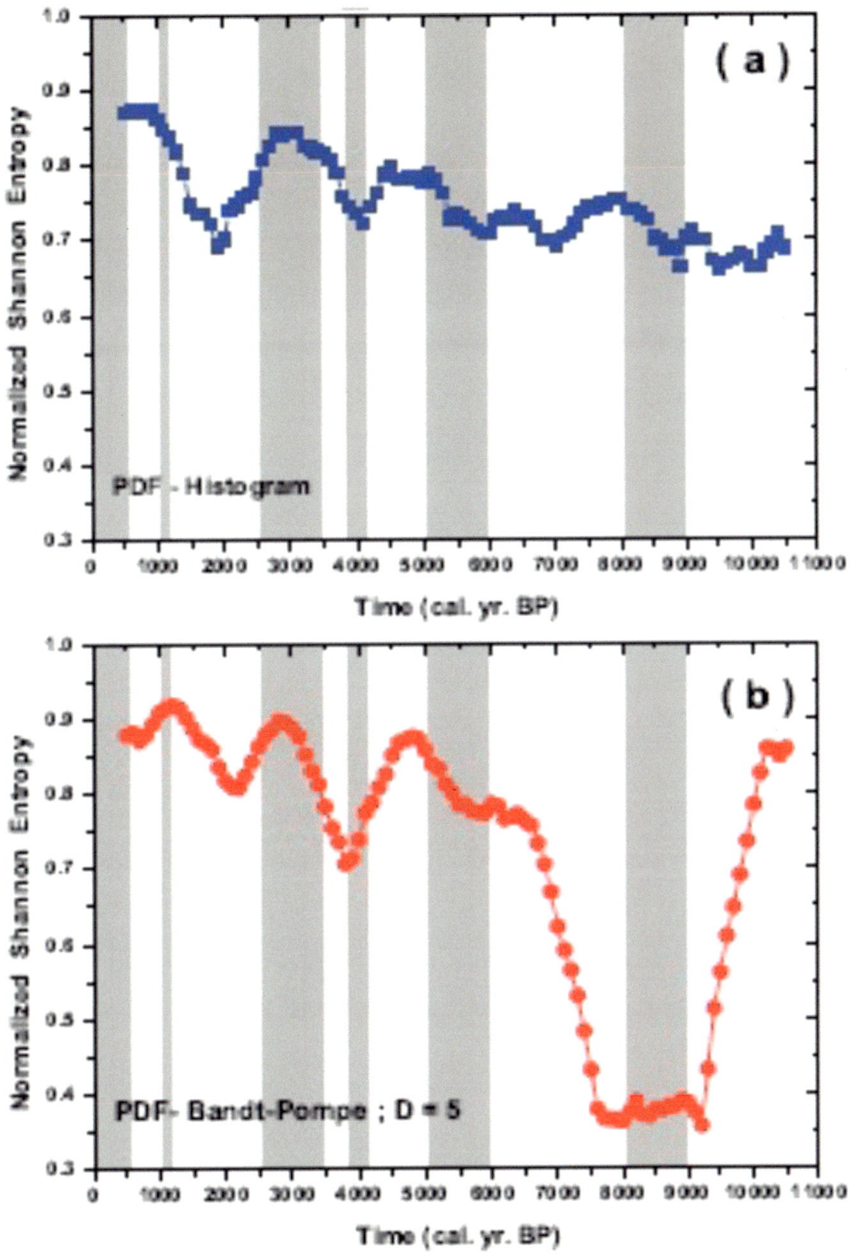

Figure 2: Normalized Shannon entropy for the ENSO proxy time series using a sliding time window of 1000 data values and a lag of 100 data values between successive windows. *a)* For PDF-histogram we consider $N_{bin} = 100$ in the interval $[38.06, 243.06]$ and in the case of *b)* For PDF-Bandt and Pompe we take the ordinal pattern length of $D = 5$. The grey bands correspond to the major periods of Holocene rapid climate change (RCC).

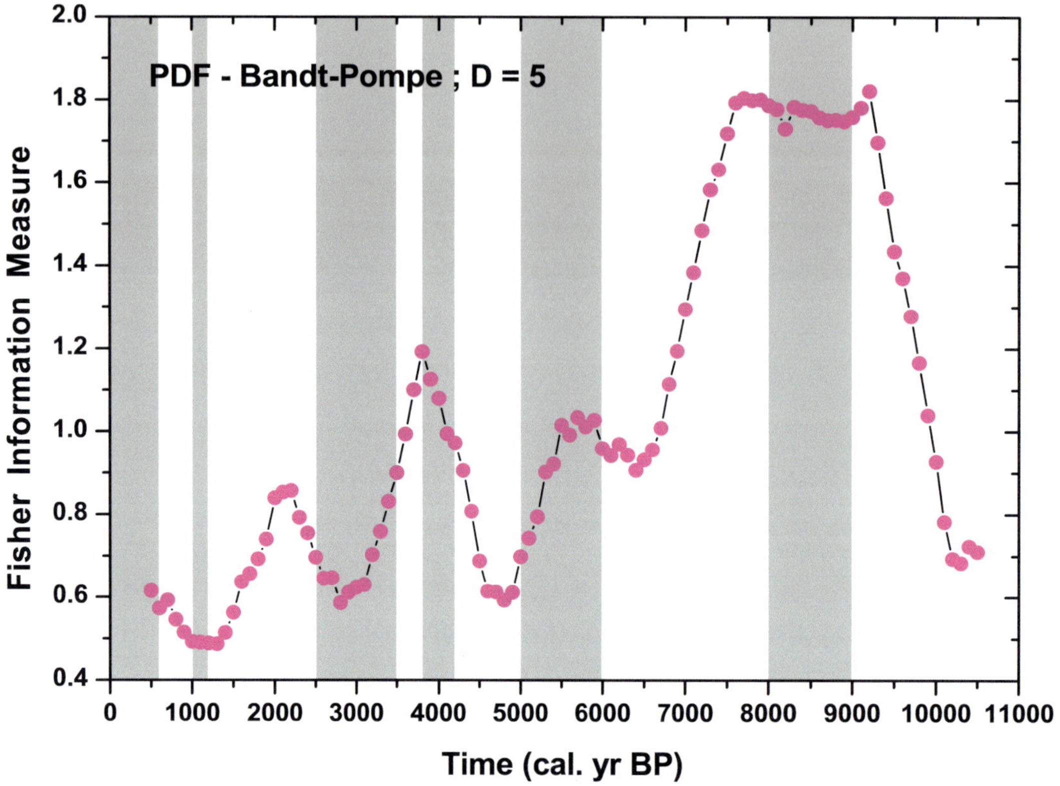

Figure 3: Information Fisher measure time evolution for the ENSO proxy time series using a sliding time window of 1000 data values and a lag of 100 data values between successive windows. PDF-Bandt and Pompe was used in it evaluation with ordinal pattern length of $D = 5$. The grey bands correspond to the major periods of Holocene rapid climate change (RCC).

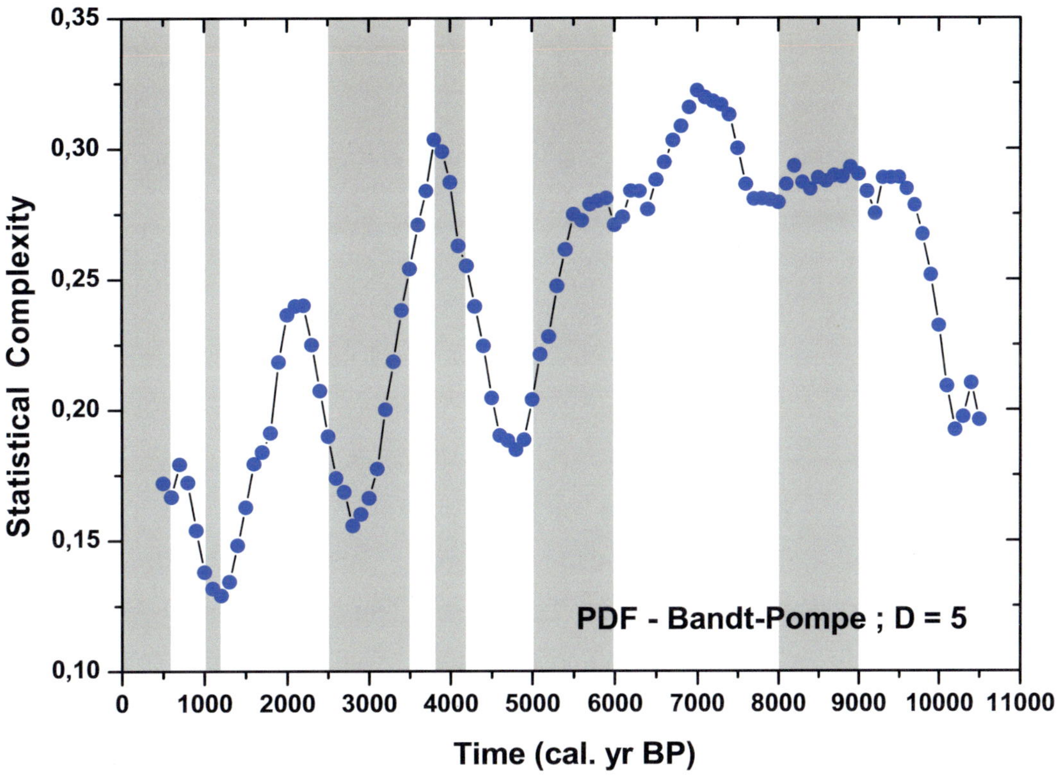

Figure 4: Similar to Fig. 3 for the Shannon-Jensen statistical complexity.

5 Results and Discussion

The evolution of the quantifiers based on Information Theory (normalized Shannon entropy, Fisher information measure and Shannon-Jensen statistical complexity) corresponding to the proxy ENSO data (Fig. 1) were evaluated using both amplitude histograms and the Bandt-Pompe methodology to obtain the corresponding PDFs [74]. We computed the evolution of above quantifiers through time using a sliding time window of 1000 data (1000 $years$) values with a lag of 100 data (100 $years$) values between successive windows. A sliding window was used to allow for a better localization of changes in the dynamics of ENSO through time. The amplitude histogram was evaluated using $N_{bins} = 100$ in the interval $[38.06, 243.06]$ (corresponding to the minimum and maximum amplitude values of the all time series).

ENSO is a climate pattern that occurs across the tropical pacific ocean on average every five years, but over a period which varies from three to seven years. Then for the estimation of the Bandt & Pompe PDF, we used a length of patterns (embedding dimension) of $D = 5$. This is the same value used by Tsonis *et al.* for the embedding dimension in the characterization of this ENSO proxy time series using nonlinear dynamics metrics tools [13, 14].

The pattern and window lengths were selected to obtain a good temporal localization of changes in the dynamics (up to a millennial time scales) without compromising the data needs to obtain reliable statistics for the permutation entropy quantifier. A similar procedure was used for the evaluation of the Fisher information measure and Shannon-Jensen statistical complexity. We should point out that the results obtained when we considered $D = 4$ and $D = 6$, were similar to those shown below, however, in the last case, it was necessary to increase the window length in order to satisfied the condition $M \gg D!$.

As explained in [74], the main difference between both entropy measures (and also in relation with the others quantifiers) is that, unlike $\mathcal{H}^{(Hist)}$, $\mathcal{H}^{(BP)}$ is invariant with respect to strictly monotonous distortions of the data. We should point out that the present proxy data is based on "red color intensity" of sedimentary data and related to rainfall by a function which is unknown but very likely to be nonlinear. However, due to its invariance, $\mathcal{H}^{(BP)}$ applied to any other data series that is strictly monotonously related to "red color intensity" would yield the same results. Therefore, in general, ordinal data processing tools based on ranked numbers would lead to results that are more robust (and useful) for proxy data analysis than those obtained using tools based on metric properties.

The results of normalized Shannon entropy, evaluated with PDF histograms and PDF Bandt-Pompe, are displayed in Fig. 2. We also show in this figure, as grey bands, the major periods of Holocene rapid climate change (RCC) defined by [15] from glacier fluctuation records (tuned to high-resolution GISP2 record). RCCs periods were identified by [15] at 9000–8000 BP, 6000–5000 BP, 4200–3800 BP, 3500–2500 BP, 1200–1000 BP, and since 600 BP. These periods have been frequently used as a framework for the examination of Holocene climate variability [16].

Normalized Shannon entropy evaluated with PDF-histogram, $\mathcal{H}^{(Hist)}$ (Fig. 2.a), shows an overall increasing trend for the whole record (11000 BP to present) that can be interpreted as an increase in uncertainty. This result is in agreement with previous results from Tsonis [14], who found a decrease in predictability between the first (11000-5000 BP) and second period (5000 BP-present). However, the results from Fig. 2.a also show that the increase in $\mathcal{H}^{(Hist)}$ (or decrease in predictability) is not gradual throughout the period and there are substantial fluctuations or cycles with shorter periods of increasing and decreasing entropy. These cycles have an approximate period of 2000 years (9000 to 7000 BP, 6000 to 4000 BP, 4000 to 2000 BP). Within cycles of $\mathcal{H}^{(Hist)}$, there is tendency for local entropy maxima (or minimum predictability) to coincide with periods of rapid climate change (RCC). Both the local maxima and the amplitude of the "cycles" in $\mathcal{H}^{(Hist)}$ also show an increasing trend towards the present. An interesting observation is that most RCC periods seem to be associated to high rates of entropy ($\mathcal{H}^{(Hist)}$) gain, suggesting that variability and uncertainty of ENSO increased during these RCCs.

As seen in Fig. 2.b the normalized Shannon entropy evaluated with PDF-Bandt-Pompe, $\mathcal{H}^{(BP)}$, points out a quite different ENSO dynamics between the early (11000–6500 BP) and middle-to-late Holocene (6500 BP to present) that is not captured by the $\mathcal{H}^{(Hist)}$ quantifier. The period 11000-6500 BP presents the lowest variability, see Fig. 1, as well as an obvious trend around 9500-8800 BP (upward) and 8800-8000 BP (downward). As mentioned before, the Bandt & Pompe methodology is based on ordinal data analysis, and unlike $H^{(Hist)}$, displays robustness against trends present in the data.

The initial values of $\mathcal{H}^{(BP)}$ at the early Holocene are quite high ($\mathcal{H}^{(BP)} = 0.85$). These values indicate that the ordinal patterns (describing the relative magnitude of five successive events) are quite unpredictable. Soon after 10000 BP, a rapid decrease in permutation

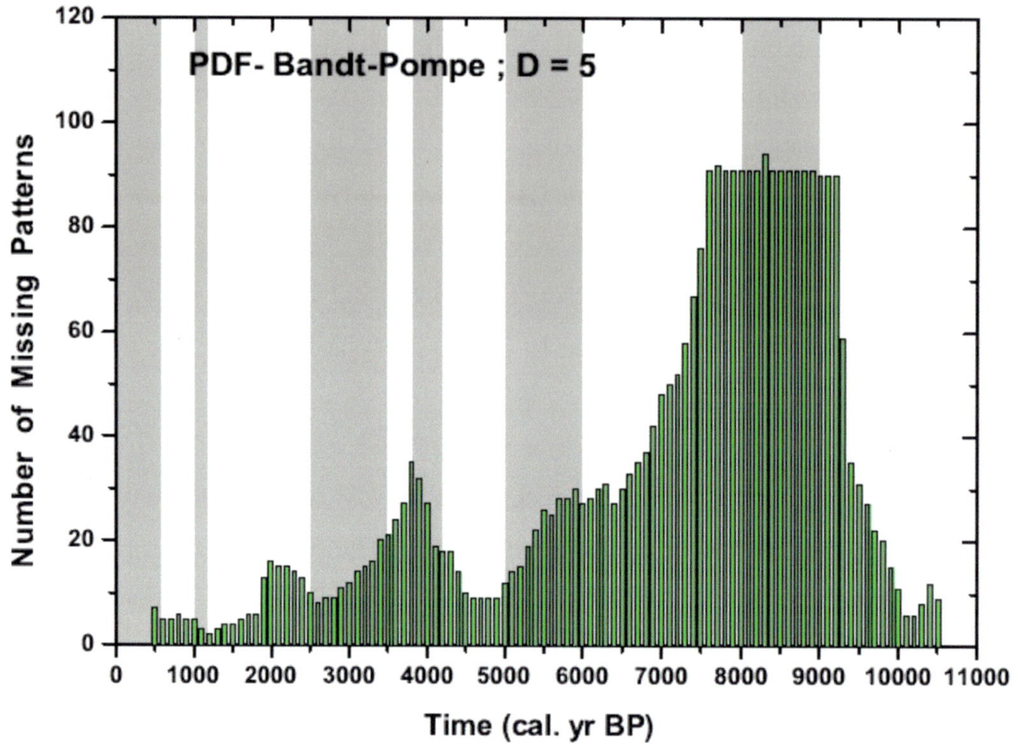

Figure 5: Missing ordinal patterns obtained using the Bandt and Pompe methodology on the ENSO proxy time series. We used a sliding time window of 1000 data values, a lag of 100 data values between successive windows, and an ordinal pattern length of $D = 5$. The grey bands correspond to the major periods of Holocene rapid climate change (RCC).

entropy occurs over a period of less than 1000 years reaching a minimum $\mathcal{H}^{(BP)}$ at around 9200 BP. This is followed by an interval of remarkable quasi steady entropy occurring over more than 1500 years, with a significantly low entropy value $\mathcal{H}^{(BP)} \sim 0.37$. The dramatic decrease in the $\mathcal{H}^{(BP)}$ quantifier suggests a remarkable shift in dynamics towards a more organized or predictable condition which lasts for more than a millennia.

As discussed in [74], further inspection of the Bandt and Pompe ordinal patterns during this period (9500–7500 BP), show the existence of numerous missing ordinal patterns (Fig. 5). After this time interval, a decreasing trend in the number of missing patterns is observed, with exception of two peaks centered around 4000 BP and 2000 BP. As explained in the previous section, missing ordinal patterns tend to appear in more persistent time series (evident by visual inspection of Fig. 1) from either chaotic dynamics or random be-

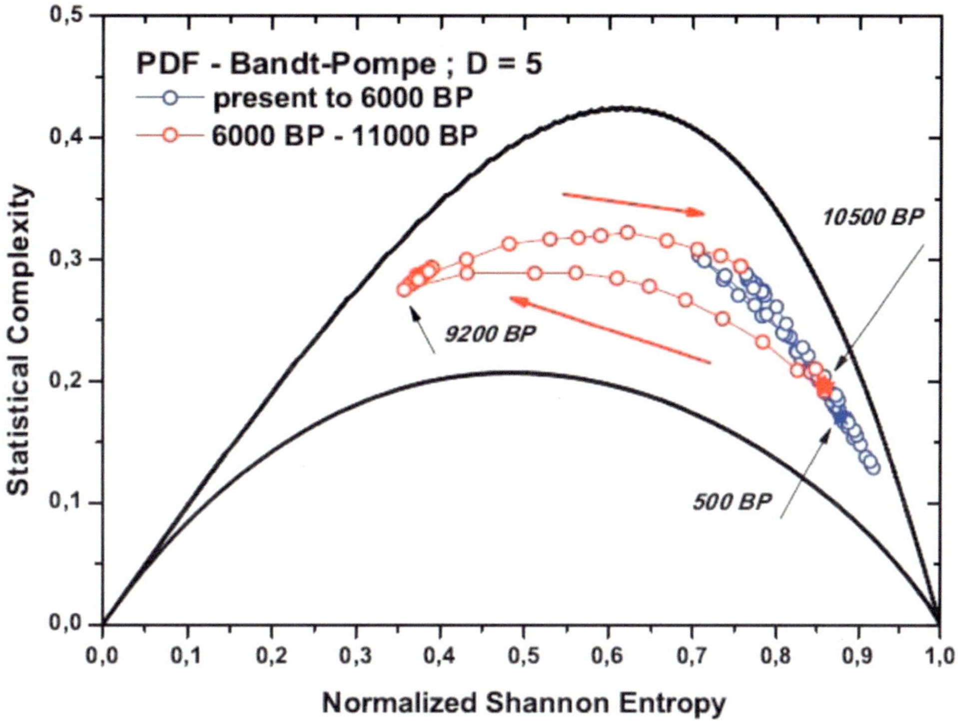

Figure 6: The time evolution of the the ENSO proxy time series in the causality plane $\mathcal{H}^{(BP)} \times \mathcal{C}^{(BP)}$. The maximum ($\mathcal{C}_{max}$) and minimum complexity ($\mathcal{C}_{min}$) are drawn (continuous line) as a function of the normalized Shannon entropy $\mathcal{H}^{(BP)}$. PDF-Bandt and Pompe with ordinal pattern length of $D = 5$ ($N = 120$) was considered.

havior with long term correlations [72]. We should note that centered within this interval of quasi steady low $\mathcal{H}^{(BP)}$ with numerous missing ordinal patterns, is the 9000-8000 BP RCC period. Therefore, these results suggest that during this RCC period, the ENSO record has a dynamic behavior that is distinctively different from that during the others RCCs of the Holocene. Following the end of this first Holocene RCC period, at around 7500 BP, there is another important change in dynamics that results in an abrupt increase in $\mathcal{H}^{(BP)}$ (during a 1000 year period) to a value slightly higher than 0.75 at 6500 BP.

As opposed to the early Holocene, during the period starting in 6500 BP to the present, the permutation entropy quantifier, $\mathcal{H}^{(BP)}$, follows a cyclic behavior similar to that of $\mathcal{H}^{(Hist)}$. During the mid-to-late Holocene period, $\mathcal{H}^{(BP)}$ displays its highest regularity and predictability at 4000 BP (RCC).

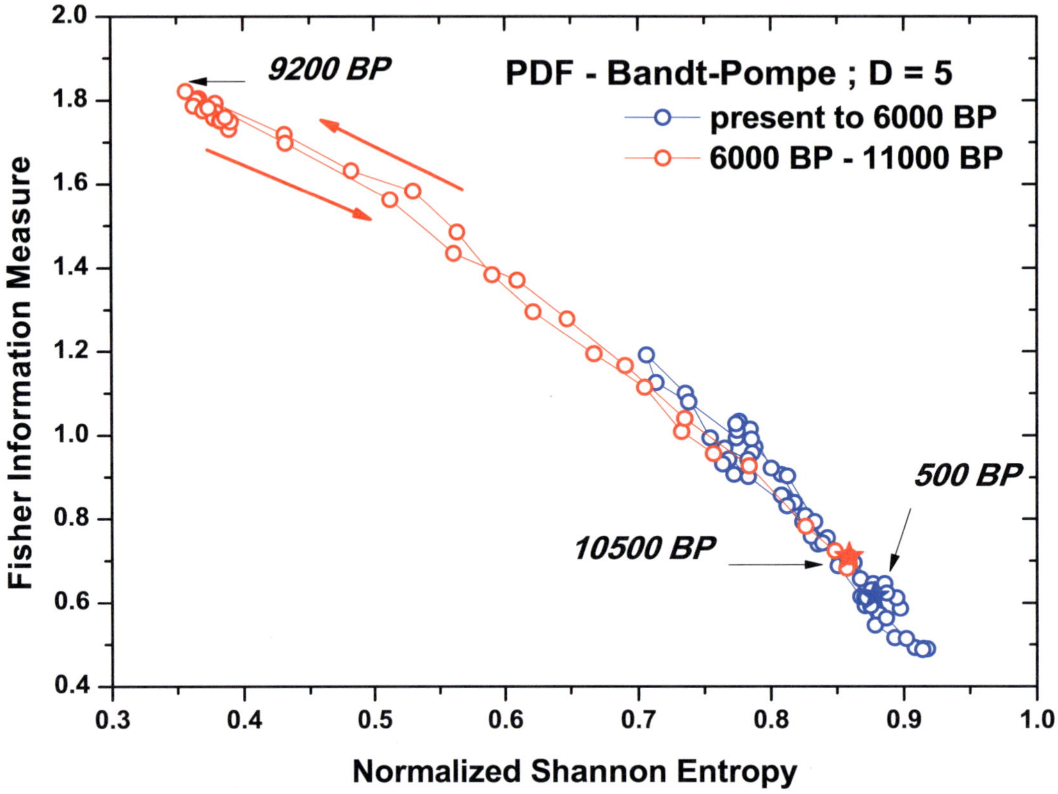

Figure 7: The time evolution of the the ENSO proxy time series in the causality plane $\mathcal{H}^{(BP)} \times \mathcal{F}^{(BP)}$. PDF-Bandt and Pompe with ordinal pattern length of $D = 5$ was considered.

The temporal evolution over the Holocene of the Fisher information measure and Shannon-Jensen statistical complexity evaluated with PDF-Bandt-Pompe, $\mathcal{F}^{(BP)}$ and $\mathcal{C}^{(BP)}$ are displayed in Figs. 3 and 4 respectively. The Fisher information measure $\mathcal{F}^{(BP)}$ (see Fig. 3) presents maximum values centered during the periods 9500–7500 BP and 6000–5500 BP, as well as around 3800 BP and 2200 BP. This shows a good agreement between the peaks of Fisher information and three of the RCC periods. Clearly, a different behavior is observed before and after 6500 BP, with oscillatory characteristics in the middle and late Holocene. Similar trends are observed for the Shannon-Jensen statistical complexity $\mathcal{C}^{(BP)}$ (see Fig. 4), with a quite stable complexity ($\mathcal{C}^{(BP)} \sim 0.28$) in the early to middle Holocene (9500–6000 BP).

The entropy quantifiers based on PDF-Histogram and PDF-Bandt-Pompe point out to the existence of cycles (periods of increasing and decreasing entropy) with a period close

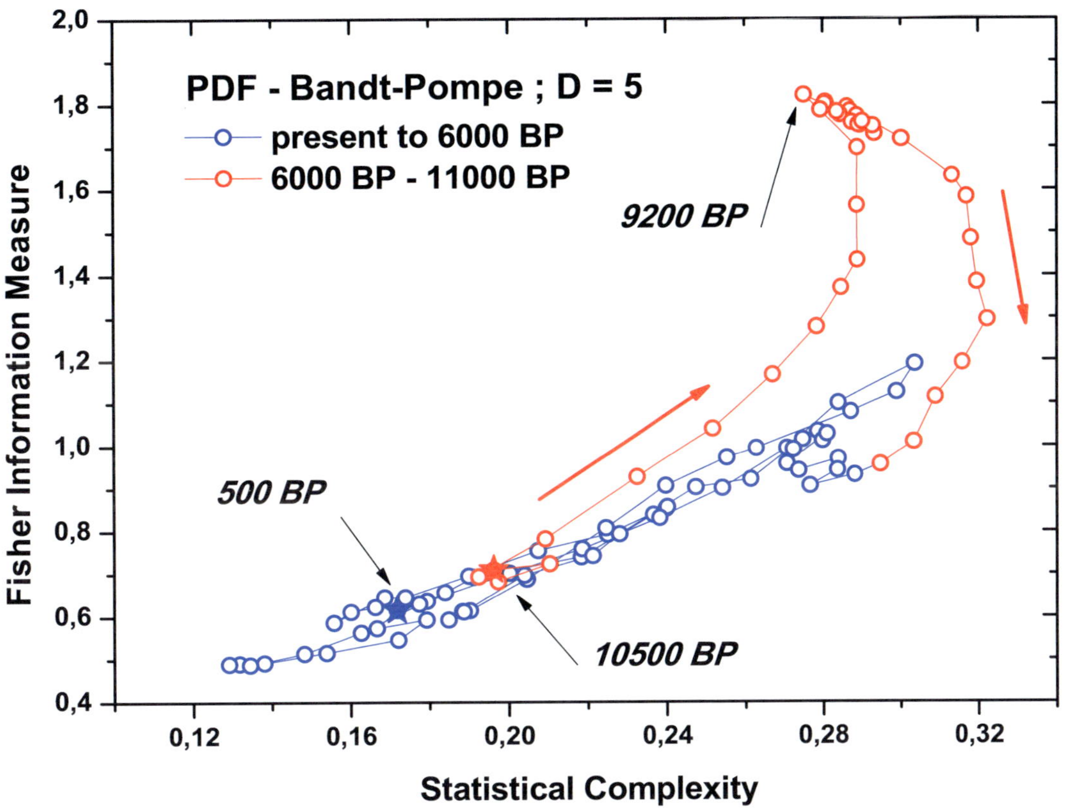

Figure 8: The time evolution of the the ENSO proxy time series in the causality plane $\mathcal{C}^{(BP)} \times \mathcal{F}^{(BP)}$. PDF-Bandt and Pompe with ordinal pattern length of $D = 5$ was considered.

to 2000 years during the mid-to-late Holocene ([74]). Here we find that a similar behavior is observed in the Fisher information measure, as well as in the Shannon-Jensen statistical complexity. These cyclic dynamics is consistent with that observed by Moy and co-workers [11] using wavelet analysis.

An interesting observation, is that within these two cycles there is a tendency for high rates of entropy gain to occur during the two longer RCC periods of the mid-to-late Holocene (6000-5000 and 3500-2500 BP). This indicates that both amplitude variability, measured by $\mathcal{H}^{(Hist)}$, and pattern variability, measured by $\mathcal{H}^{(BP)}$, of ENSO tend to increase (predictability decreases) during these the two longer RCCs. As opposed to the other minor RCCs, the major RCC events (6000-5000 and 3500-2500 BP) coincide with maxima in the $\Delta^{14}C$ and ^{10}Be events that suggest a decline in solar output at these times which might be responsible by the increased in aridity the South American tropics suggested by several proxy records [16].

The temporal evolution of the ENSO proxy time series (Fig. 1) in $\mathcal{H}^{(BP)} \times \mathcal{C}^{(BP)}$ phase space is shown in Fig. 6. The plot also displays the lines of maximum and minimum (possible) values for the complexity $\mathcal{C}_{max}$ and $\mathcal{C}_{min}$ as a function of $\mathcal{H}^{(BP)}$ [41]. It becomes clear from this figure that the ENSO proxy data displays a different dynamical behavior for time intervals 11000–6000 BP and 6000 BP to present. The first time interval (11000–6000 BP) is characterized by a loop in the entropy-complexity causality plane. The ENSO "trajectory/orbit" tends to remain close to $\mathcal{C}_{min}$ during the very early Holocene (10500–8400 BP), while it tends to approach the line of maximum complexity $\mathcal{C}_{max}$ as it evolves towards the end of the interval (8400–6000 BP).

As can be see in Fig. 6, the points corresponding to the first RCC (9000–8000 BP) are all located close to the left corner of the "orbit/trajectory" loop. Note that the point corresponding to 9200 BP corresponds to the point of return in the loop. These results are of an instructive character, notwithstanding the fact that a clear frontier between the two phases is not always easily discernible from the ENSO data series (see Fig. 1). An oscillatory behavior is observed for the interval 6000 to present. The ENSO trajectory is almost equidistant from minimum and maximum complexity path. For a given entropy, we found that the values of complexity of the ENSO proxy are higher that those corresponding to correlated pure noises, like fBm and fGn (see Rosso *et al.* [40] and Zunino *et al.* [68]). In consequence, we can hypothesize a mixed dynamics of determinism and correlated noise for the ENSO record.

Moreover, in order to visually analyze the dynamics of the ENSO proxy, we can get important complementary insights *via* the Vignat-Bercher's Shannon-Fisher $\mathcal{H}^{(BP)} \times \mathcal{F}^{(BP)}$ plane [75, 76, 77, 78, 79] and also by introducing a new complexity-Fisher plane, $\mathcal{C}^{(BP)} \times \mathcal{F}^{(BP)}$, that we will employ to characterize our different time ENSO dynamic-regimes (the time parameter does not explicitly appear in the graph, of course). Figs. 7 and 8 show these two causality planes for the ENSO dynamic evolution. It is clear from these figures that two different regimes are observed in the ENSO dynamics, one corresponding to 11000 – 6000 BP and other from 6000 BP to present. From these results we can see that a wealth of geographic information regarding plane-locations for different kinds of dynamics becomes easily identifiable.

6 Conclusions

The present work investigate the dynamics of the Holocene proxy of El Niño-Southern Oscillation (ENSO) record corresponding to Pallcacocha Lake sedimentary data [11, 12]. As can be seen in the ENSO proxy time series depicted in Fig. 1, in which higher values of red colour intensity represent strong El Niño events, a clear switch in the dynamics take place in the last $5,000$ showing more frequent and stronger El Niño event. In this work we are particularly interested in using several Information Theory quantifiers in order to get new insights into the ENSO dynamics during the Holocene period. For that, the original Pallcacocha Lake time series has been interpolated to a sample of one year using Hermite polynomial resulting in a new time series containing $11,000$ data points. Then, the normalized Shannon entropy, the Fisher information measure and Shannon-Jensen statistical complexity were computed using both amplitude histograms and the Bandt and Pompe methodology to obtain the corresponding probability distribution function. As can be seen in the previous sections, these Information Theory quantifiers point out to varying aspects of the dynamics of the ENSO proxy during the Holocene. Important details in the ENSO dynamics can be easily identified using the causality planes entropy-complexity, entropy-Fisher and complexity-Fisher. For example, all quantifiers reveal the existence of cycles of a period close to $2,000$ years during the Holocene, that is consistent with previous results by Moy [11]. It is also found that the dynamics of the ENSO proxy record during the RCC interval 9000-8000 BP is remarkably different from that of the others RCCs of the Holocene. As explained by Mayewski *et al.* [16], this RCC interval is the only one that coincides with a significant increase in volcanic aerosol production, and it occurred when bipolar ice sheet dynamics still had the potential for substantial effects on global climate. Both entropy quantifiers, $\mathcal{H}^{(Hist)}$ and $\mathcal{H}^{(BP)}$, are low during this period and the number of missing patterns is very high suggesting higher predictability of ENSO. In agreement with this observed behavior, the Fisher information measure $\mathcal{F}^{(BP)}$ presents the highest values in this interval, and the Shannon-Jensen statistical complexity $\mathcal{C}^{(BP)}$ presents high stable values. As can be observed, several important details regarding the variability of ENSO dynamics during the Holocene emerge with the use of Information Theory quantifiers, specially in combination with the Bandt and Pompe methodology that naturally includes temporal causality.

Conflict of interest: None.

Acknowledgments

We wish to thank Dr. Martín Gómez Ravetti, Dr. Carlos Riveros and Dr. José Rodriguez for very useful discussions and comments on the current manuscript. This research has been partially supported by a scholarship from The University of Newcastle awarded to L. C. Carpi . A. Figliola and O. A. Rosso acknowledge partial support from the Consejo Nacional de Investigaciones Científicas y Técnicas (CONICET), Argentina. OAR gratefully acknowledges support from CAPES, PVE fellowship, Brazil.

References

[1] Maraun, D. *What Can We Learn from Climate Data? Methods for Fluctuation, Time/Scale and Phase Analysis*, PhD Thesis, University of Potsdam, Potsdam, Germany, p. 62, 2006.

[2] Butzer, K.W. *Early hydraulic civilization in Egypt*, Chicago University Press, Chicago, IL, 1976.

[3] Carpenter, R. *Discontinuity in Greek civilization*, Cambridge University Press, London, p. 88, 1966.

[4] Stanley, J.D.; Krom, M.D.; Cliff, R.A.; Woodward, J.C. Short Contribution: Nile Flow Failure at the End of the Old Kingdom, Egypt: Strontium Isotopic and Petrologic Evidence, *Geoarchaeology* **18** e3 (2003).

[5] Armento, B.J. *A Message of Ancient Days*, Houghton-Mifflin, Boston, p. 85, 1991.

[6] Haug, G.H.; Gunther, D.; Peterson, L.C.; Sigman, D.M.; Hughen, K.A.; Aeschlimann, B. Climate and Collapse of the Maya Civilization, *Science* **299** 1731–1735 (2003).

[7] Nogués-Bravo, D.; Rodríguez, D,J.; Hortal, J.; Batra, P.; Araújo, M.B. Climate Change, Humans, and the Extinction of the Woolly Mammoth, *PLoS Biol.* **6** e79 (2008).

[8] Das, M.; Parthasarathy, S. Anomaly Detection and Spatio-Temporal Analysis of Global Climate System, *Proceedings of the Third International Workshop on Knowledge Discovery from Sensor Data*, Paris, 142–150 (2009).

[9] Cane, M.A. The evolution of El Niño, past and future, *Earth an Planetary Science Letters* **164** 1–10 (2004).

[10] Bjerknes, J. Atmospheric teleconnections from the equatorial Pacific, *Mon. Wea. Rev.* **97** 163–172 (1969).

[11] Moy, C.M.; Seltzer, G.O.; Rodbell, D.T.; Anderson, D.M. "Variability of El Niño Southern Oscillation activity at millennial time scales during the Holocene epoch", *Nature*, **420** 162–165 (2002).

[12] Moy, C.M.; Seltzer, G.O.; Rodbell, D.T.; Anderson, D.M. Laguna Pallcacocha Sediment Color Intensity Data, *IGBP PAGES/World Data Center for Paleoclimatology Data Contribution Series 2002-76.* NOAA/NCDC Paleoclimatology Program, Boulder CO, 2002. Data available online at: http://www.ngdc.noaa.gov/paleo/pubs/moy2002.

[13] Wang, G.; Tsonis, A.A. On the variability of ENSO at millennial time scales, *Geophys. Res. Lett.* **35** L17702 (2008).

[14] Tsonis, A.A. Dynamical changes in the ENSO system in the last 11,000 years, *Clymate Dynamics* **33** 1069–1074 (2008).

[15] Denton, G.H. Karlén, W. Holocene climatic variations: their pattern and possible cause, Quaternary Research **3** 155–205 (1973).

[16] Mayewski, P.A.; Rohling, E.E.; Stanger, J.C.; Karlén, W.; Maasch, K.A.; Meeker, L.D.; Meyerson, E.A.; Gasse, F.; van Kreveld, S.; Holmgren, K.; Lee-Thorp, J.; Rosqvist, G.; Rack, F.; Staubwasser, M.; Schneider, R.R.; Steig, E.J. Holocene climate variability, *Quaternary Research* **62** 243–255 (2004); and references therein.

[17] . Rodbell, D.T.; Seltzer, G.O.; Anderson, D.M.; Abbott, M.B.; Enfield, D.B.; Newman, J.H.; An ~15,000-year record of El Niño-driven alluviation in southwestern Ecuador, *Science* **283** 516–520 (1999).

[18] Moy, C.M. *A Continuous Record of Late-Quaternary El Nino-Southern Oscillation from the Southern Ecuadorian Andes*, M.S. Thesis, Syracuse University, USA, 2000.

[19] Shannon, C.E. A mathematical theory of communication, *Bell Syst. Technol. J.* **27** 379-423, 623-56 (1948).

[20] Fisher, R.A. On the mathematical foundations of theoretical statistics, *Philos. Trans. R. Soc. Lond. Ser. A* **222** 309–368 (1922).

[21] Frieden, B.R. *Physics from Fisher Information: A Unification*, Cambridge University Press, Cambridge, 1998.

[22] Mayer, A.L.; Pawlowski, C.W.; Cabezas, H. Fisher Information and dinamic regime changes in ecological systems, *Ecological Modeling* **195** 72–82 (2006).

[23] Hall, M.J.W. Quantum properties of classical Fisher information, *Phys. Rev. A* **62** 012107 (2000).

[24] Ferri, G.I.; Pennini, F.; Plastino, A. LMC-complexity and various chaotic regimes, *Phys. Lett. A* **373** 2210–2214 (2009).

[25] Zografos, K.; Ferentinos, K.; Papaioannou, T. Discrete approximations to the Csiszár, Renyi, and Fisher measures of information, *Canad. J. Stat.* **14** 355–366 (1986).

[26] Pardo, L.; Morales, D.; Ferentinos, K.; Zografos, K. Discretization problems on generalized entropies and R-divergences, *Kybernetika* **30** 445–460 (1994).

[27] Madiman, M.; Johnson, O.; Kontoyiannis, I. Fisher Information, compound Poisson approximation, and the Poisson channel, *IEEE Int. Symp. Inform. Theory, ISIT 2007* 976–980 (2007).

[28] López-Ruiz, R.; Mancini, H.L.; Calbet, X. A statistical measure of complexity, *Phys. Lett. A* **209** 321–326 (1995).

[29] Grassberger, P. Toward a quantitative theory of self-generated complexity, *Int. J. Theo. Phys.* **25** 907–938 (1986).

[30] Grassberger, P. How to measure self-generated complexity, *Physica A* **140** 319–325 (1986).

[31] Crutchfield, J.P.; Young, K. Inferring statistical complexity, *Phys. Rev. Lett.* **63** 105–108 (1989).

[32] Wackerbauer, R.; Witt, A.; Altmanspacher, H.; Kurths, J.; Scheingraber, H. Comparative classification of complexity-measures, *Chaos, Solitons & Fractals* **4** 133–173 (1994).

[33] Feldman, D.P. Crutchfield, J.P. Measures of statistical complexity: Why?, *Phys. Lett. A* **238** 244–252 (1998).

[34] Shiner, J.S.; Davison, M. Landsberg, P.T. Simple measure for complexity, *Phys. Rev. E* **59** 1459 (1999).

[35] Kantz, H.; Kurths, J.; Meyer-Kress, G. *Nonlinear Analysis of Physiological Data*, Springer, Berlin, 1998.

[36] Martín, M.T.;Plastino, A.; Rosso, O.A. Statistical complexity and disequilibrium, *Phys. Lett. A* **311** 126–132 (2003).

[37] Lamberti, P.W.; Martín, M.T.; Plastino, A.; Rosso, O.A. Intensive entropic nontriviality measure, *Physica A* **334** 119–131 (2004).

[38] Wootters, W.K. Statistical distance and Hilbert space, *Phys. Rev. D* **23** 357–362 (1981).

[39] Rosso, O.A.; Martín, M.T.; Figliola, A.; Keller, K.; Plastino, A. EEG analysis using wavelet-based information tools, *J. Neurosc. Meth.* **153** 163–182 (2006).

[40] Rosso, O.A.; Larrondo, H.A.; Martín, M.T.; Plastino A.; Fuentes, M.A. Distinguishing noise from chaos, *Phys. Rev. Lett.* **99** 154102 (2007).

[41] Martín, M.T.; Plastino, A.; Rosso, O.A. Generalized statistical complexity measures: Geometrical and analytical properties, *Physica A* **369** 439–462 (2006).

[42] Feldman, D.P.; McTague, C.S.; Crutchfield, J.P. The organization of intrinsic computation: Complexity-entropy diagrams and the diversity of natural information processing, *Chaos* **18** 043106 (2008).

[43] Plastino, A.R.; Plastino, A. Symmetries of the FokkerPlank equation and FisherFrieden arrow of time, *Phys. Rev. E* **54** 4423–4326 (1996).

[44] Calbet, X.; López-Ruiz, R. Tendency towards maximum complexity in a nonequilibrium isolated system, *Phys. Rev. E* **63** 066116 (2001).

[45] Rosso, O.A.; Martín, M.T.; Plastino, A. Brain electrical activity analysis using wavelet based informational tools (II): Tsallis non-extensivity and complexity measurements, *Physica A* **320** 497–511 (2003).

[46] Kowalski, A.M.; Martín, M.T.; Plastino, A.; Proto, A.N.; Rosso, O.A. Wavelet statistical complexity analysis of classical limit, *Phys. Lett. A* **311** 180–191 (2003).

[47] Rosso, O.A.; Martín, M.T.; Plastino, A. Evidence of self-organization in brain electrical activity using wavelet based informational tools, *Physica A* **347** 444–464 (2005).

[48] Plastino, A.;Rosso, O.A. Entropy and statistical complexity in brain activity, *Europhysics News* **36** 224–228 (2005). Special Number on: "Nonextensive Statistical Mechanics - Theory and Applications", Boon, J.P.; Tsallis, C. (Guest Editors).

[49] Kowalski, A.M.; Martín, M.T.; Plastino, A.; Rosso, O.A. Entropic non-triviality, the classical limit, and geometry-dynamics correlations, *Int. J. of Modern Phys. B* **14** 2273–2285 (2005).

[50] Kowalski, A.M.; Martín, M.T.; Plastino, A.; Rosso, O.A. Bandt-Pompe approach to the classical-quantum transition, *Physica D* **233** 21–31 (2007).

[51] Zunino, L.; Pérez, D.G. Martín, M.T.; Plastino, A.; Garavaglia, M.; Rosso, O.A. Characterization of gaussian self-similar stochastic processes using wavelet-based informational tools, *Phys. Rev. E* **75** 021115 (2007).

[52] Rosso, O.A.; Masoller, C. Detecting and quantifying temporal correlations in stochastic resonances by Information Theory measures, *European Physics Journal B* **69** 37–43 (2009).

[53] Rosso, O.A.; Masoller, C. Detecting and quantifying stochastic and coherence resonances via Information Theory complexity measurements, *Phys. Rev. E* **79** 040106(R) (2009).

[54] Rosso, O.A.; Craig, H.; Moscato, P. Shakespeare and other English renaissance authors as characterized by Information Theory complexity quantifiers, *Physica A* **388** 916–926 (2009).

[55] Zunino, L.; Zanin, M.; Tabak, B.M.; Pérez, D.G.; Rosso, O.A. Complexity-entropy causality plane: a useful approach to quantify the stock market inefficiency, *Physica A* **389** 1891–1901 (2010).

[56] Rosso, O.A.; de Micco, L.; Larrondo, H.A.; Martín, M.T.; Plastino, A. Generalized statistical complexity measure, *Int. J. Bif. and Chaos* **20** 775–785 (2010).

[57] De Micco, L.; González, C.M.; Larrondo, H.A.; Martín, M.T.; Plastino, A.; Rosso, O.A. Randomizing nonlinear maps via symbolic dynamics, *Physica A* **387** 3373–3383 (2008).

[58] Mischaikow, K.; Mrozek, M.; Reiss, J.; Szymczak, A. Construction of symbolic dynamics from experimental time series, *Phys. Rev. Lett.* **82** 1114–1147 (1999).

[59] Powell, G.E.; Percival, I.C. A spectral entropy method for distinguishing regular and irregular motion of hamiltonian systems, *J. Phys. A: Math. Gen.* **12** 2053–2071 (1979).

[60] Rosso, O.A.; Blanco, S.; Jordanova, J.; Kolev, V.; Figliola, A.; Schürmann, M.; Başar, E. Wavelet entropy: a new tool for analysis of short duration brain electrical signals, *J. Neurosc. Meth.* **105** 65–75 (2001).

[61] Rosso, O.A.; Mairal, M.L. Characterization of time dynamical evolution of electroencephalographic records, *Physica A* **312** 469-504 (2002).

[62] Bandt, C.; Pompe, B. Permutation entropy: a natural complexity measure for time series, *Phys. Rev. Lett.* **88** 174102 (2002).

[63] Keller, K.; Sinn, M. Ordinal analysis of time series, *Physica A* **356** 114–120 (2005) .

[64] Larrondo, H.A.; González, C.M.; Martín, M.T.; Plastino, A.; Rosso, O.A. Intensive statistical complexity measure of pseudorandom number generators, *Physica A* **356** 133–138 (2005).

[65] Larrondo, H.A.; Martín, M.T.; González, C.M.; Plastino, A.; Rosso, O.A. Random number generators and causality, *Phys. Lett. A* **352** 421–425 (2006).

[66] O. A. Rosso, L. Zunino, D. G. Pérez, A. Figliola, H. A. Larrondo, M. Garavaglia, M. T. Martín, and A. Plastino, "Extracting features of Gaussian self-similar stochastic processes via the Bandt and Pompe approach", *Phys. Rev. E*, **76** 061114 (2007).

[67] Rosso, O.A.; Vicente, R.; Mirasso, C.R. Encryption test of pseudo-aleatory messages embedded on chaotic laser signals: an information theory approach, *Phys. Lett. A* **372** 1018–1023 (2008).

[68] Zunino, L.; Pérez, D.G.; Martín, M.T.; Garavaglia, M.; Plastino, A.; Rosso, O.A. Permutation entropy of fractional brownian motion and fractional gaussian noise, *Phys. Lett. A* **372** 4768–4774 (2008).

[69] Zunino, L.; Pérez, D.G.; Kowalski, A.; Martín, M.T.; Garavaglia, M.; Plastino, A.; Rosso, O.A. Fractional Brownian motion, fractional Gaussian noise, and Tsallis permutation entropy, *Physica A* **387** 6057–6068 (2008).

[70] De Micco, L.; Larrondo, H.A.; Plastino, A.; Rosso, O.A. Quantifiers for stochasticity of chaotic pseudo random number generators, *Phil. Trans. Royal Soc. A* **367** 3281–3296 (2009).

[71] Zunino, L.; Zanin, M.; Tabak, B.M.; Pérez, D.G.; Rosso, O.A. Forbidden patterns, permutation entropy and stock market inefficiency, *Physica A* **388** 2854–2864 (2009).

[72] Carpi, L.C.; Saco, P.M.; Rosso, O.A. Missing Ordinal Patterns in Correlated Noises, *Physica A* **389** 2020–2029 (2010).

[73] Amigó, J.M.; Zambrano, S.; Sanjuán, M.A.F. True and false forbidden patterns in deterministic and random dynamics, *Europhys. Lett.* **79** 50001 (2007).

[74] Saco, P.M.; Carpi, L.C.; Figliola, A.; Serrano, E.; Rosso, O.A. Entropy Analysis of the Dynamics of EL Niño/Southern Oscillation during the Holocene, *Physica A* **389** 5022–5027 (2010).

[75] Vignat, C.; Bercher, J.F. Analysis of signals in the FisherShannon information plane, *Phys. Lett. A* **312** 27–33 (2003).

[76] Romera, E.; Dehesa, J.D. The Fisher-Shannon information plane an electron correlation tool, *J. of Chem. Phys.* **120** 8906–8912 (2004).

[77] Sen, K.D.; Antolin, J.; Angulo, J.C. Fisher-Shannon analysis of ionization processes and isoelectron series, *Phys. Rev. A* **76** 032502 (2007).

[78] Szabó, J.B.; Sen, K.D.; Nagy, A. Fisher-Shannon information plane for atoms, *Phys. Lett. A* **372** 2428–2430 (2007).

[79] Angulo, J.C.; Antonin, J.; Sen, K.D. Fisher-Shannon plane and statistical complexity for atoms, *Phys. Lett. A* **372** 670–674 (2008).

Concepts and Recent Advances in Generalized Information Measures and Statistics, 2013, 341-374

CHAPTER 14

Erythrocytes Viscoelasticity Under the lens of Wavelet–Information Theory Quantifiers

Ana María Korol [a,b,*], María Teresa Martín [c,f], Bibiana Riquelme [a,b], Mabel D'Arrigo [a], Osvaldo A. Rosso [d,e,f]

[a] Facultad de Ciencias Bioquímicas y Farmacéuticas, Universidad Nacional de Rosario, Suipacha 531, 2000 Rosario, Argentina

[b] Applied Optical Group, Instituto de Física Rosario, CERIDER-CONICET, 2000 Rosario, Argentina

[c] Instituto de Física, Facultad de Ciencias Exactas, Universidad Nacional de La Plata, C.C. 727, 1900 La Plata, Argentina

[d] Departamento de Física, Instituto de Ciências Exatas, Universidade Federal de Minas Gerais, Av. Antônio Carlos, 6627 - Campus Pampulha, 31270-901 Belo Horizonte - MG, Brazil

[e] Chaos & Biology Group, Instituto de Cálculo, Facultad de Ciencias Exactas y Naturales, Universidad de Buenos Aires, Pabellón II, Ciudad Universitaria, 1428 Ciudad Autónoma de Buenos Aires, Argentina

[f] Consejo Nacional de Investigaciones Científicas y Técnicas (CONICET), Argentina

Abstract

We present an application of wavelet-based Information Theory quantifiers (Relative Wavelet Energy, Normalized Total Wavelet Entropy, Wavelet MPR-Statistical Complexity and Entropy-Complexity plane) on red blood cells membrane viscoelasticity characterization. These quantifiers exhibit important localization advantages provided by the Wavelet Theory. The present approach produces a clear characterization of this dynamical system. We have found out an evident manifestation of a random process on the red cell samples of healthy individuals, which is weakly less when studying glucose incubated erythrocytes, while there is a sharp reduction of randomness on analyzing a human haematological disease, such as β-thalassemia minor.

* Corresponding author. E-mail: korol@ifir-conicet.gov.ar

Andres M. Kowalski, Raul D. Rossignoli and Evaldo M. F. Curado (Eds)

Keywords: Time series; Hemorheology; Wavelet analysis; Entropy; Generalized Statistical Complexity; Thalassemia; Glycosylation.

1 Introduction

Why should anyone deal with mathematical physics analysis of a physiological system? The answer could be that drawn by Eugene Wigner [1], one of the important physicists of the twenty's century: "that the enormous usefulness of mathematics in natural sciences is something bordering on the mysterious and that there is no rational explanation for it", at least it serves well to show the big power of mathematics in describing physical reality. Subjected to rigorous testing, a mathematical description of a system helps to put our knowledge of that system into a rigorous quantitative form. As a matter of fact, anyone reading this book probably does not need to be convinced of the value of mathematical approach to physiological analysis.

Theoretical physics has long possessed a general measure of the uncertainty associated with the behavior of a probabilistic process: the Shannon entropy of the underlying probability distribution [2], which was originally introduced by Boltzmann over 100 years ago. In the fifty's Kolmogorov and Sinai [3, 4] adapted Shannon's information theory to the study of dynamical systems. This work constitutes the foundation for the statistical characterization of the deterministic origin of the apparent randomness, the chaos theory, born in the late sixty's through early eighty's. The effort on describing the unpredictability of non-linear dynamical systems was largely successful. The metric entropy, Lyapunov exponents and fractal dimensions provide applicable quantities that can be used to detect not only the presence of a deterministic chaotic behavior, but also to quantify it [5, 6]. However, it has become better understood that measuring the randomness and unpredictability of a system fails to capture the correlation structure in its behaviour. It is well known the larger and intricate are the correlations between the system's constituents the more structured the underlying distribution.

These ideas led to a considerable effort to develop a general theory that quantifies the degree of structure present in the process, and the quantities proposed as general measures are usually called complexity measures. We call as "statistical complexity measures" (see Chapter 8) those proposed in the framework of Information Theory. Information Theory has been applied with great success to many different biological systems, and many of them have been found to be either random or chaotic [7, 8, 9, 10, 11, 12, 13, 14, 15, 16, 17, 18, 19, 20, 21, 22, 23, 24, 25]. The present chapter on erythrocytes (red blood cells) viscoelastic properties is another example of the appropriate uses of Information Theory in a biological system.

Human red blood cells (RBC) are viscoelastic, showing viscous and elastic components integrated in a complex viscoelasticity when undergo fluid shear stress. This dynamic behaviour has a physiological significance since *in vivo* the erythrocytes continually change their shape and adapt passively to capillary circulation. Then the clinical interpretation of erythrocyte deformation attempts to link pathological features with the visual microscopy inspection of erythrocytes samples. It is well known that on healthy donors the erythrocytes are discocytes which is the dynamical expression of the best capability of the cells to be able to adapt in flow.

Complexity refers to certain systems quality of being intricate and hardly predictable, ranging from turbulent flows of the atmosphere to cells behaviour. For succeeding in the plan of explaining complex behaviour, biologist, physicists, mathematics and engineers have been looking for simplified models mathematically tractable and able to catch the essence of complexity. We would like to emphasize the dynamics, not the tools. Analysis based on Fourier Transform of red blood cells deformation photometrically recorded time series has been performed previously (see in example [8]). However, this kind of time series exhibit a some degree of non-stationarity and present a small length, making in way difficult the application of this technique in order to obtain reliable results. Wavelet transform do not require stationary condition of the signal under analysis and can be used with small amount of data.

Our preliminary studies on red blood cells (RBCs) complex behaviour [7, 8, 9, 10, 11], have shown the competition between stochastic ordinal Brownian motion (oBm) and fractional Brownian motion (fBm), and make a crossover from one relaxation scenario when the erythrocytes are at rest to the other one, when the erythrocytes are subjected to shear stress. On this basis we applied quantifiers based on Information Theory in the study of the manifestation of this behaviour.

The present chapter is organized as follow. In Section 2, the materials and methods used in the present work are given. Wavelet based Information Theory quantifiers are described in Section 3. In Section 4 we present our results and discuss it. Finally, in Section 5 some conclusions are given.

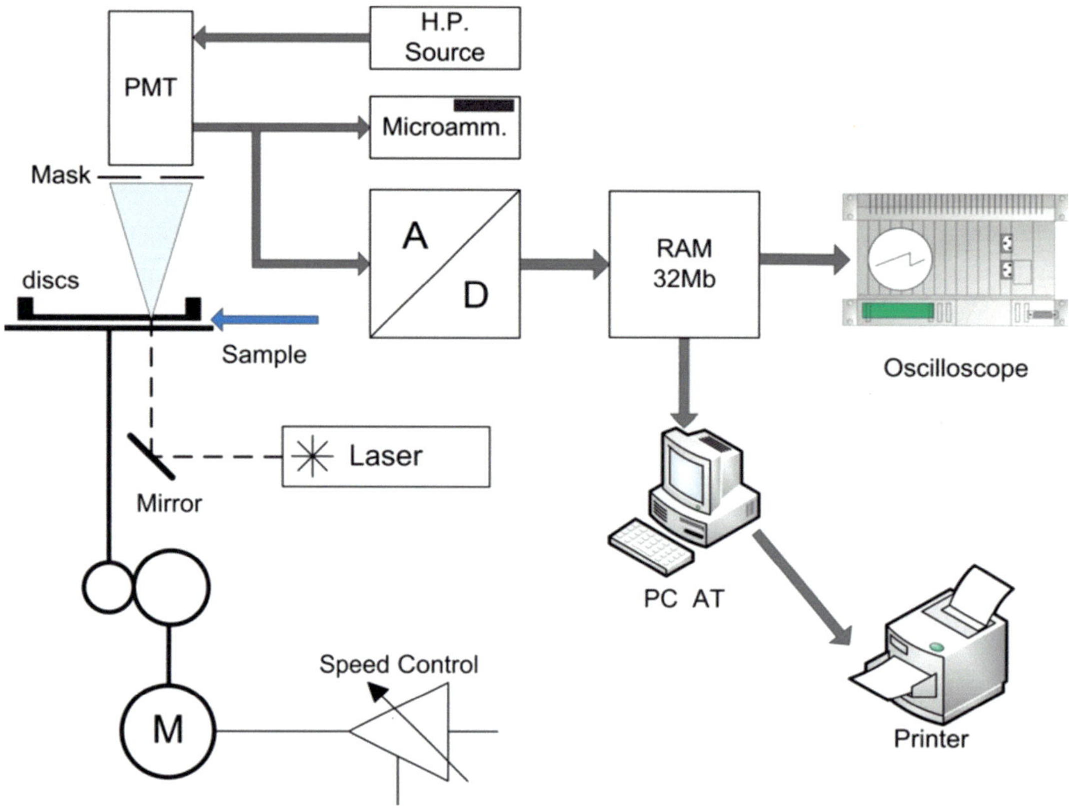

Figure 1: Scheme of the home-made erythrocytes deformer divice. Figure take from Ref. [12].

2 Materials and Methods

Erythrocytes (red blood cells) are one of the main constituents of blood. Normal erythrocytes [26] are biconcave discs having a mean diameter of about 7.8 μm and a thickness of 2.5 μm at the thickest point and in the center ≤ 1 μm. The average volume of the erythrocytes is $90 - 95$ μm^3. The shape of the erythrocytes can change noticeably as the cells pass through capillaries which in some cases have 3 μm of diameter.

2.1 β-thalassemia minor

Thalassemias are a diverse group of genetic blood diseases characterized by absent or decreased production of normal hemoglobin, resulting in a microcytic anemia of varying degree

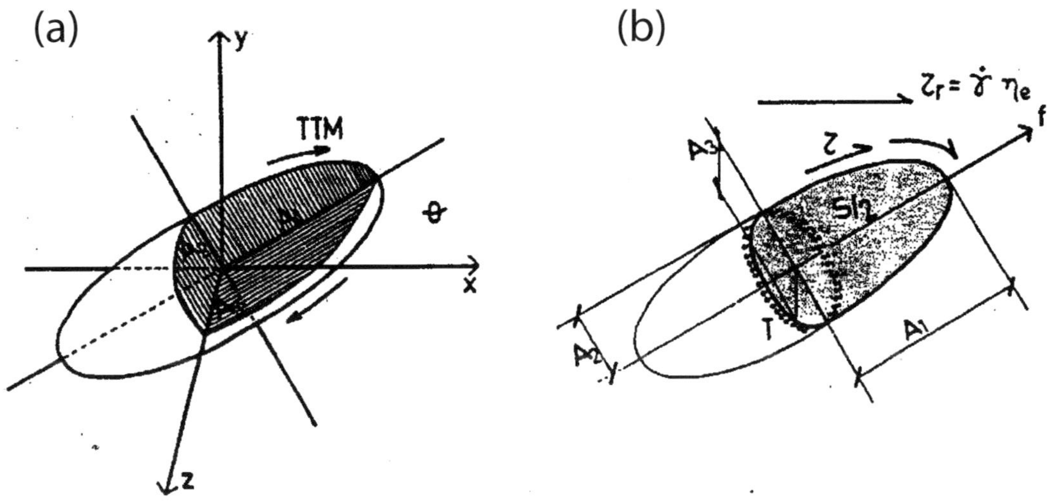

Figure 2: Scheme of a tank treated deformed cell. *a)* Principal radii and inclination of ellipsoidal cell. *b)* Stress conditions and equilibrium state at the coss section of the cell deformed by shear stress τ_r. Figure take from Ref. [37].

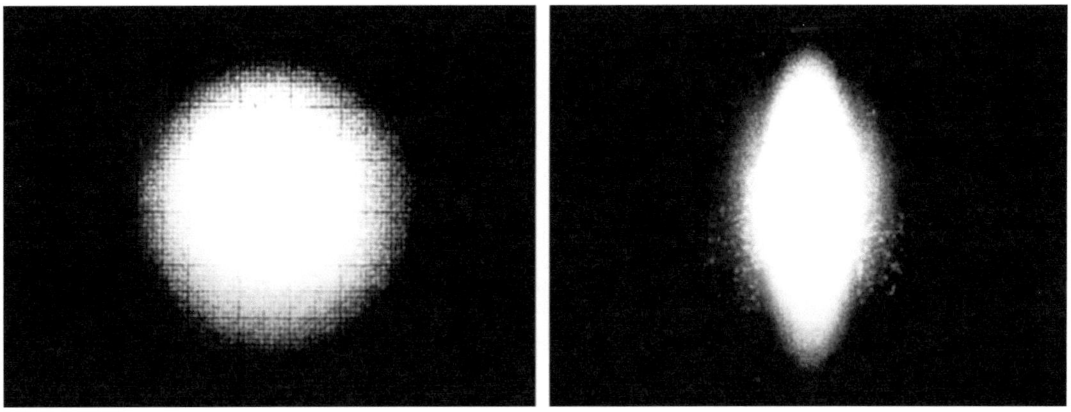

Figure 3: Diffraction pattern of the erythrocyte: circular shape at rest, elliptical shape under well defined shear stress. Figure take from Ref. [12].

[27]. Thalassemia occurs most frequently in people of Italian, Greek, Middle Eastern, Southern Asian and African ancestry. The two main types are called alpha and beta thalassemias, depending on which part of an oxygen-carrying protein (called hemoglobin) is lacking in the red blood cells. The genes controlling globin production are on chromosome #16 (α globin genes) and chromosome #11 (β, γ and δ genes). In the thalassemia patient, a mutation or deletion of the genes that control globin production occurs. This leads to a decreased

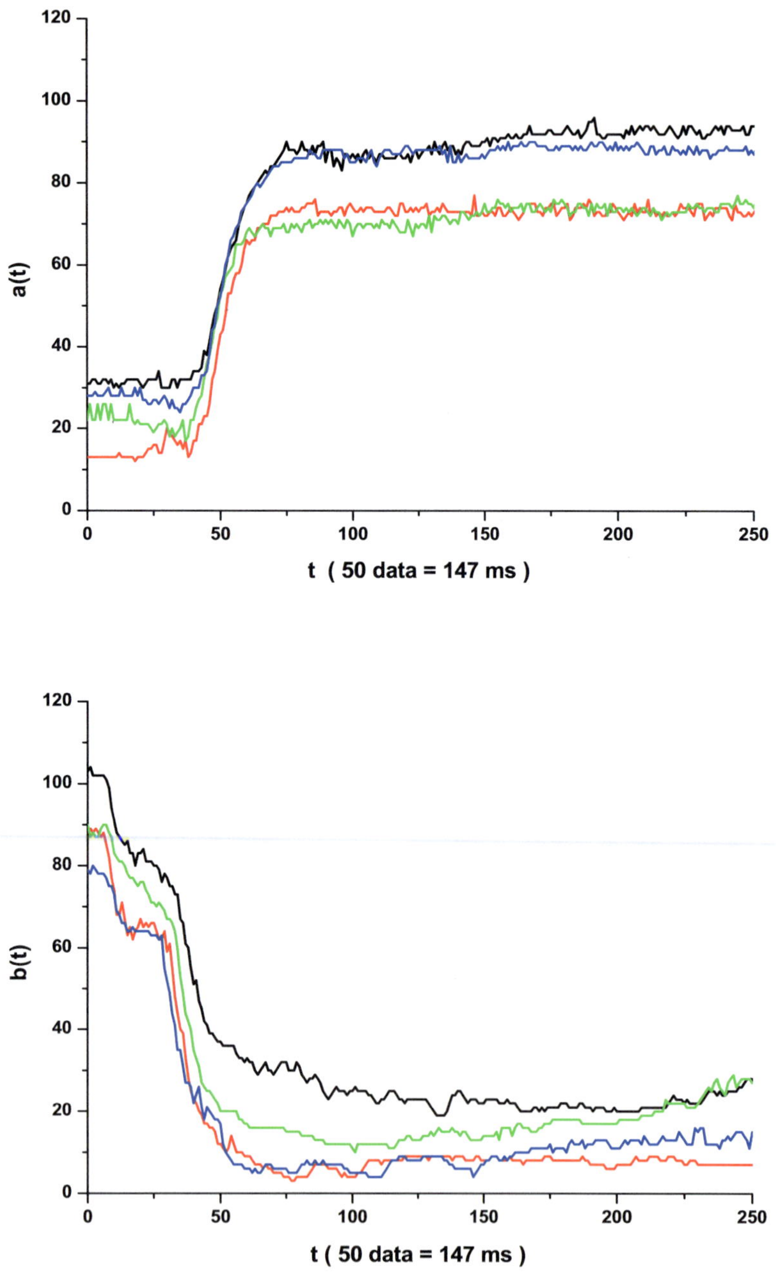

Figure 4: Typical creep $a(t)$ and recovery $b(t)$ curves of diffracted intensity measured along the major axis of the elliptical pattern files recorded by the erythrodeformeter. Note that the first 56 data points represents the transitory regime. Figure take from Ref. [10].

production of the corresponding globin chains and an abnormal hemoglobin ratio (α/non-α). This abnormal ratio leads to decreased synthesis of hemoglobin and the expression of thalassemia. The globin that is produced in normal amounts winds up in excess and forms red cell aggregates or inclusions. These aggregates become oxidized and damage the cell membrane, leading either to hemolysis, ineffective erythropoiesis, or both. The quantity and properties of these globin chain aggregates determine the characteristics and severity of the thalassemia.

β-thalassemia results in an excess of alpha globins, which leads to the formation of alpha globin tetramers that accumulate in the erythroblast (immature red blood cell). These aggregates are very insoluble and precipitation interferes with erythropoiesis, cell maturation and cell membrane function, leading to ineffective erythropoiesis and anemia. α-thalassemia results in an excess of beta globins, which leads to the formation of beta globin tetramers. These tetramers are more stable and soluble, but under special circumstances can lead to hemolysis, generally shortening the life span of the red cell.

A person with this condition, carries the genetic trait for β-thalassemia and will usually experience no health problems other than a mild anemia, sometimes incorrectly prescribed with iron supplements that have no effect on the anemia. In the case both parents carry β-thalassemia minor, that is the β-thalassemia trait, there is 25% chance with each pregnancy their child will inherit the two normal β globin genes, so will be a normal individual; there is a 50% chance with each pregnancy their child will inherit one normal β globin gene, so will be a β-thalassemia minor individual; and also there is 25% chance with each pregnancy their child will inherit the two abnormal β globin genes. In the last possibility will be β-thalassemia in its most severe form, which requires regular blood transfusions and extensive ongoing medical care. Due to the wide range in severity, it is important for each individual the diagnosis of β-thalassemia minor, despite this does not cause problems in the normal life. Rheological studies in thalassemia carriers suggest the existence of decreased red cell deformability [28].

2.2 Glucose incubated erythrocytes

A very important concern of a blood transfusion centre is how to maintain the functional properties of erythrocytes during their storage. Storage time, based on the post transfusion survival of the erythrocytes, has grown constantly in the last decades, along with the progress made in the field of preservation solutions. The anticoagulant preservation solutions contain enough glucose to support continuing adenosine triphosphate generation during storage, resulting in improved viability.

Elevated levels of glucose in blood, which is frequently observed in diabetes mellitus, are known to cause damage and death of erythrocytes [29]. Modification in the hemorheological properties of cells, which may be associated with glucose induced auto oxidation and glycosylation of erythrocytes proteins in hyperglycemia, can adversely affect the fluidity and oxygen delivery function of the erythrocytes [30, 31].

The normal glucose level ranges from 0.7 to 1.05 g/l (3.89–5.83 mmol/l) and in diabetic patients it is increased attaining to 3.6 g/l (20 mmol/l) value, which can produce *in vivo* glycosylation affecting the cellular function. Even though the biomechanical mechanisms of damage of cell membranes at elevated levels of glucose are not fully understood, evidence of molecular changes in erythrocytes constituents, resulting in increased calcium concentration and therefore altered erythrocyte viscoelasticity, has been extensively studied [32].

2.3 Blood samples

Two groups laboratory essays were considered in the present work: 1) β thalassemia minor and 2) Glucose incubated erythrocytes. Rheological studies were carried out within the first four hours after withdrawal to avoid cell membranes damage. All anticoagulated blood samples were kept at $5°C$. Patients and healthy controls, in all cases, gave their informed consent to participate in the study.

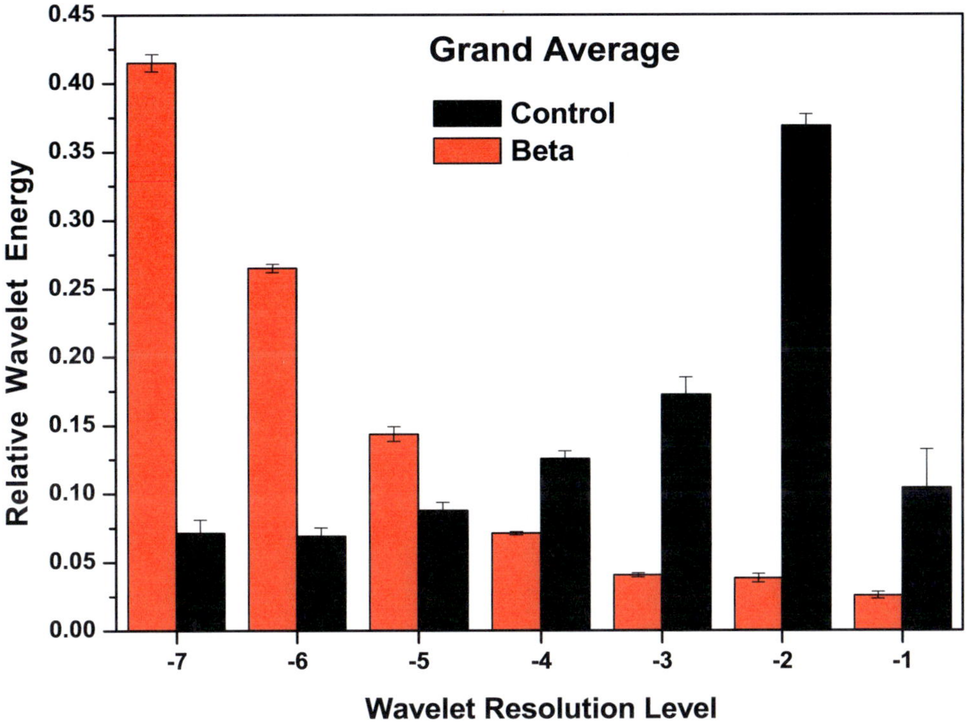

Figure 5: Grand average corresponding to Relative Wavelet Energy (RWE) for 5 healthy controls and 14 β-thalassaemic samples. Figure take from Ref. [12].

First Group — β thalassemia minor:

Samples were obtained by cubital venipuncture with disposable syringes and large gauge needles to avoid mechanical cell damage [33], using as anticoagulant EDTA NA_2, in 19 individuals: 14 β-thalassaemia patients and 5 healthy controls. Healthy controls were non-smokers, non-alcoholic individuals and were not receiving any therapeutic drug. Thalassemia patients were diagnosed as β-thalassemia minor by conventional methods and classified by molecular biological ones (PCR-ARMS) as β^0 (β^0 39).

Second Group — Glucose incubated erythrocytes:

Human venous blood samples from 8 healthy volunteers, red cell type 0, were anticoagulated with EDTA NA_2 and maintained at $4°C$ until they were processed. Whole blood was centrifuged at $800 \times g$ during 10 min, plasma and buffer coat were then removed. The remaining red blood cells (RBCs) were washed three times with phosphate buffer saline (PBS: pH 7.4) at $25°C$. Each packet of RBCs was divided into four aliquots. One aliquot was left untreated (w.t.) as a control sample. The other three aliquots were incubated

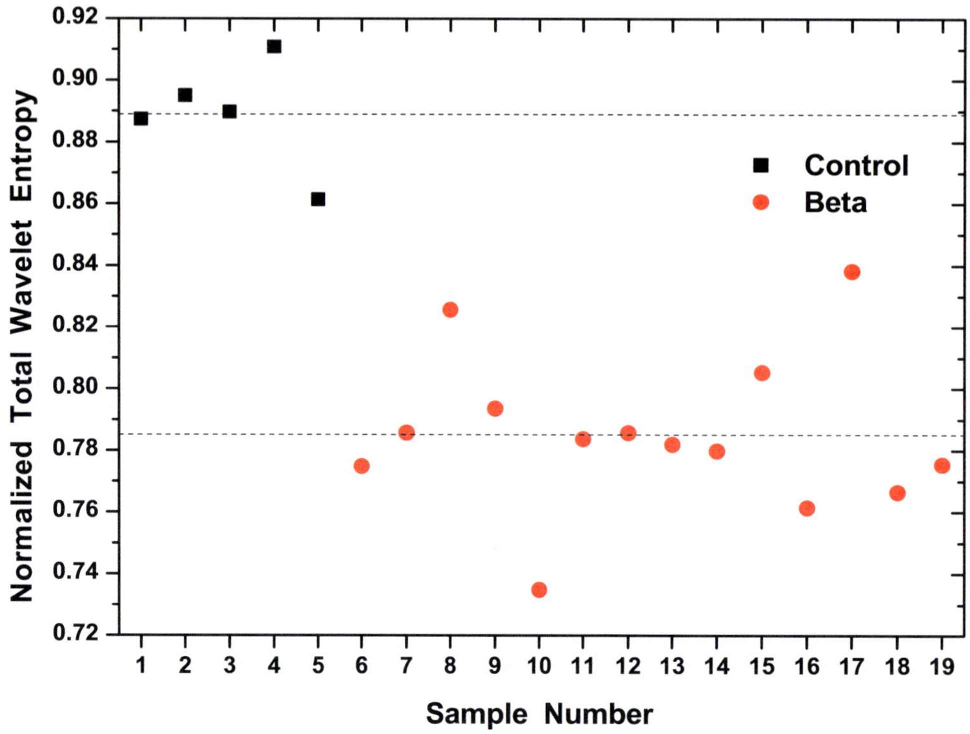

Figure 6: Two clearly different clouds for each population: healthy control donors (square), and β-thalassemia patients (circles). The horizontal lines correspond to the average Normalized Total Wavelet Entropy ($\mathcal{H}$) values. Figure take from Ref. [12].

vitro with three glucose concentrations solutions at $37°C$ during 2 h. The hematocrit during incubation was 50%. These glucose incubating media were prepared by solving 2%, 5%, and 10% (w/v) of glucose in PBS, corresponding to 20, 50 and 100 g/l as recommended by Bourdon *et al.* [34].

An *in vitro* non physiological 10% glucose concentration was chosen to accelerate the non enzymatic glycosylation that is known to depend on time and exposure to glucose [35]. After that, the cells were washed three times with PBS at $25°C$ to remove any remnants of glucose. Each aliquot of RBCs was resuspended in autologous plasma to obtain the required hematocrit for each rheological determination. The erythrocytes suspensions were obtained according to the experimental procedure described above and following the International Committee for Standardization on Haematology [33].

We could observe the presence of haemolysis during the post glucose washing with PBS at higher glucose concentration than 10%. The erythrocyte osmotic resistance and the

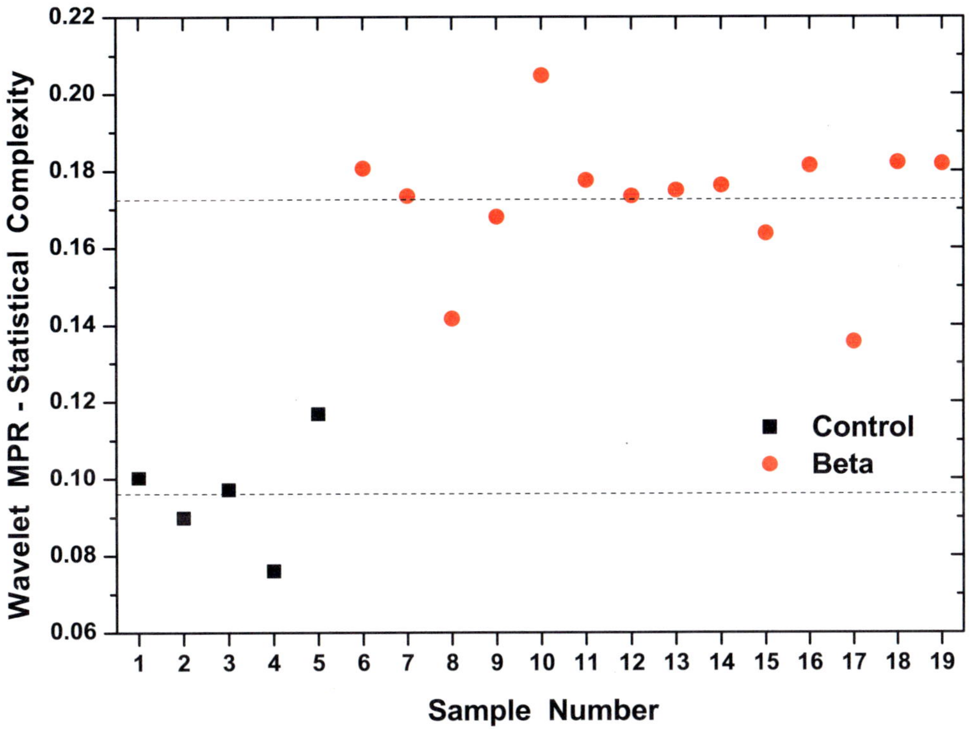

Figure 7: Two clearly different clouds for each population: healthy control donors (square), and β-thalassemia patients (circles). The horizontal lines correspond to the average Wavelet MPR-Statistical Complexity ($\mathcal{C}$) values. Figure take from Ref. [12].

erythrocytes viscoelastic parameters were measured in a previous work [36], evaluating the increasing in the fragility and rheological alterations of red blood cells with glucose level in the incubating media.

2.4 Red blood cells suspension and medium

In order to prepare the samples, 100 μl of anti coagulated whole blood were suspended into 4 ml of isotonic viscous medium. 5% w/v polyvinylpyrrolidone (PVP 360, Sigma, MW 360 kDa) in phosphate buffered saline (PBS: 0.150M NaCl, and 0.005 M (K2PO4 + KPO4H); pH 7.4 ± 0.05; 295 ± 8 mOsmol/kg). Medium viscosity adjusted at 22 ± 0.5 cP at $25°C$.

A scheme of the home-made erythrodeformeter device is presented in Fig. 1. For its validation and reliability we refer the reader to the original work of Rasia [37]. This device allows us to tests the ability of red blood cells to change their circular shape into an ellipsoidal one, when are subjected to a well controlled fluid shear stress. The erythrodeformeter

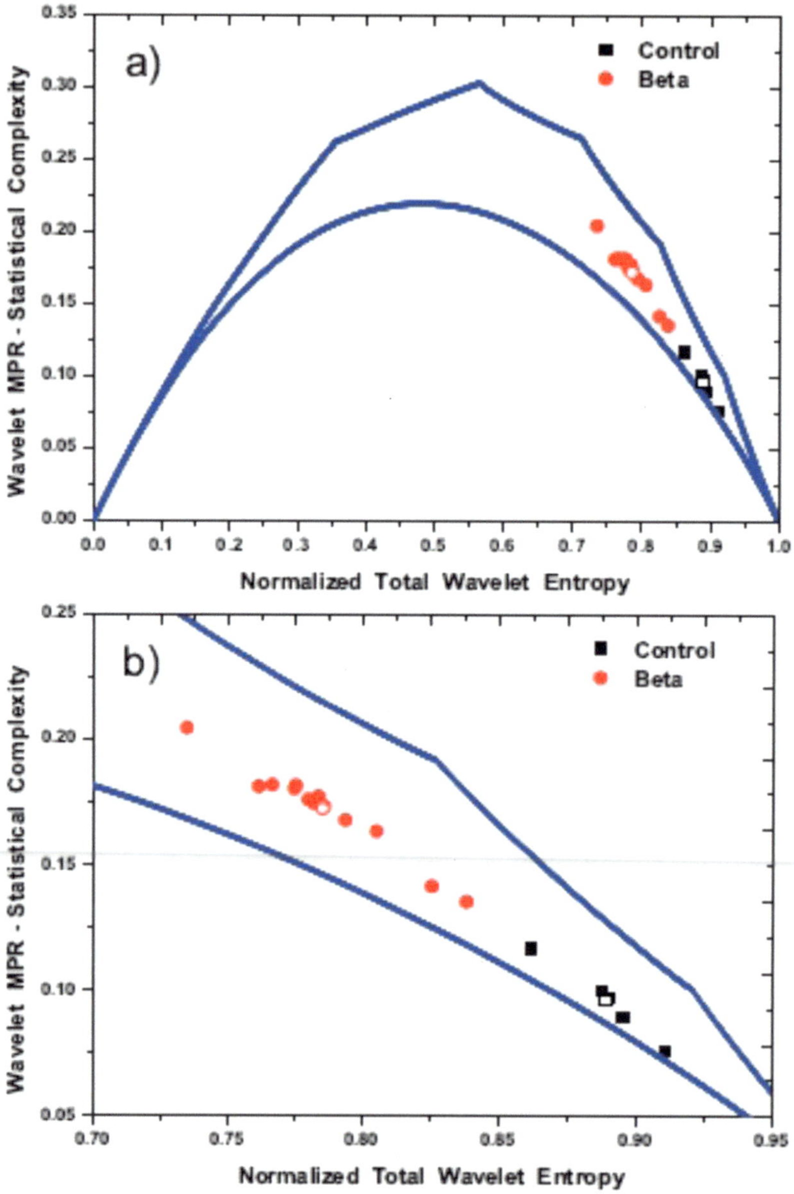

Figure 8: *a)* Entropy-Complexity plane $\mathcal{H} \times \mathcal{C}$ for $N_J = 7$. The continuous lines display the bound curves $\mathcal{C}_{min}$ and $\mathcal{C}_{max}$ associated to the maximum and minimum values of wavelet MPR-Complexity for a fixed value of the Normalized Total Wavelet Entropy. Also, different marks are used to distinguish the numerical calculated values for controls and β-thalassaemia signals. The open symbols correspond to the average values (see Fig. 6 and Fig. 7). *b)* Detail of the Entropy-Complexity plane. Figure take from Ref. [12].

device has two plane disks, both of them of flint glass, superposed, coaxials, parallels and horizontals. The driving motor is coupled to the lower disk by two helicoidal gears that provide a great rigidity to the transmission system. This allows the rotational disk to start or stop rotation in a short time (< 1 ms), so can be considered as instantaneous.

The light source is a 5 mW HeNe laser. When the layer of diffracting cells is perpendicularly transversed by the laser beam, the diffraction pattern can be observed on a ground glass screen. Diffracted intensity corresponding to each principal diameter of the elliptical diffraction pattern falls onto a masked photomultiplier tube (PMT), after passing through a thin straight slot in the mask placed exactly on the corresponding axis of the elliptical pattern.

All units, except the oscilloscope and the PC, are homemade equipments. The PMT signal output is either relayed to the microammeter or to the oscilloscope. When the PMT is connected to the microammeter, two values of electrical current can be measured. They correspond to the long and short axis of the elliptical pattern. On the other hand, the PMT is connected to the oscilloscope through an 8-bit A/D converter having a memory divided into 32 sectors with 256 bytes each. The conversion delay is $0 : 1$ ms and the converter board has two important controls which must be adjusted for optimal performance: one selects the memory sector to be filled during the data acquisition, the other selects the lapse of time during which a set of 256 *data* points will be acquired and stored [7, 8].

2.5 Deformability Index

Viscoelasticity of RBC membrane and viscosity of cytoplasmic fluid play a key role in the blood flow behaviour. So techniques capable of assessing the rheological properties of the membranes over large number of cells become of great interest. It can provide important information about the state of the membrane structure in normal or abnormal health conditions. Also, an important advantage of this procedure is that the results are averaged over several millions of cells and that the whole process requires no more than 30 minutes.

RBC suspended in a viscous continuous medium (Newtonian fluid), subjected to a well controlled shear stress, assumes an ellipsoidal shape. The ellipsoid takes a Φ inclination with respect to the flow direction while its membrane rotates steadily, this movement is the so called tank treading motion. In Fig. 2, it is represented the deformed cell by a triaxial ellipsoid ($A_1 \neq A_2 \neq A_3$), where the coordinate system $Oxyz$ is placed in the elongated cell center, remaining the cell volume constant, while the erythrocytes take a position in the flow field being θ the angle with respect to the x-axis. We assume that the ellipsoid tends to a prolate spheroid ($A_1 > A_2 = A_3$) with a circular cross section. The light intensity measurements taken along the major and minor axes of the ellipsoidal diffraction pattern, is produced by an ellipsoidal Franhoufer diffractive particle which is inversely proportional to the original particle. A true value of the deformability index (DI) can be obtained by using calculated values of A_1 and A_3: $DI = (A_1 - A_3)/(A_1 + A_3)$.

2.6 Data acquisition

The erythrocyte suspension is placed between the two flint glass disks, and the driving motor allows the lower disk rotation. Normal erythrocytes being at rest can be considered as a monodisperse population having discoidal shape with almost the same size. Light diffraction under Fraunhofer theory conditions may be applied to obtain quantitative information of diffracting particles such as suspended erythrocytes. The cells in dilute suspension, under shear stress take a three axial ellipsoidal shape having the major axis oriented towards the shear field direction. The laser beam transverses perpendicularly a thin layer of the erythrocytes suspension and it is diffracted producing a Fraunhofer diffraction pattern that is either circular when the cells are at rest or elliptical when they are deformed by a shear stress field. The typical patterns are shown in Fig. 3.

The start of data acquisition is externally triggered by the erythrodeformeter motor switch. Once the driven motor is switched on, the PMT signal output is read, converted and stored in a preselected memory sector. Each time the A/D converter completes a conversion and, as soon as, the creep time series has been stored, it returns to the initial position ready to start a new cycle of conversion and storage of another set of data into a new memory sector. A similar process is carried out when the motor is switched off to store a recovery curve (measured while the cells recover their circular shape from the ellipsoidal one) as well as it was done with the creep one. The A/D converter can receive data from the memory and plot them repeatedly on the oscilloscope screen. A 25-pin connector allows the transfer of data from the A/D converter to the PC by an interfaced bus. Data can be stored in the PC hard disk for being numerically processed offline.

Typical creep and recover time series are shown in Fig. 4 ($50\ data\ points = 147\ ms$).

These time series correspond to diffracted intensity measured in the major axis of the elliptical pattern. From these figures it is clear that the signals present non-stationary characteristics.

3 Wavelet based Information Theory quantifiers

3.1 Wavelet transform

The wavelet representation provides precise measurements regarding *(a) when* and *to what degree* transient events take place in a signal and *(b) when* and *how* the frequency contents of the signal changes over time [38, 39, 40]. The wavelet is a simple and quickly vanishing oscillating function. Unlike sine and cosine of Fourier analysis, which are precisely localized in frequency but extend infinitely in time, wavelets are relatively localized in both time and frequency. Furthermore, wavelet are band-limited; they are composed of not one but a relatively limited range of several frequencies.

A wavelet family $\Psi_{a,b}$ is the set of elemental functions generated by *(i)* scaling (stretching or shrinking the basic wavelet) and *(ii)* translation (moving the basic wavelet to different time positions at any scale without changing its shape) of a unique admissible mother wavelet $\Psi(t)$:

$$\Psi_{a,b}(t) = |a|^{-1/2}\Psi\Big(\frac{t-b}{a}\Big) , \qquad (1)$$

where $a, b \in \mathbb{R}$, $a \neq 0$, are the scale and translation parameters, respectively, and t is the time.

The *continuous wavelet transform* (CWT) of a signal $\mathcal{S}(t) \in L^2(\mathbb{R})$ (the space of real square summable functions) is defined as the correlation between the function (signal) $\mathcal{S}(t)$ with the family wavelet $\Psi_{a,b}$ for each a and b [38, 39, 40]:

$$(W_\Psi \mathcal{S})(a,b) = |a|^{-1/2} \int_{-\infty}^{\infty} \mathcal{S}(t)\ \Psi^*\Big(\frac{t-b}{a}\Big)\ dt = <\mathcal{S}, \Psi_{a,b}> \qquad (2)$$

where $*$ means complex conjugation. This equation can be inverted in order to archive perfect reconstruction of the original signal. $<\mathcal{S}, \Psi_{a,b}>$ represent the wavelet coefficients. In principle the CWT produce an infinite number of coefficients. The information it display at closely space scales or at closely spaced time points is highly correlated. In consequence, the CWT provides a redundant representation of the signal under analysis. It is also time consuming to compute directly.

The *discrete wavelet transform* (DWT) provides a non-redundant, highly efficient wavelet representation that can be implemented with a simple recursive filter scheme and the original signal reconstruction can be obtained by an inverse filtering operation. For *(i)* a special election of the mother wavelet function $\Psi(t)$ and *(ii)* the discrete set of parameters $a_j = 2^{-j}$ and $b_{j,k} = 2^{-j}k$, with $j, k \in \mathbb{Z}$ (the set of integers), the family $\Psi_{j,k}(t) = 2^{j/2}\Psi(2^j t - k)$ constitutes an orthonormal basis of $L^2(\mathbb{R})$, also called the Hilbert space, consisting of finite-energy signals. The correlated DWT provides a non-redundant representation of the signal $\mathcal{S}$ and its values constitute the coefficients in a wavelet series, $< \mathcal{S}, \Psi_{j,k} >= C_j(k)$. The DWT produces only as many coefficients as there are samples within the signal under analysis $\mathcal{S}$, without the loss of any information at all. These wavelet coefficients provide full information in a simple way and a direct estimation of local energies at the different relevant scales. Moreover, the information can be organized into a multiresolution manner, yielding a hierarchical scheme of nested subspaces in $L^2(\mathbb{R})$ [38, 39, 40].

If the wavelet is required to form a proper orthogonal basis, it has the advantage that any arbitrary function can be uniquely decomposed and the decomposition can be inverted. Orthogonal cubic spline functions were used here as mother wavelets and the time-frequency information was organized in a hierarchical scheme [38, 39, 40]. Among several alternatives, symmetric and orthogonal, cubic spline functions were used. They combine in suitable proportion smoothness, on the one hand, with numerical advantages, on the other one (for a complete discussion, see [41, 42]).

In the following, we assume that the signal is given by the sampled values $\mathcal{S} = \{s_0(n); n = 1, \cdots, M\}$, which correspond to an uniform time grid with sampling time t_s and M the total number of samples. For simplicity we suppose that the sampling rate is $t_s = 1$. If the decomposition is carried out over all resolutions levels, $N_J = \log_2(M)$, the wavelet expansion reads

$$\mathcal{S}(t) = \sum_{j=-N_j}^{-1} \sum_{k} C_j(k)\, \Psi_{j,k}(t) = \sum_{j=-N_j}^{-1} r_j(t) \ , \tag{3}$$

where the wavelet coefficients $C_j(k)$ can be interpreted as the local residual errors between successive signal approximations at scales j and $j-1$, respectively, and $r_j(t)$ is the detail signal at scale j, that contains information of the signal $\mathcal{S}(t)$ corresponding with the frequencies $2^{j-1}\omega_s \leq |\omega| \leq 2^j\omega_s$.

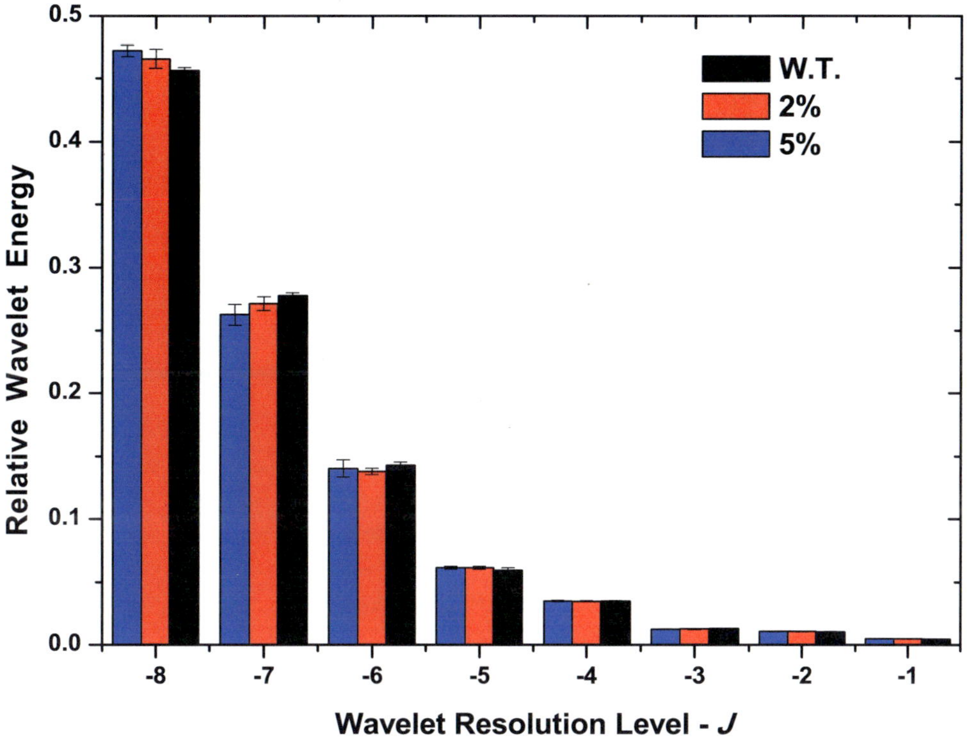

Figure 9: Grand average corresponding to relative Wavelet Energy (RWE) for eight samples without treatment, eight samples 2% glucose incubated and six samples 5% glucose incubated. Figure take from Ref. [13].

3.2 Relative Wavelet Energy

Since the family $\Psi_{j,k}(t)$ is an orthonormal basis for $L^2(\mathbb{R})$, the concept of energy is linked to the usual notions derived from Fourier's theory. The wavelet coefficients are given by $C_j(k) =< \mathcal{S}, \Psi_{j,k} >$ and the energy, at each resolution level $j = -1, \cdots, -N_J$, will be the energy of the detail signal

$$E_j \; = \; ||r_j||^2 \; = \; \sum_k \; |C_j(k)|^2 \; . \tag{4}$$

The total energy can be obtained in the fashion

$$E_{tot} \; = \; ||\mathcal{S}||^2 \; = \; \sum_{j=-N_J}^{-1} \sum_k |C_j(k)|^2 \; = \; \sum_{j=-N_J}^{-1} \; E_j \; . \tag{5}$$

Finally, we define the normalized ρ_j-values, which represent the *relative wavelet energy* (RWE)

$$\rho_j \; = \; E_j / E_{tot} \tag{6}$$

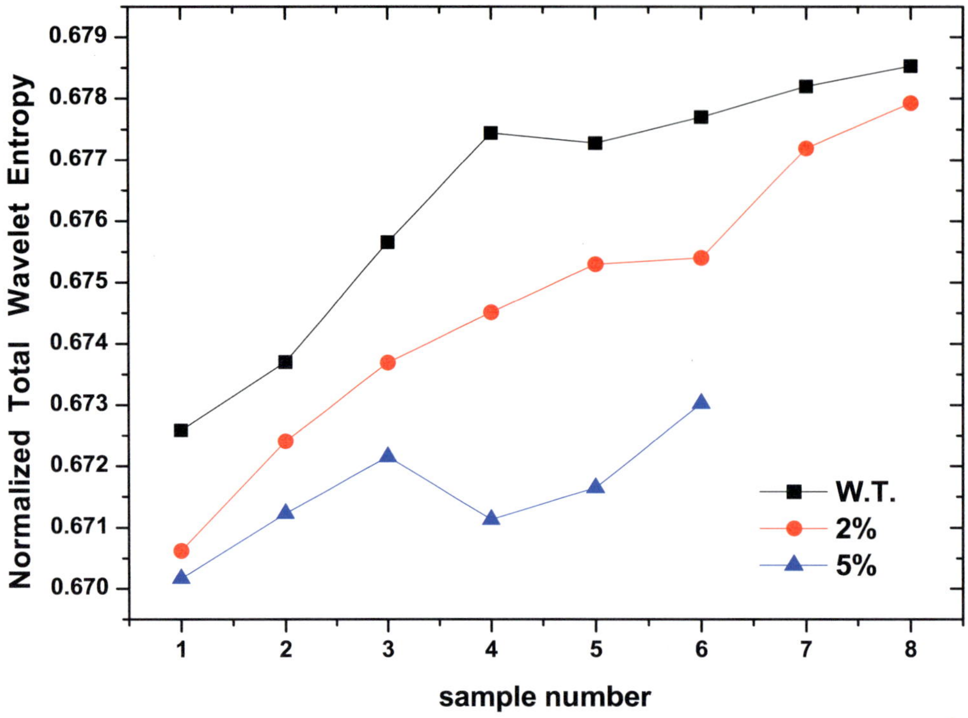

Figure 10: Normalized Total Wavelet Entropy (NTWS) for eight samples without treatment (square), eight samples 2% glucose incubated (circle) and six samples 5% glucose incubated (triangle). Figure take from Ref. [13].

for the resolution levels $j = -1, -2, \cdots, -N_J$.

The ρ_j yield, at different scales, the probability distribution for the energy. Clearly,

$$\sum_{j=-N_J}^{-1} \rho_j = 1 . \tag{7}$$

The distribution $P^{(W)} = \{\rho_j; j = -1, \cdots, -N_J\}$ can be considered as a time-scale probability density that constitutes a suitable tool for detecting and characterizing specific phenomena in both the time and the frequency planes [14, 23, 43].

3.3 Normalized Total Wavelet Entropy

The Shannon entropy [2] gives a useful criterion for analyzing and comparing probability distributions. It provides a measure of the information contained in any distribution. We

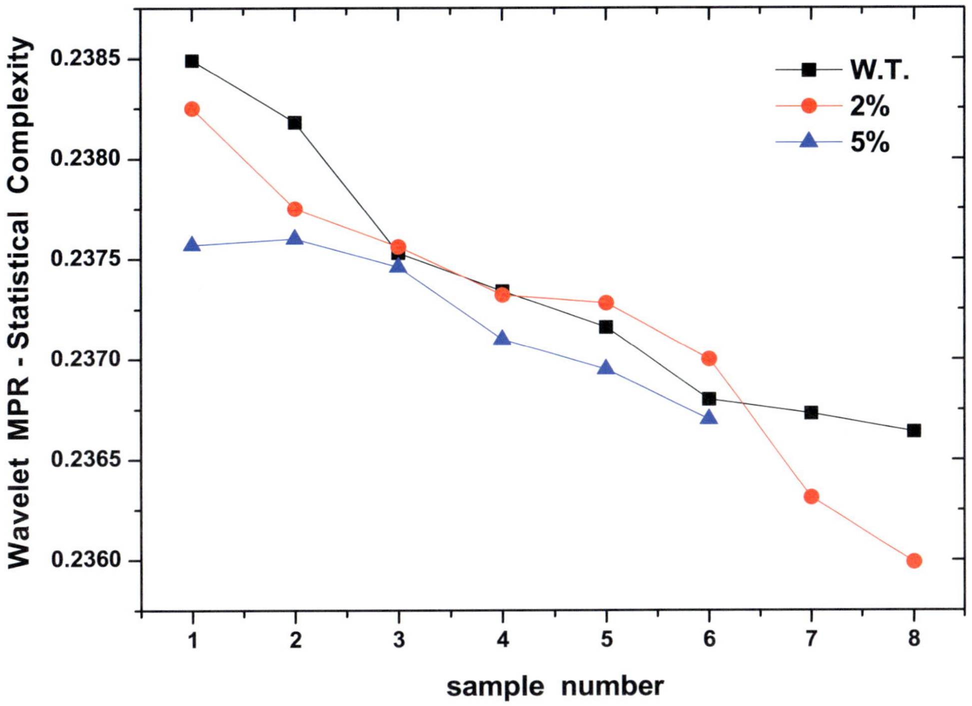

Figure 11: Wavelet MPR-Statistical Complexity for eight samples without treatment (square), eight samples 2% glucose incubated (circle) and six samples 5% glucose incubated (triangle). Figure take from Ref. [13].

define the *Shannon Wavelet Entropy* (SWS) [14, 23, 43] as

$$\mathrm{S}[P^{(W)}] \;=\; -\sum_{j<0} \rho_j \cdot \ln[\rho_j] \ , \tag{8}$$

where $P^{(W)}$ denotes the probability energy wavelet distribution given by Eq. (6). Thus, the SWS appears as a measure of the degree of order/disorder of the signal, giving useful information about the underlying dynamical process associated with the signal [16, 24]. Indeed, a very ordered process can be represented by a periodic mono-frequency signal (signal with a narrow band spectrum). A wavelet representation of such a signal will be resolved at one unique wavelet resolution level, *i.e.*, all RWE will be (almost) zero except at the wavelet resolution level which includes the representative signal frequency. For this special level the RWE will (in our chosen energy units) almost equal unity. As a consequence, the SWS will acquire a very small, vanishing value. A signal generated by a totally random process can be taken as representative of a very disordered behaviour. This kind of signal will have a wavelet representation with significant contributions coming from

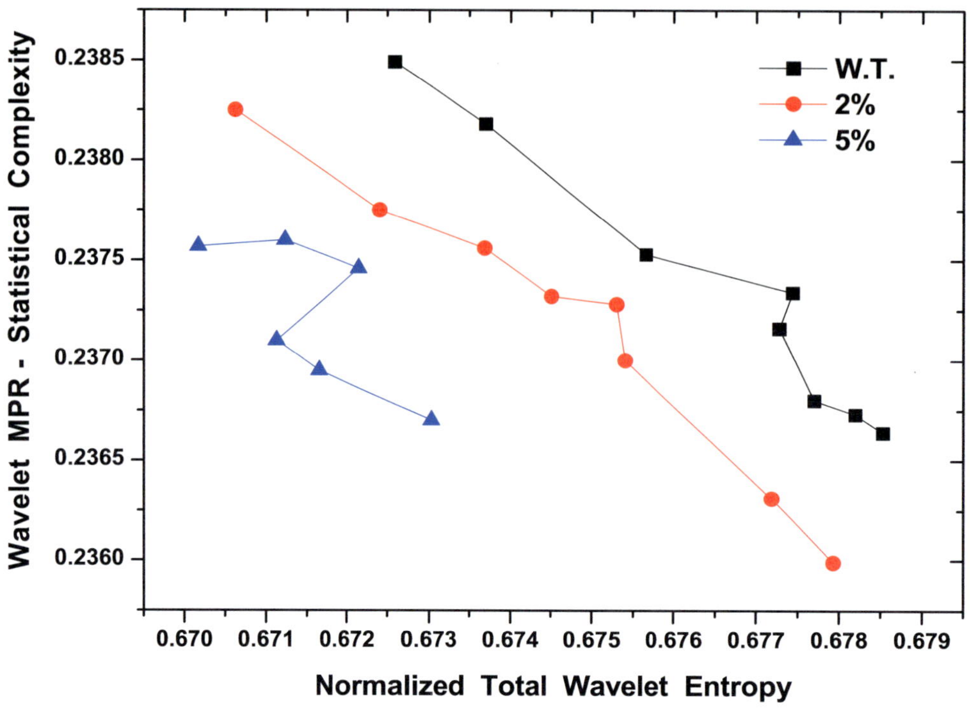

Figure 12: Entropy-Complexity plane $\mathcal{H}\times\mathcal{C}$ for $N_J = 8$ for eight samples without treatment (square), eight samples 2% glucose incubated (circle) and six samples 5% glucose incubated (triangle). Figure take from Ref. [13].

all frequency bands. Moreover, one could expect that all contributions will be of the same order. Consequently, the RWE will be almost equal at all resolutions levels, and the SWS will acquire its maximum possible value. We define, the *Normalized Total Wavelet Entropy*, as the normalized SWS given by

$$\mathcal{H}[P^{(W)}] = \mathrm{S}[P^{(W)}] \;/\; \mathrm{S}_{max} \tag{9}$$

where $\mathrm{S}_{max} = \mathrm{S}[P_e^{(W)}] = \ln N_J$, being P_e the uniform the uniform distribution $P_e^{(W)} = \{1/N_J, \cdots, 1/N_J\}$.

3.4 Wavelet MPR-Statistical Complexity

It is widely known that an entropy measure does not quantify the degree of structure or patterns present in a process [44]. Moreover, it was recently shown that measures of statistical or structural complexity are necessary because they capture the property of organization [45]. This kind of information is not discriminated by randomness measures.

The opposite extremes of perfect order and maximal randomness (a periodic sequence and a fair coin toss, for example) possess no complex structure, and then these systems are too simple and should have zero statistical complexity. At a given distance from these extremes, a wide range of possible degrees of physical structure exists, which should be quantified by the statistical complexity measure (SCM).

Lamberti, Martín, Plastino and Rosso [46] (see also Chapter 8) introduced an effective statistical complexity measure. This statistical complexity version is: *a)* an intensive quantity (in the thermodynamical sense), *b)* able to grasp the essential details of the erythrocytes dynamics, *c)* capable of distinguishing between different degrees of periodicity and chaos. It also provides important additional information regarding the peculiarities of the underlying probability distribution, not already detected by the entropy, in particular in the present case on the probability distribution associated with the photometrically recorded time series. This statistical complexity measure is defined, following the intuitive notion advanced by López-Ruiz *et al.* [47] (see also Chapter 7), through the product between the normalized Shannon entropy ($\mathcal{H}$) and the so colled "disequilibrium" ($\mathcal{Q}_J$).

Then for the wavelet energy based PDF, $P^{(W)}$, we define the *Wavelet MPR-Statistical Complexity*[1] as

$$\mathcal{C}[P^{(W)}] = \mathcal{H}[P^{(W)}] \cdot \mathcal{Q}_J[P^{(W)}] \, , \tag{10}$$

where $\mathcal{H}$, the normalized Shannon entropy, is given by Eq. (9). The disequilibrium $\mathcal{Q}_J$ is defined in terms of the extensive (in the thermodynamical sense) Jensen-Shannon divergence. That is,

$$\mathcal{Q}_J[P^{(W)}] = Q_0 \cdot \mathcal{J}_S[P^{(W)}, P_e^{(W)}] \, , \tag{11}$$

with

$$\mathcal{J}_S[P^{(W)}, P_e^{(W)}] = \mathrm{S}\left[\frac{P^{(W)} + P_e^{(W)}}{2}\right] - \frac{1}{2}\mathrm{S}[P^{(W)}] - \frac{1}{2}\mathrm{S}[P_e^{(W)}] \tag{12}$$

the above-mentioned Jensen-Shannon divergence and Q_0 a normalization constant ($0 \leq \mathcal{Q} \leq 1$) equal the inverse of the maximum possible value of $\mathcal{J}_S[P^{(W)}, P_e^{(W)}]$. This value is obtained when one of the components of $P^{(W)}$, say p_m, is equal to 1 and the remaining p_i are equal to zero. The Jensen-Shannon divergence, which quantifies the difference between two

[1] that is the Shannon Jensen Statistical Complexity defined in Chapter 8, but now evaluated with a PDF based on wavelet transform of the signal.

(or more) probability distributions, is especially useful to compare the symbol composition between different sequences [48].

The complexity measure constructed in this way is intensive, as many thermodynamic quantities [46]. We stress the fact that the above SCM is not a trivial function of the entropy because it depends on two different probabilities distributions, the one associated to the system under analysis ($P^{(W)}$ in the present case), and the uniform distribution, $P_e^{(W)}$. Furthermore, it was shown that for a given $\mathcal{H}$ value, there exists a range of possible SCM values [49]. Thus, it is clear that $\mathcal{C}$ carries important additional information (related to the correlational structure between the components of the physical system) that is not contained in the entropic functional [50, 51].

A detailed analysis of the $\mathcal{C}$-behaviour demonstrates the existence of bounds to $\mathcal{C}$ that we called $\mathcal{C}_{max}$ and $\mathcal{C}_{min}$. These bounds can be systematically evaluated by recourse to a careful geometric analysis performed in the space of probabilities Ω [49]. The corresponding values of $\mathcal{C}_{max}$ and $\mathcal{C}_{min}$ depend only on the probability space's dimension and, of course, on the functional form adopted by the amount of disorder $\mathcal{H}$ and the disequilibrium $\mathcal{Q}$. $\mathcal{H} \times \mathcal{C}$ diagrams, the so-called entropy-complexity plane, are important and yield system's information independently of the values that the different control parameters may adopt (see, for instance, Refs. [23, 47, 52, 53], and references therein). The bounds also provide us with relevant information that depends on the system's particular characteristics, as for instance, the existence of global extrema, or the peculiarities of the system's configuration for which such extrema is obtained.

4 Results and Discussion

4.1 β-thalassemia minor samples

The analyzed time series correspond to the fluctuations of the circular pattern of the erythrocytes for healthy donors (control samples) and heterozygous β-thalassemia patients. The erythrocytes suspensions were obtained according to the experimental procedure described above and following the International Committee for Standarization on Haemathology [33]. The photometrically recorded time series have a length of $M = 256$ *data* and the corresponding discrete orthogonal wavelet transform (using a spline cubic mother wavelet) were obtained for resolution levels $j = -1, \cdots, -7$. Taken the wavelet coefficients set $\{C_j(k)\}$, the energy wavelet based probability distribution $P^{(W)} = \{\rho_j\}$, corresponding to controls and β-thalassemia were evaluated.

The grand average of the relative wavelet energy for controls and β-thalassemia are presented in Fig. 5. The relative energy bands appear better distributed in the β-thalassemia patients, than in control samples, which could be a signature of the appearance of chaos on the first ones. The relative wavelet energy on healthy donor samples is greater on levels $j = -1, -2, -3$, while on β-thalassemia patient samples the relative wavelet energy is concentrated on levels $j = -5, -6, -7$. In particular on level $j = -7$ the relative wavelet energy value on β-thalassemia patient samples, is important and this behaviour could correspond to a minor entropy.

We show the Normalized Total Wavelet Entropy ($\mathcal{H}$) values obtained for the 19 different samples studied in Fig. 6. We observed that the wavelet entropy averaged values for control samples is $< \mathcal{H}^{(c)} >= 0.889$, this value could be related to homogeneous discocytes samples which is the dynamical expression of the cell capability of changing their shape in flow. On the other hand the entropy values for β-thalassemia patients have an averaged value of $< \mathcal{H}^{(\beta)} >= 0.785$. This decrease in the mean value could be caused by the loss of viscoelasticity in some cells and could be associated with acanthocyte and dacrocyte cells in the samples. This is compatible with the fact that all the patients studied were β-thalassemia minor, so they do not need to receive blood transfusion and their population of erythrocytes is constituted by healthy and damaged cells. We have to remark that all the entropy values, $\mathcal{H}^{(c)}$, for controls are higher than the corresponding entropy values, $\mathcal{H}^{(\beta)}$, corresponding to β-thalassemia patients; there are two clouds of points for each population of samples; and they are clearly differentiated.

The wavelet MPR-Statistical Complexity ($\mathcal{C}$) values obtained for the different samples: healthy donors, and β-thalassemia patients analyzed are presented in Fig. 7. Note that the averaged complexity values for β-thalassemia patients is $< \mathcal{C}^{(\beta)} >= 0.172$, while for control samples population is $< \mathcal{C}^{(c)} >= 0.096$, this reduction on the wavelet MPR-Statistical Complexity reflects the interplay between its disequilibrium and the amount of information stored in the red blood cells deformability dynamics. This information measure quantifies aspects of the intricate structures hidden in the photometrically recorded series of red blood cells. In the group of control samples, the erythrocytes are constituted by healthy cells, while in the group of β-thalassemia patients, the erythrocytes are constituted by healthy and damaged cells. Moreover, $< \mathcal{C}^{(\beta)} >$ is not only higher than $< \mathcal{C}^{(c)} >$ but also all the values obtained for control samples are smaller than the ones obtained for β-thalassemia patients.

In Fig. 8 we display our results in the entropy-complexity plane $\mathcal{H} \times \mathcal{C}$. As mentioned previously (see Sec. 3.4), the corresponding complexity values are bounded by $\mathcal{C}_{min}$ and $\mathcal{C}_{max}$ curves, evaluated in this case for $N_J = 7$. Their values are displayed in this figure using continuous lines. Only values in the interior of the region defined by these two limit curves can be numerically obtained. All the photometrically recorded series in the present entropy-complexity planes present values that are almost equidistant from the two limit curves; however the localization of the clouds of points corresponding to controls and β-thalassemia patients are in well differentiated regions.

It is clear from Fig. 8 that our introduced wavelet based informational tools are able to distinguish between the two sample populations. The control samples are more disordered. The dynamics corresponds to white noise, where the maximum values of entropy are reached together with minimum values of complexity. While, the β-thalassemia patients samples exhibit decreasing entropy values and increasing complexity values, that could correspond to short memory fractional Brownian motion; results consistent with our previous works [7, 8, 10, 54].

For the population of control samples we found that in our entropy-complexity plane the pertinent point lies in the high entropy interval, $(0.860, 0.911)$ and low complexity values $(0.075, 0.125)$. In fact this behaviour is quite close to the total random expected values corresponding to $\mathcal{H} \approx 1$ and $\mathcal{C} \approx 0$. In consequence one can associate an almost random behaviour with low correlation structures presented in the time series. All the photometrically recorded series from β thalassemia patients are localized in the entropy-complexity plane in the entropy region lying in the interval $(0.735, 0.838)$ and complexity interval $(0.125, 0.225)$. This decrease in entropy values reduced random behaviour and increase in the complexity values can be associated with correlated structures immersed in the time series.

From inspection of Figs. 6 to 8 it is clear that our introduced wavelet-based informational tools are able to distinguish between the two types of RBC under study. In particular, on Fig. 8, β-thalassaemia patients are almost equidistant between the curves of maximum and minimum complexity, while the control samples are closer to the curve of minimum complexity. On comparing the two groups of samples both have strong stochastic behaviour, however the control ones became almost ideal noise with $\mathcal{H} = 1$ and $\mathcal{C} = 0$, while disease samples appear more correlated and consequently, entropy decreases and complexity increases.

β-thalassemia minor is a silent disease, usually incorrectly prescribed with iron. This fact could explain the difference observed in the wavelet based Information Theory quantifiers in relation with the controls, even when they do not differ excessively. In fact, little information exists to explain the erythrocytes viscoelastic properties. The present results suggest that the percentage of discocytes and also acontocytes, leptocytes and target cells, which are present in a sample are reflected in the corresponding photometrically recorded series of β-thalassaemia patients affecting the corresponding stochastic dynamics with an increase of the degree of correlation on them.

4.2 Glucose incubated erythrocytes samples

In a previous work [36], was observed that the erythrocyte Deformability Index (DI), which is a measure of the erythrocyte elongation (see Sec. 2.5), decreased as the glucose concentration increased. Beyond a glucose concentration of 2%, the DI is rapidly decreased (for samples without treatment DI = 0.65; for a 2%, 5% and 10% glucose incubated samples DI =0.64, 0.62, and 0.52 respectively).

We observed the presence of haemolysis during post glucose washing with PBS at all the samples incubated at 10% glucose concentration, and in two samples incubated at 5% glucose concentration. That is the reason for not including in our analysis the samples incubated at 10%, as well as those samples with less incubated percentage of glucose concentration in which after being washed with PBS appeared the presence of haemolysis. This fact implied that oxidative stress due to high glucose concentrations, which caused damage to the erythrocyte membrane proteins even in a relative short exposure time, affected the erythrocyte rheological characteristics.

The analyzed time series correspond to the fluctuations of the elliptical long axes pattern from erythrocytes without treatment, 2% and 5% glucose incubated RBCs. The photometrically recorded time series have a length of 512 *data* points and the corresponding discrete orthogonal wavelet transform (using a spline cubic mother wavelet) were obtained for resolution levels $j = -1, \cdots, -8$. Taken the wavelet coefficients set $\{C_j(k)\}$, the relative wavelet energy based probability distribution $P^{(W)} = \{\rho_j\}$, corresponding to the photometrically recorded time series were evaluated.

We show in Fig. 9 the realtive wavelet energies (RWE) values for each one of the eight wavelet frequency band considered, averaged over the samples corresponding to the same case. The RWE for all the samples studied is greater on level $j = -8$, and decreased on increasing j. By comparing the RWE, for the eight wavelet levels considered, for 2%, 5% and without treatment samples, we can not make tentative distinctions between the different populations. To make progress on the study, one needs more insights of the dynamics going through the Normalized Total Wavelet Entropy $\mathcal{H}$, and wavelet MPR-Statistical Complexity $\mathcal{C}$ values.

In Fig. 10, we show the Normalized Total Wavelet Entropy $\mathcal{H}$ values obtained for the different samples studied in the present work. We observed that the value for the wavelet entropy averaged over all the without treatment samples is $< \mathcal{H}^{(wt)} >= 0.67639 \pm 0.00219$, this value could be related to homogeneous discocytes in the samples which is the dynamical expression of the cell capability of changing their shape in flow. On the other hand the average entropy values for the samples that have been incubated with 2% and 5% glucose are $< \mathcal{H}^{(2\%)} >= 0.67463 \pm 0.0024$ and $< \mathcal{H}^{(5\%)} >= 0.67156 \pm 0.00097$ respectively. This decrease in the mean value could be caused by the alterations of viscoelasticity in some cells, as a result of the glucose action.

The wavelet MPR-Statistical Complexity measure $\mathcal{C}$ values obtained for the different samples analyzed are presented in Fig. 11. The averaged complexity values results: for samples without treatment $< \mathcal{C}^{(wt)} >= 0.23736 \pm 0.00068$, for 2% and 5% glucose $< \mathcal{C}^{(2\%)} >= 0.23718 \pm 0.00074$ and $< \mathcal{C}^{(2\%)} >= 0.23723 \pm 0.00037$, respectively. As we mention, the variation on the wavelet MPR-Statistical Complexity values reflects the interplay between its disequilibrium and the amount of information stored in the red blood cells deformability dynamics, as we mentioned before. This information measure quantifies aspects of the intricate structures hidden in the photometrically recorded series of red blood cells.

In Fig. 12 we display our results in the plane $\mathcal{H} \times \mathcal{C}$. As mentioned previously (see Sec. 3.4), the corresponding complexity values are bounded by $\mathcal{C}_{min}$ and $\mathcal{C}_{max}$ curves evaluated, in this case for $N_J = 8$. All the photometrically recorded series in the present entropy-complexity plane present values that are almost equidistant from each one of these two curves $\mathcal{C}_{min}$ and $\mathcal{C}_{max}$ (not shown in the figure due to scale values used). However the localization of points corresponding to the different conditions is in well differentiated regions.

It is clear that our introduced wavelet based informational tools are able to distinguish between the different samples populations (see Fig. 12). The RBCs without treatment samples are more disordered (bigger entropy values) than the 2% glucose and also these ones are more disorder than the 5% glucose treatment. The dynamics for RBCs without treatment samples could correspond to fractional Gaussian noise (fGn), results consistent with a previous work [54], while, the 2% and 5% glucose incubated samples exhibit decreasing entropy values and increasing complexity values, that could correspond to short memory fractional Brownian motion (fBm); results consistent with our previous works [8, 55].

From inspection of Figs. 10 to 12 it is clear that our introduced wavelet-based Information Theory tools are able to distinguish between the different samples of RBC under study. Particularly, in Fig. 12, the different sample populations are almost equidistant between the curves of maximum and minimum complexity, even thought this general characteristic we can say that while the without treatment samples are closer to the curve of maximum complexity bound, the 2% and 5% glucose-incubated samples are closer to the minimum complexity bound.

5 Conclusions

For succeeding in the plan of explaining complex behaviour from first principles, researchers have been looking for simplified models, mathematically tractable and able to catch the essence of complexity. The present review applies Information Theory tools derived from the orthogonal discrete wavelet transform and their application to the analysis and characterization of red blood cells deformation.

Now, the natural question that arises is the following one: which are the mechanical conditions the RBCs must fulfill to guarantee that the statistical mechanics calculations would predict, with an adequate degree of accuracy, the averaged quantities obtained from our laboratory experiment. And when trying to answer such an apparently simple question, one discovers that even the simplest system can give place to very intricate behaviour.

As we mention above, our preliminary studies on red blood cells (RBCs) complex behaviour [7, 8, 9, 10, 11], have shown the competition between stochastic ordinal Brownian motion (oBm) and fractional Brownian motion (fBm), and make a crossover from one relaxation scenario when the erythrocytes are at rest to the other one, when the erythrocytes are subjected to shear stress. On this basis we applied quantifiers based on Information Theory in combination with orthogonal wavelets transform in order to study the manifestations of this behaviour.

In the present chapter we reviewed the application of wavelet-based Information Theory quantifiers (Normalized Total Shannon Entropy, intensive Statistical Complexity and Entropy-Complexity plane (see Chapter 8)) on red blood cells membrane viscoelasticity characterization. These quantifiers exhibit important localization advantages provided by the Wavelet Theory. The present approach produces a clear characterization of this dynamical system, finding out an evident manifestation of a random process on the red cell samples of healthy individuals, and its sharp reduction of randomness on analyzing a human haematological disease, such as β-thalassaemia minor [12]. We also review the application to the characterization on the effects produced by an increased concentration of glucose in suspending medium on erithrocytes [13]. The results show that the increase in glucose concentration does not produce significant changes on the Relative Wavelet Entropy, while significant ones on the Normalized Total Wavelet Entropy, which combined with Wavelet MPR-Statistical Complexity values allow to identify different behaviour for all the different populations in the entropy-complexity plane. Modification in the hemorheological properties of cells could be clearly detected with these Wavelet-based Information Theory quantifiers

Taken into account the results presented in the previous section, we can conclude that the wavelet based Information Theory quantifiers as well as the representation entropy-complexity plane, $\mathcal{H} \times \mathcal{C}$, could be a useful tool when dealing with biological data, which always have a stochastic component due to omnipresent dynamical noise, and could be used to classify different degrees of random behaviour. Moreover, this representation plane is an effective tool for revealing the sometimes hidden difference between chaos and noise as was recently shown [52]. The entropy-complexity plane yields additional information about the dynamics of the erythrocytes deformability. The use of the proposed entropy complexity plane, based on data wavelet transform, can contribute to the analysis of erythrocytes membrane viscoelastic properties, and may also lead to a better understanding of their dynamics. Certainly the use of this plane is not intended to replace conventional analysis; instead, it aims to provide further insights into the underlying dynamics.

Conflict of interest: None.

Acknowledgment

O. A. Rosso acknowledges partial support from CONICET, Argentina, and CAPES, PVE fellowship, Brazil. M. T. Martín acknowledges partial support from CONICET, Argentina.

References

[1] Wigner, E. The unreasonable effectiveness of mathematics in the natural sciences. *Communications in Pure and Applied Mathematics* **13** 1–14 (1960).

[2] Shannon, C.E. A mathematical theory of communication. *Bell Syst. Technol. J.* **27** 379–423, 623–656 (1948).

[3] Kolmogorov, A.N. A new metric invariant of transitive dynamic system and automorphysms in Lebesgue spaces. *Dokl. Akad. Nauk. SSSR* **119** 861–864 (1958).

[4] Sinai, Y.G. On the concept of entropy of dynamical system. *Dokl. Akad. Nauk. SSSR* **124** 768–771 (1959).

[5] Ott, E.; Sauer, T.; Yorke, J.A. *Coping with Chaos*, Wiley, New York, 1994.

[6] Sprott, J.C. *Chaos and Time Series Analysis*, Oxford University Press, Oxford, 2004.

[7] Korol, A.M.; Valverde, J.R.; Rasia, R.J. Viscoelasticity: fractal parameters studied on mammalian erythrocytes under shear stress. *Experimental Mechanics* **42**, 172–177 (2002).

[8] Korol, A.M.; Rasia, R.J. Signatures of deterministic chaos in dyslipidemic erythrocytes under shear stress. *Chaos* **13** 87–93 (2003).

[9] Korol, A.M.; Rasia, R.J.; Rosso, O.A. Alterations of thalassemic erythrocytes detected by wavelet entropy. *Physica A* **375** 257–264 (2007).

[10] Korol, A.M.; Foresto, P.; Rosso, O.A. Self-organizing dynamics of human erythrocytes under shear stress. *Physica A* **386** 770–775 (2007).

[11] Korol, A.M.; Foresto, P.; Darrigo, M.; Rosso, O.A. Diabetic erythrocytes test by correlation coefficient *The Open Medical Informatics Journal* **2** 105–111 (2008).

[12] Korol, A.M.; Foresto, P.; Pérez, S.; Martín, M.T.; Rosso, O.A. Preliminary characterization of erythrocytes deformability on the entropy–complexity plane. *Open Medical Informatics Journal* **4** 164–170 (2010).

[13] Korol, A.M.; Rosso, O.A.; Martín, M.T.; D'Arrigo, M.; Riquelme B.D. Impairment of erythrocytes incubated in glucose medium: a wavelet information theory analysis. *Cell Biochemistry and Biophysics* **60** 329–334 (2011).

[14] Rosso, O.A.; Blanco, S.; Jordanova, J.; Kolev, V.; Figliola, A.; Schürmann, M.; Başar, E. Wavelet entropy: a new tool for analysis of short duration brain electrical signals. *J. Neurosc. Meth.* **105** 65–75 (2001).

[15] Rosso, O.A.; Martín, M.T.; Plastino, A. Brain electrical activity analysis using wavelet based informational tools. *Physica A* **313** 587–608 (2002).

[16] Rosso, O.A.; Mairal, M.L. Characterization of time dynamical evolution of electroencephalographic records. *Physica A* **312** 469–504 (2002).

[17] Rosso, O.A.; Martín, M.T.; Plastino, A. Brain electrical activity analysis using wavelet based informational tools (II): Tsallis non-extensivity and complexity measurements. *Physica A* **320** 497–511 (2003).

[18] Rosso, O.A.; Blanco, S.; Rabinowicz, A. Wavelet analysis of generalized Tonic-Clonic epileptic seizures. *Signal Processing* **83** 1275–1289 (2003).

[19] Plastino, A.; Martín, M.T.; Rosso, O.A. Generalized information measures and the analysis of brain electrical signals. In *Nonextesive Entropy - Interdisciplinary applications.* Edited by: Gell-Mann, M.; Tsallis, C. Santa Fe Institute Studies in the Sciences of Complexity, 261–293 (2004).

[20] Rosso, O.A.; Martín, M.T.; Plastino, A. Evidence of self-organization in brain electrical activity using wavelet based informational tools. *Physica A* **347** 444–464 (2005).

[21] Rosso, O.A.; Hyslop, W.; Gerlach, E.; Smith, R.L.L.; Rostas, J.; Hunter, M. Quantitative EEG analysis of the maturational changes associated with childhood absence epilepsy. *Physica A* **356** 184–189 (2005).

[22] Plastino, A.; Rosso, O.A. Entropy and statistical complexity in brain activity. *European Physical Society - Europhysics News* **36** 224–228 (2005). Special Number on: "Nonextensive Statistical Mechanics - Theory and Applications", Boon, J.P.; Tsallis, C. (Guest Editors).

[23] Rosso, O.A.; Martín, M.T.; Figliola, A.; Keller, K.; Plastino, A. EEG analysis using wavelet-based informational tools. *J. Neurosc. Meth.* **153** 163–182 (2006).

[24] Rosso, O.A. Entropy changes in brain function. *Int. Journal of Psychophysiology* **64** 75–80 (2007).

[25] Pereyra, M.; Lamberti, P.W.; Rosso, O.A. Wavelet Jensen-Shannon divergence as a tool for studying the dynamics of frequency band components in EEG epileptic seizures. *Physica A* **379** 122–132 (2007) .

[26] Guyton, A.C.; Hall, J.E. *Textbook of Medical Physiology*, W. B. Sanders Company, Philadelphia, USA, 1996.

[27] Weatheral, D.J.; Clegg, J.B. *Thalassemia Syndromes*, Blackwell Scientific Publication, London, 1981.

[28] Pérez, S.; Acosta, I.; Milani, A.; Rasia, R.J. New evidence of red blood cell rheological disorders in beta-thalassaemia minor. *Clin. Hemorheol. Microcirc.* **31** 89–96 (2004).

[29] Brown, C.; Ghali, H.; Zhao, Z.; Thomas, L.; Friedman, E. Association of reduced red blood cell deformability with diabetic nephropathy. *Kidney Int.* **67** 295–300 (2005).

[30] McMillan, D.E.; Utterback, N.G.; La Puma, J. Reduced erythrocytes deformability in diabetes. *Diabetes* **27** 895–901 (1978).

[31] McMillan, D.E.; Utterback N.G.; Mitchell, T.P. Doublet formation of diabetic erythrocytes as a model of impaired membrane viscous deformation. *Microvascular Research* **26** 205–220 (1983).

[32] Szelachowska, W.; Gries, A.F.; Kinalska, I. Activity of Ca/Mg ATPase in erythrocyte membrane of women with diabetes mellitus type I. *Endocrynol. Pology* **43** 23–29 (1992).

[33] International Committee of Standarization Haemathology. (Expert Panel on Blood Rheology). Guidelines for measurements of blood viscosity and erythrocyte deformability. *Clin Hemorheol. Microcirc.* **6** 439–453 (1996).

[34] Bourdon, E.; Loreau N.; Blanche, D. Glucose and free radicals impair the antioxidant properties of serum albumin. *The FASEB J.* **13** 233–244 (1999).

[35] Lapolla, A.; Gerhardiger C.; Dal Frá, M.; Franchin, A.; Fedele, D.; Crepaldi, G. Glycated erythrocyte membrane proteins and hemorheological parameters in insulin dependen diabetic subjects. *Climical Hemorheology* **11** 405–415 (1991).

[36] Riquelme, B.; Foresto, F.; Darrigo, M.; Valverde J.; Rasia, R. A dynamic and stationary rheological study of erythrocytes incubated in a glucose medium. *J. Biochem. Biophys. Methods* **62** 131–141 (2005).

[37] Rasia, R. Quantitative evaluation of erythrocyte viscoelastic properties from difractometric data: applications to hereditary spherocitosis and hemoglobinopathies. *Clinical Hemorheology* **15** 177–189 (1995).

[38] Daubechies, I. *Ten lectures on wavelets*, SIAM, Philadelphia, 1992.

[39] Mallat, S. *A wavelet tour of signal processing*, Academic Press, San Diego, 1999.

[40] Samar, V.; Bopardikar, A.; Rao R.; Swartz, K. Wavelet analysis of neuroelectric waveforms: a conceptual tutorial. *Brain Lang.* **66** 7–60 (1999).

[41] Unser, M. Spline: a perfect fit for signal and image processing. *IEEE Signal Process. Mag.* **16** 22–38 (1999).

[42] Thevenaz, P.; Blue, T.; Unser, M. Interpolation revisited. *IEEE Trans. Med. Imaging* **19** 739–758 (2000).

[43] Blanco, S.; Figliola, F.; Quian Quiroga, R.; Rosso, O.A.; Serrano, E. Time-frequency analysis of electroencephalogram series (III): information transfer function and wavelets packets. *Phys. Rev. E* **57** 932–940 (1998).

[44] Feldman, D.P.; Crutchfield, J.P. Measures of statistical complexity: Why?. *Phys. Lett. A* **238** 244–252 (1998).

[45] Feldman, D.P.; McTague, C.S.; Crutchfield, J.P. The organization of intrinsic computation: Complexity-entropy diagrams and the diversity of natural information processing *Chaos* **18** 043106 (2008).

[46] Lamberti, P.W.; Martín, M.T.; Plastino, A.; Rosso, O.A. Intensive entropic nontriviality measure. *Physica A* **334** 119–131 (2004).

[47] López-Ruiz, L.; Mancini, H.; Calbet, X. A statistical measure of complexity. *Phys. Lett. A* **209** 321–326 (1995).

[48] Grosse, I.; Bernaola-Galván, P.; Carpena, P.; Román-Rolán, R.; Oliver, J.; Stanley, H.E. Analysis of symbolic sequences using the Jensen-Shannon divergence. *Phys. Rev. E* **65** 041905 (2002).

[49] Martín, M.T.; Plastino, A.; Rosso, O.A. Generalized statistical complexity measures: Geometrical and analytical properties. *Physica A* **369** 439–462 (2006).

[50] Rosso, O. A.; Masoller, C. Detecting and Quantifying Temporal Correlations in Stochastic Resonances by Information Theory Measure. *Eur. Phys. J. B* **69** 37–43 (2009).

[51] Rosso, O.A.; Masoller, C. Detecting and quantifying stochastic and coherence resonances via information theory complexity measurements. *Phys. Rev. E* **79** 040106(R) (2009).

[52] Rosso, O.A.; Larrondo, H.A.; Martín, M.T.; Plastino, A.; Fuentes, M.A. Distinguishing noise from chaos. *Phys. Rev. Lett.* **99** 154102 (2007).

[53] Rosso, O.A.; De Micco, L.; Larrondo, H.A.; Martín, M.T.; Plastino, A. Generalized statistical complexity measure. *Int. J. Bif. and Chaos* **20** 775–785 (2010).

[54] Zunino, L.; Pérez, D.G.; Martín, M.T.; Plastino, A.; Garavaglia, M.; Rosso, O.A. Characterization of gaussian self-similar stochastic processes using wavelet-based informational tools. *Phys. Rev. E* **75** 021115 (2007).

[55] Korol, A.; Rasia, R.J. Correlated random walk: a fractal approach to erythrocyte viscoelastic properties. *Clinical Hemorheology and Microcirculation* **20** 97–103 (1999).

Concepts and Recent Advances in Generalized Information Measures and Statistics, 2013, 375-407

CHAPTER 15

Information-Theoretic Analysis of the Role of Correlations in Neural Spike Trains

Fernando Montani [a,*], **Simon R. Schultz** [b]

[a] Instituto de Física La Plata, CONICET and Universidad Nacional de La Plata 115 y 49 C.C: 67 1900 La Plata Argentina

[b] Department of Bioengineering, Imperial College London, South Kensington, London SW7 2AZ

Abstract

We have applied an information theoretic approach to gain insights of the role of spike correlations in the neuronal code. First, we illustrate and compare the different methods used in the literature to remove sample size dependent bias from information estimations. Then, we use a modified version of the information components breakdown to quantify the contribution of individual members of the population, the interaction between them, and the overall information encoded by the ensemble of neurons making an especial emphasis of the separation between contributions due to the noise and signal spike correlations. This formalism is applied to a set of multi-neuronal spike data with different stimuli configurations.

Keywords: Information Theory; Neural Code; Sampling bias correction; Spike correlations; Sensory encoding.

1 Introduction

Early in the twentieth century, [1], using the afferent nerve of stretch receptor in the frog muscle, showed that it was possible by amplification to record the electrical impulse in single nerve fibers. They summarized their results in this single sentence: "The impulses set up by a single end-organ occur with a regular rhythm at a frequency which increases with the load on the muscle and decreases with the length of time for which the load has been applied". This observation that the firing rate of peripheral touch receptors coded for the pressure applied to a patch of skin began a long period of investigation of "the neural code" by neuroscientists.

* Corresponding author. E-mail: fmontani@gmail.com

Andres M. Kowalski, Raul D. Rossignoli and Evaldo M. F. Curado (Eds)

However, understanding brain function requires interdisciplinary approaches involving many levels of study: from the molecular level through the cellular level (individual neurons), to the level of relatively small assemblies of neurons (for example cortical columns); to the case of larger subsystems such as the one which subserves visual perception, and up to the level of large systems including the cerebral cortex, the cerebellum, and the nervous system as a whole. Neuroscience encompasses approaches ranging from molecular and cellular studies to human psychophysics and psychology. Moreover, theoretical analysis of physiological data and computational modeling are important tools for characterizing what nervous systems do, determining how they function, and understanding why they operate in particular ways.

Unfortunately, it is extremely difficult to understand exactly how sensory information is processed in the mammalian brain. For example, the influx of sensory information to a single mammalian retina is detected by an array of millions of receptors, each capable of detecting the difference between hundreds of gray levels. They have integration times which allow them to detect new signals in tens to hundreds of milliseconds [2]. These inputs are then processed by hundreds of millions of other neurons, and many of them interact with each other in complex ways. Moreover, ongoing (recurrent) activity in the brain, which very likely reflects information processing functions in the brain, may modify the sensory representation dynamically [3]. With this huge level of complexity, the experimental tools which have been available during the last thirty years were insufficient to account for all the involved variables.

One of the main experimental challenges in neuroscience during the last thirty years was to demonstrate that distributed neural populations in the visual cortex process information in a cooperative way. The visual cortex is composed of a large number of areas, which contain neurons that are tuned to different visual features. Correlations between the time-series of activity of pairs of individual neurons in the visual cortex were investigated in many laboratories beginning in the early 1980s [4, 5, 6, 7, 8, 9, 10, 11, 12, 13], most often with the motivation of revealing structural coupling between cells. This approach to functional anatomy had been methodologically and conceptually outlined by [14, 15, 16, 17, 18, 19], and successfully applied to the invertebrate nervous system [20]. The results of these different studies showed the presence of temporally precise synchronization (between the spike trains of different cells) within the cat and macaque visual cortex. But, these correlations occurred between cells showing thc samc feature preference. Because the peaks in the correlograms were centered around zero the obtained correlations were attributed to common input. No attempt was made at this stage to determine whether response synchronization depend on the stimulus configuration.

Cortical neurons receive 3,000-10,000 synaptic inputs, 85 % of which are excitatory [21, 22]. Nearly half of the excitatory input to any one neuron comes from nearby neurons that fall within a cylinder of 100-200 μ m radius, arranged as a column, sometimes termed a mini-column [23, 24, 25, 26]. This suggests that cortical neurons receive abundant excitatory input and are embedded in a network of highly convergent signals. The networks have a recurrent nature, it is likely therefore that most excitatory neurons receive similar inputs and emits spikes under similar conditions. This means that the conditions that lead to a response of any of one neuron in mini-column are likely to involve considerable activity from a large number of its inputs and its outputs. It is therefore reasonable to expect that many input spikes will arrive in synchrony within a very small time window. In addition, neurons operating under *in vivo* conditions, during which they receive a barrage of synaptic input, have been shown to have much shorter effective membrane integration times than they do *in vitro*, indicating that input spikes coordinated on a millisecond timescale may be especially effective in driving output spiking [27, 28, 29].

Understanding the role of correlated neuronal activity is key to understanding the neural code, and thus neural information processing. A better understanding of the role of correlations will have widespread implications, not only for our basic understanding of how the brain operates, but also in our understanding of neuronal dysfunctions [30].

However, the matter of how the coding of sensory information is affected by spike correlations across neurons has been fraught with dissension, and the debate has become polarized into two different postures that co-exist in the neuroscientific community. One of the positions, supported by many authors [31, 32, 33, 34, 35, 36, 37, 38] is that correlations are important in the coding of sensory information, whereas others have proposed that they are either epiphenomenal [39, 40, 41, 42, 43, 44, 45, 46, 47, 48], or are involved in other aspect of information processing, such as attention [49].

This dissension relates partly to different methods used to assess the role of synchronized spike firing. For instance, one of the most used methods consisted in looking for stimulus-dependent changes in cross-correlograms [31, 32, 33, 35, 36]. However, the firing rate can alter the shape of cross-correlograms making it difficult to separate information carried by firing rates from information carried by correlations. Moreover, cross-correlograms account only for near synchronous spikes. This is, they do not account for correlations which occur on a longer time scales and among patterns of spikes.

Information Theory provides us with a measure of the uncertainty in a message (see chapter 2 for a review of Information Entropy Measures). It was invented as a general framework for quantifying the ability of a coding scheme or a communication channel (such an optic nerve) to convey information [50]. It is assumed that the code involves a number of symbols (such as different neuronal responses), and that the coding and transmission process are stochastic and noisy. Communication channels, if they are noisy, have only limited capacities to convey information. The techniques of information theory are used to evaluate these limits and find coding schemes that saturate them. Moreover, information theory allows us to address how much the neuronal responses tell us about the stimulus. We can use them to ask what forms of neuronal response are optimal for conveying information about the neuronal stimuli. Therefore many brain structures can be considered as communication channels, and an appropriate mathematical framework is the Information Theory. Inherent in the application of Information Theory to neuroscience is the assumption that the representation of information in the nervous system is *efficient.*

2 Estimating entropy and information

At the heart of almost all applications of Information Theory to neural spike trains is the estimation of entropy. In this section we will review some of the basic problems of entropy and mutual information estimations for discrete variables and give a brief classification of different approaches to the problem.

It is well known that the naive or "plugin" estimator of entropy tends to underestimate it - it is biased [51]. This is can be understood in the framework of the Jensen inequality [52]. The entropy is a concave function and therefore a "plugin" estimator of entropy tends to underestimate the real value of the entropy (a rigorous proof of this statement can be found in [53]). The usual approach has been to attempt to avoid this by the use of a perturbative expansion in the asymptotic regime [54]. An alternative idea, that of calculating entropy by counting coincidences, was proposed a long time ago by Ma for physical systems in the micro-canonical ensemble, where a uniform distribution of entropy corresponds to states of fixed energy [55]. In a neuroscience context, this corresponds to the assumption that the probability distribution of response words at a fixed spike count is close to uniform [56, 57]. The Bayesian approach proposed by Nemenman, Shafee and Bialek [58], building upon ideas formulated by [59] and [60], extends this idea to arbitrarily complex distributions. The goal of the NSB method is to construct a Bayesian prior which generates a nearly uniform distribution of entropies in order to correct sample size dependent bias at its source.

In the following we will review the basic ideas of the NSB entropy estimator method [61], compare the effectiveness of NSB in comparison with the computation of the entropy *via* the naive "plugin" approach, and with an analytical approximation for the finite sampling bias [54]. We will then suggest some practical practical ways that information estimates can be improved when dealing with real data.

Consider the problem of estimating the Shannon entropy for a given probability distribution $\mathbf{p} = \{p_i\}$,

$$H = -\sum_{i=1}^{K} p_i \log_2 p_i \tag{1}$$

where the index i runs over K possibilities. For instance, $\mathbf{p}$ might be the distribution of spike counts observed to be fired by a neuron – in which case $i$would represent "number of spikes + 1" (as zero spikes fired is also a possible response). Consider N samples (trials) which were obtained from a given experiment, where each possibility i occurred n_i times: if N is much bigger than K we can approximate $p_i \approx f_i = n_i/N$, and therefore the entropy can be expressed in terms of the observed frequencies as

$$H_{naive} = -\sum_{i=1}^{K} f_i \log f_i \tag{2}$$

The problem is that this "plugin" approach tends to underestimate the entropy. Neurophysiological experiments can only collect a finite number of trials, and therefore the true stimulus-response probabilities cannot be estimated exactly. The estimated probabilities fluctuate around their true values. Several attempts to solve this problem were made by [62], who made asymptotic bias corrections by adding a term of order $O(K/N)$. This approach was developed further by [54], and amounts to (in brief)

$$H = H_{naive} + Bias[H(R)] \tag{3}$$

where

$$Bias[H(R)] \approx -\frac{1}{2Nln2}\sum_{s}\hat{R}_s \tag{4}$$

and $\hat{R}_s$ denotes the number of relevant bins for for a trial with stimulus s, *i.e.*, the response bins in which the occupancy probability $p(i|s)$ (at given s) is non-zero. Note that if a bin is observed never to be occupied, it is ambiguous whether that is because it has a true zero occupancy probability, or because responses simply have not been observed for long enough. Choice of $\hat{R}_s$ is a subtle issue, and is to some extent dependent upon data characteristics. [54] made use of a Bayesian prior for the number of relevant bins, and iteratively re-estimated it. This procedure has been found to be effective for a number of types of single-unit electrophysiology data where dimensionality was not excessive (*e.g.*, [63, 64]), but after much detailed analysis, we have found its use to be inadequate for the analysis of many datasets to which we have had access, where luckily other methods could be made to work. Note that this does not suggest that this analytical estimate is of no value: indeed, it is extremely useful as a simple "rule of thumb" to determine what kind of sampling regime the dataset is in.

In 1998 [56] suggested a time discretization technique, and in order to correct sampling size bias of the information estimations, they proposed an asymptotic approximation slightly different from [54]. They extrapolated the value of information for an hypothetical infinite sample (Quadratic Extrapolation method - ***QE***)

$$H_N{}^{plug-in} = H_{true} + \frac{a}{N} + \frac{b}{N^2}, \tag{5}$$

where the free parameters a and b, depend on the characteristic of the stimulus-response probabilities. The parameters a and b are estimated from the data-set by simply selecting the best fit criteria, plotting the value of the 'plug-in information' *versus* the sampling size. Thus H_{true} is estimated extrapolating the interpolating curve at the limit point of infinite sampling that is $\frac{1}{N} \longrightarrow 0$.

Several approaches for estimating entropies without using such an asymptotic expansion approach have been presented [61, 53]. We will follow the approach of Nemenman *et al.* [61].

Recall our examination of the probability distribution $\mathbf{p}$ from equation (2). Bayes' rule tells us that we can express the posterior probability of $\mathbf{p}$ $(\equiv \{p_i\}\ ,\ i = 1..K$ discrete random variable), given that we have just observed i to have occurred n_i times.

$$P(p|n_i) = \frac{P(n_i|p)P(p)}{P(n_i)} \tag{6}$$

Note that the number of times we observe each response value i to occur must add up to the total number of experimental trials N.

$$\sum_{i=1}^{K} n_i = N \tag{7}$$

In equation (6), the "prior" distribution is $P(\mathbf{p})$ – in principle, we could choose a prior such that our estimator of the entropy of $P(\mathbf{p})$ does not depend upon the number of trials. This is obviously a desirable quality! To do this, we can express $P(\mathbf{p})$ in terms of the Dirichlet family of priors [61]. This should allow us to construct a prior which does not depend on inverse powers of N (which even the naive approximation does implicitly, as $f_i = \frac{n_i}{N}$. The Dirichlet family are characterised by a parameter β; they can be written as

$$P_\beta(p) = \frac{1}{Z}\delta(1 - \sum_{i=1}^{K} p_i)\prod_{i=1}^{K} p_i^{\beta-1} \tag{8}$$

where

$$Z = \frac{\Gamma^K(\beta)}{\Gamma(K\beta)} \tag{9}$$

δ and Z are functions which enforce the normalization of $\mathbf{p}$ and P_β respectively (δ being the Dirac delta function, and Γ stands for Euler's Γ-function). Maximum likelihood estimation corresponds to Bayesian estimation with this prior in the limit $\beta \to 0$, while a uniform prior is implemented by β =1 ([58]) observed that fixing a particular value of β (and thus fixing the prior) specifies the entropy almost exactly. For an "incorrect" prior, the estimate of entropy is thus dominated by the prior, rather than determined by our actual knowledge – that is to say, it is biased.

Ideally, we would like to compute the whole *a priori* distribution of entropies

$$P_\beta(H) = \int dp_1 dp_2 ... dp_K P_\beta(\{p_i\})\delta(H + \sum_{i=1}^{K} p_i \log_2 p_i) \tag{10}$$

But this is quite difficult to achieve. In order to get an entropy estimate with small bias one could simply fix a flat prior distribution of entropy $P(H)$. One way of doing so is by defining

$$P(H) = 1 = \int \delta(H - \xi) d\xi \tag{11}$$

where ξ is the expected entropy. If we could find a family of priors $P_\beta(\mathbf{p})$ which result in δ functions over H, and if by changing β we move the peak across the whole range of entropies uniformly, then we will effectively be choosing the proper prior for each entropy. We will thus come close to the objective of removing bias at its source. Because the entropy of distributions chosen for P_β is sharply defined and monotonically dependent on the parameter β we can effect this goal by averaging over β. The main idea of the NSB approach is thus to construct a prior

$$P_\beta(p) = \frac{1}{Z}\delta(1 - \sum_{i=1}^{K} p_i) \prod_{i=1}^{K} p_i^{\beta-1} \frac{d\xi(\beta)}{d\beta} P(\beta) \tag{12}$$

which will perform this task even when N is small. In this equation Z is again a normalizing coefficient and $\frac{d\xi(\beta)}{d\beta}$ ensures the uniformity for *a priori* expected entropy ξ.

This Dirichlet priors allow all the Kdimensional integrals to be calculated analytically (see [59], giving for the moments of the entropy

$$(H^{NSB})^m = \frac{\int d\xi \rho(\xi, n) \langle H^m(n)\rangle_{\beta(\xi)}}{\int d\xi \rho(\xi, n)} \tag{13}$$

where $\mathbf{n}=\{n_i\}$, and m=1,2 correspond to the entropy and its second moment. $\langle H^m[n_i]\rangle_{\beta(\xi)}$is the expectation value of the mth entropy moment at a fixed β [59] and the posterior density is a function of the proposed Dirichlet prior,

$$\rho(\xi|n) = P_\beta(\xi)\frac{\Gamma(\kappa(\xi))}{\Gamma(\beta(\xi))} \prod_{i=1}^{K} \frac{\Gamma(n_i + \beta(\xi))}{\Gamma(\beta(\xi))} \tag{14}$$

Summarizing, the main idea of NSB is to construct a Bayesian prior, which generates a nearly uniform distribution of entropies in order to avoid bias at its origin. These results were obtained by [58, 61]; we have tried here to give, in as simplistic terms as possible, an explanation of how the method works.

2.1 Shannon Mutual Information

The Shannon mutual information transmitted by the population response about the whole set of stimuli is defined as

$$I(S;\mathbf{R}) = \sum_{s\in S}\sum_{\mathbf{r}\in\mathbf{R}} P(s)P(\mathbf{r}|s)\log_2\frac{P(\mathbf{r}|s)}{P(\mathbf{r})} \tag{15}$$

where $P(\mathbf{r}|s)$ is the probability of observing response $\mathbf{r}$ when stimulus s is presented, $P(s)$ is the probability of presentation of stimulus s, and $P(\mathbf{r})$ is the probability of observing $\mathbf{r}$ in response to any stimulus. Notice that the neural population response is denoted by $\mathbf{r}$ (drawn from a response space R*)* and a sensory stimulus from a given stimulus set S will be denoted by s.

Notice that Shannon mutual is a distance-like measure (the Kullback-Leibler divergence) between two probability distributions: the joint probability of responses and stimuli, $P(\mathbf{r},\mathbf{s})$, and joint probability distribution that would apply if responses did not depend at all upon stimuli, $P(\mathbf{r})P(s)$. It can be alternatively expressed as,

$$I(S;\mathbf{R}) = H(R) - H(R|S) \tag{16}$$

where $H(R)$ is the total response entropy and $H(R|S)$ is the noise entropy. These two quantities are defined as

$$H(R) = -\sum_{\mathbf{r}\in R} P(\mathbf{r})\log_2 P(\mathbf{r}), \tag{17}$$

and

$$H(R|S) = -\sum_{s\in S} P(s)\sum_{r\in R} P(\mathbf{r}|s)\log_2 P(\mathbf{r}|s), \tag{18}$$

The mutual Information $I(S;\mathbf{R})$ thus quantifies the reduction of uncertainty about the stimulus that can, on average, be gained by observing a single trial of the neural response.

We can expect to have noise correlations if $P(\mathbf{r}|s)$ at fixed stimulus is different from the independent response probability, $P_{ind}(\mathbf{r}|s) = \prod_{c=1}^{C}\prod_{t=1}^{T} P(r_{ct}|s)$, where c denotes the label of each neuron and t denotes time bin. In the case in which the responses were independent stochastic variable, we can factorize $P_{ind}(\mathbf{r}\,|s)$ in terms of the product of the single bin marginals $P(\mathbf{r}|s)$. The noise entropy of $P_{ind}(\mathbf{r}\,|s\,)$ reads

$$H_{ind}(R) = -\sum_{\mathbf{r},\mathbf{s}} P_{ind}(\mathbf{r}|s)\log_2 P_{ind}(\mathbf{r}|s), \tag{19}$$

and $P_{ind}(\mathbf{r}|s)$ is much better sampled than $P(\mathbf{r}|s)$ since they depend only on the easier to sample single bin parameters. Thus, $Hind(R|S)$ is much more straightforward to estimate than the 'correlated' noise entropy $H(R|S)$.

The asymptotic value of the shuffled conditional entropy responses with respect to the stimuli, $H_{sh}(R|S)$, is the same as $H_{ind}(R|S)$ while its bias is very similar to the bias of $H(R|S)$ [65]. This observation leads to the idea of estimating the information using the shuffled $H_{sh}(R|S)$ in order to cancel the bias of $H(R|S)$. Thus, in order to correct bias as much as possible, we can subtract the value of the independent $H_{ind}(R|S)$, that is almost unbiased. This strategy allow us to define the Shuffled Mutual Information as

$$I_{Shuffled}(R;S) = H(R) - H(R\,|S) + H_{sh}(R\,|S) - H_{ind}(R\,|S)\,, \tag{20}$$

This method can be understood as using the difference $H_{ind}(R|S) - H_{sh}(R|S)$ as the estimation of the bias for the plain information term.

2.1.1 The information component breakdown

The huge degree of complexity of the different biophysics process involved in the mammalian brain, makes completely unlikely for us to describe and predict the timing of each spike deterministically. Instead, a complete description of the stochastic relationship between a stimulus and a response would require us to know the probabilities corresponding to every sequence of spikes that can be evoked by the stimulus. Thus, the neural responses and their relationship to the stimuli are completely characterized by the probability distribution of spike times as function of the stimulus.

The 'neural code' refers to the neural representation of information. The independent-spike code would corresponds to the case in which the generation of each spike is independent of all other spikes in the train. In contrast, if the individual spikes do not encode independently of each other, we can call the code a correlation code, because correlations between spike times may carry additional information. Certainly, information is likely to be carried by both individual spikes and through correlations. Thus we must establish some criterion to characterize the neural code.

The dependence between responses and stimuli may take a number of forms – firing rate dependence, pairwise correlation dependence, *etc.* Information component breakdown method (see Figure 1) allow us to quantify the effect of the contribution of individual members of the population, the interaction between them, and the overall information encoded by the ensemble of neurons evaluating separately contributions due to the noise and signal spike correlations. This allows us to assess the different ways in which the correlations contribute to the neural code.

It is usually referred as synergy (redundancy) when the information conveyed by an ensemble of neurons is larger (smaller) than the sum information conveyed by each neuron independently. If synergistic or redundant interactions between cells are observed, the information components can help us understand the mechanisms from which they arise. More importantly, the exact breakdown of Information into its coding mechanisms allows a precise quantification of all the modalities with which correlations contribute to the neural code.

The technical approach we present here is a modification of a previously developed technique [57, 63, 66] in which a Taylor series expansion was used to break the Shannon information into its components. [67] generalised this approach by substituting correlation functions with probability functions, resulting in an exact information component breakdown. The main disadvantage of the Pola approach is that the correlation components cannot be explicitly written in terms of entropies of probability distributions. Thus, advanced methods to remove bias deviations from entropy estimations cannot be used for all components.

However, we have found an alternative way to rewrite the correlational components as *approximate* entropies (*i.e.*, an entropies plus additional terms which are very small and relatively immune to bias), allowing us to use the NSB method for estimating entropy from a limited number of experimental samples
[61]. Let's consider a time period of duration T unit time bins in which the activity of a

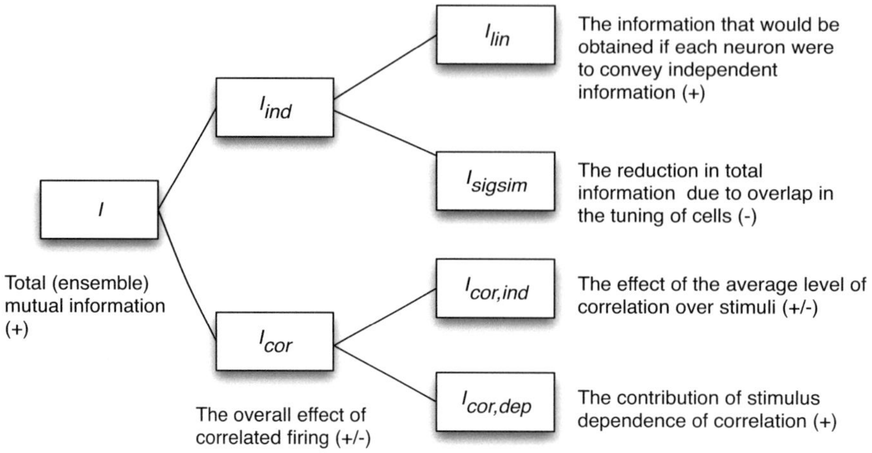

Figure 1: Schematic depiction of the Information component breakdown. The ensemble mutual information can first be broken down into the amount of information that would have been present if responses (cells and/or time bins, depending on the application) were independent, and an additional component due to the effect of noise correlations. Each of these components can be broken down further to dissect out the effects of tuning curve similarity, stimulus-dependence of correlation, *etc.*

given pair of cells is observed. We recall that the mutual information is given by equation (16).

In order to understand the meaning of correlation, we bring back to our minds that probability of obtaining independent population responses is given by $P_{ind}(\mathbf{r}\,|s) = \prod_{c=1}^{C}\prod_{t=1}^{T} P(r_{ct}\,|s)$, which averaged over stimuli reads

$$P_{ind}(\mathbf{r}) = \langle P_{ind}(\mathbf{r}\,|s)\rangle_s\,, \tag{21}$$

where c is the label for each cell (up to C=2 cells in the pair for the analysis presented here, although the formalism is quite general in this respect), t indexes the time bin up to a maximum value of L corresponding to duration T, and the stimulus average $\langle x\rangle_s = \sum_{s=1}^{S} P(s)x$. The presence of noise correlation (correlation in the response variability for a fixed stimulus) or signal correlation (correlation in the tuning of, or signal conveyed by response variables) is indicated by $P_{ind}(\mathbf{r}|s) \neq P(\mathbf{r}|s)$ and $P(\mathbf{r}) \neq P_{ind}(\mathbf{r})$, respectively. Moreover, $P(s|\boldsymbol{r})$ is the true distribution of stimuli given responses and $P_{ind}(s|\mathbf{r})$ is the distribution one would derive in absence of knowledge of correlations. In practice, we obtain $P_{ind}(\mathbf{r}|s)$ numerically by multiplying the marginal probability distributions.

The information component breakdown method allows us to write the total mutual information into a sum of components which are related to the different decoding mechanism [63, 67]:

$$I(R;S) = I_{ind} + I_{cor} = I_{lin} + I_{sig-sim} + I_{cor-ind} + I_{cor-dep}. \tag{22}$$

The first term of the information breakdown, I_{lin}, gives the total amount of information which would be conveyed if all the cells were independent:

$$I_{lin} = \sum_{c=1}^{C}\sum_{t=1}^{T}[H(R_{ct}) - H(R_{ct}|S)] \tag{23}$$

where

$$H(R_{ct}|S) = -\sum_{s \in S} P(s) \sum_{r_{ct}} P(r_{ct}|s) \log_2 P(r_{ct}|s), \tag{24}$$

and $H(R_{ct})$ is the averaged value across all the stimuli.

The signal similarity term $I_{sig-sim}$ quantifies the information loss arising from redundancy due to overlap in the tuning curves pertaining to response (cell,time) bins r_{ct} ("redundancy due to signal correlations")

$$I_{sig-sim} = H_{ind}(R) - \sum_{c=1}^{C}\sum_{t=1}^{T} H(R_{ct}), \tag{25}$$

where

$$H_{ind}(R) = -\sum_{\mathbf{r}} P_{ind}(\mathbf{r}) \log_2 P_{ind}(\mathbf{r}), \tag{26}$$

The term I_{cor} quantifies the total amount of information due to the correlated activity on the overall neural coding:

$$I_{cor} = I(R;S) - H_{ind}(R) + \sum_{c=1}^{C}\sum_{t=1}^{T} H(R_{ct}|S). \tag{27}$$

I_{cor} quantifies whether the presence of "correlation" increases or decrease the information available in the neural responses. It is important to point out that, in the systems neuroscience literature, correlations in response profiles of individual signal across different stimuli are often called "signal correlation" because they are entirely attributable to common stimulus selectivity. On the other hand, correlations which are manifested as covariation of the trial-by-trial fluctuation around the mean response to the stimulus are referred to as "noise correlations". These covariations are observed at fixed stimulus, and they are not attributable to the stimulus.

So far, it is apparent that all of the terms can be written in terms of entropies of particular distributions. However, this correlation term can be further resolved into two components, a stimulus independent component $I_{cor-ind}$ and a stimulus dependent component $I_{cor-dep}$. This last term in the [67] formalism is calculated as

$$I_{cor-dep} = I(R;S) - \chi + \sum_c H(R_c|S), \tag{28}$$

$$\tag{29}$$

where

$$\chi = -\sum_{\mathbf{r}} P(\mathbf{r}) \log_2 P_{ind}(\mathbf{r}). \tag{30}$$

Thus, the stimulus dependent correlation component is not made up entirely of entropies, and advanced entropy estimation techniques such as NSB cannot be applied. Moreover, χ can be substantially biased, which renders its calculation by previously available procedures difficult for all except very low-dimensional problems.

It is important to point out that the stimulus-dependent correlation component can be defined as the Kullback-Leiber divergence between $P(s\,|\mathbf{r})$ and $P_{ind}(s\,|\mathbf{r})$, [39, 68] – conceptually characterized by Nirenberg *et al.* as the effect of correlations upon a stimulus decoder:

$$I_{cor-dep} = D\left[P(s\,|\mathbf{r}\,)\,||P_{ind}(s\,|\mathbf{r}\,)\right] \equiv \sum_{\mathbf{r}} P(\mathbf{r}) \sum_s P(s\,|\mathbf{r}\,) \log_2 \frac{P(s\,|\mathbf{r}\,)}{P_{ind}(s\,|\mathbf{r}\,)}. \tag{31}$$

In the following, we rewrite the stimulus-dependent correlation component from "first principles" using the Kullback-Leibler, for the case in which only pairwise correlations between neurons are considered. We note that this formalism can be easily extended to a larger number of cells.

$$\begin{aligned} I_{cor-dep} &= D\left[P(s\,|r_1 r_2\,)\,||P_{ind}(s\,|r_1 r_2\,)\right] \\ &\equiv \sum_{r_1 r_2} \sum_s P(s) P(r_1 r_2\,|s\,) \log_2 \frac{P(r_1 r_2\,|s\,)}{P_{ind}(r_1 r_2\,|s\,)} - \sum_{r_1 r_2} P(r_1 r_2) \log_2 \frac{P(r_1 r_2)}{P_{ind}(r_1 r_2)} \end{aligned} \tag{32}$$

where

$$P_{ind}(r_1 r_2) = \sum_s P(s) P(r_1|s) P(r_2|s) \tag{33}$$

and

$$P(r_1r_2) = \sum_s P(s)P(r_1r_2|s) \tag{34}$$

The second term in the Kullback-Leibler divergence, $D_\Delta = \sum_{r_1r_2} P(r_1r_2)\log_2 \frac{P(r_1r_2)}{P_{ind}(r_1r_2)}$, does not change if we multiply and divide by exactly the same factor inside the logarithm,

$$D_\Delta = -\sum_{r_1r_2} P(r_1r_2)\log_2\left(\frac{P(r_1r_2)\left\langle P(r_1|s)\right\rangle_s \left\langle P(r_2|s)\right\rangle_s}{\left\langle P(r_1|s)P(r_2|s)\right\rangle_s \left\langle P(r_1|s)\right\rangle_s \left\langle P(r_2|s)\right\rangle_s}\right). \tag{35}$$

This is equivalent to saying

$$D_\Delta = \sum_{r_1r_2} P(r_1r_2)\log_2 \frac{P(r_1r_2)}{\left\langle P(r_1|s)\right\rangle_s \left\langle P(r_2|s)\right\rangle_s} - \Delta = I(r_1;r_2) - \Delta \tag{36}$$

where $\Delta = \sum_{r_1r_2} P(r_1r_2)\log_2\beta$ and

$$\beta = \frac{\left\langle P(r_1\,|s)\right\rangle_s \left\langle P(r_2\,|s)\right\rangle_s}{\left\langle P(r_1\,|s)P(r_2\,|s)\right\rangle_s}. \tag{37}$$

$I_{cor-dep}$ can therefore be rewritten as

$$I_{cor-dep} = \sum_{r_1r_2}\sum_s P(s)P(r_1r_2\,|s)\log_2\frac{P(r_1r_2\,|s)}{P_{ind}(r_1r_2\,|s)} - \tag{38}$$
$$\sum_{r_1r_2} P(r_1r_2)\log_2 \frac{P(r_1r_2)}{\left\langle P(r_1\,|s)\right\rangle_s \left\langle P(r_1\,|s)\right\rangle_s} - \Delta.$$

The stimulus dependent correlation component has thus been expressed in terms of entropy quantities with the addition of a single term Δ; the NSB estimation method can be applied to these quantities with the exception of Δ. The remaining non-entropy term (Δ) is essentially the stimulus-averaged response multiplied by a logarithmic function of the coefficient β summed over the whole space of responses. We are summing therefore in the space of responses quantities which have been averaged over the stimuli, and it is reasonable to expect a very low bias contribution from this term. Moreover, since β is the product of the stimulus-averaged response from each cell divided by the stimulus unconditional independent responses, it is reasonable to expect that it will always be close to one and therefore Δ will be close to zero. In practice, we included Δ in our calculations without bias correction; the magnitude of Δ was always observed to be small [69]. Figure 2 shows a histogram of the ratio between the true value of I_{cordep} and that obtained by setting Δ=0.

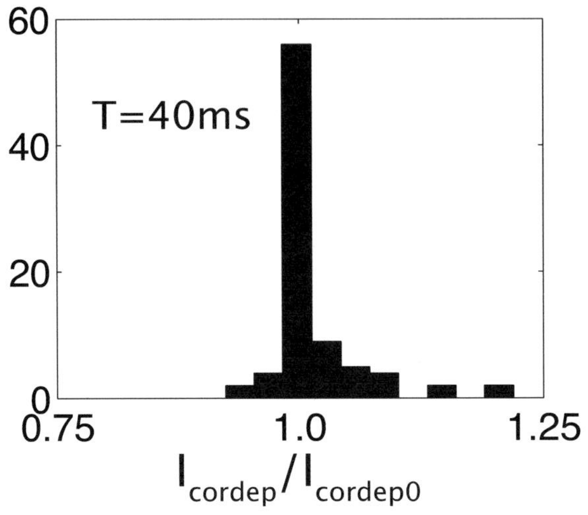

Figure 2: The effect of Δ (at $T = 40$ ms for 102 pairs of neurons). $I_{cor-dep} = I_{cor-dep0}$ corresponds to $\Delta = 0$.

We conclude from the histogram that the contribution from Δ will be reasonably small. In practice, we included Δ in our calculations (without bias correction).

Having expressed all of the quantities necessary to perform the information component analysis in terms of entropies of particular (in some cases conditional) distributions, these entropies are computed by inserting the respective distributions into the NSB entropy estimation routine. We found that, for our dataset, this procedure resulted in substantially better estimation performance than we were able to achieve with techniques that we have used previously [54]. In addition, the NSB approach provides a convenient error estimate for the entropies, which can be propagated into error estimates for the information quantities. An alternative method to estimate $I_{cor-dep}$ has been also recently developed by [65].

In the next section will will illustrate first the performance of the method NSB method with some simple examples. Then, in the following sections we will present a detailed discussion of the application of the Information breakdown method to investigate the role of correlations in neural spike train.

2.2 Example 1: Information calculation for a pair of V1 neurons

Mutual Information [50, 52] quantifies how well an ideal observer can discriminate between all members of the stimulus set based on the neuronal responses of a single trial. All our

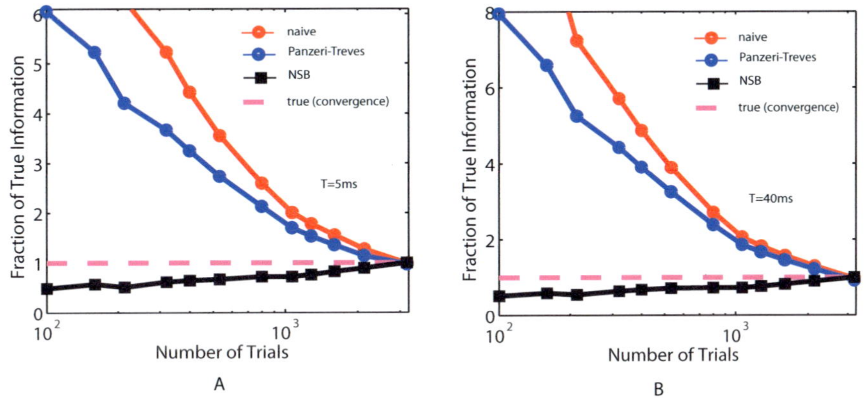

Figure 3: Relative performance of several information estimators. Two different time windows were chosen for counting spikes: **A** 5ms, resulting in cardinality $K = 10$ from the ensemble responses. **B** 40 ms, cardinality $K = 76$.

current analysis are performed through the estimation of the Mutual Information using different entropy estimators to remove bias deviations. As we briefly mentioned it can be estimated as showed in equation (16). It is important to point out that when removing sampling size bias deviation Mutual Information tends to show the opposite asymptotic behavior to entropy estimations.

In order to test the effectiveness of the entropy estimator developed by [61], we performed calculations of the total mutual information conveyed by a pair of cells in V1 ([70], we provide a more detailed description of this data set in the following sections) for which a relatively large number of trials was available, comparing the naive estimator, the Panzeri-Treves bias correction, and the NSB method. The ensemble (pair) mutual information was calculated using different subsets of the total available number of trials, which in this case was sufficient for all methods to converge to the same asymptote, which we can consider to be the true information. Figure 3 shows how the information estimate asymptotes to the true value for each approach.

The NSB approach was the only one we found to be adequate across the entire dataset, despite the relatively large number of trials available (480-3200) compared to many other information theoretic analysis in the literature. Note that this was a particularly (although not entirely unusual) hard to sample pair of cells; for some other pairs, adequate sampling was obtained by NSB at as low as 100 trials. The precise sampling behaviour does depend upon the statistics of the responses, and thus can vary from recording to recording even

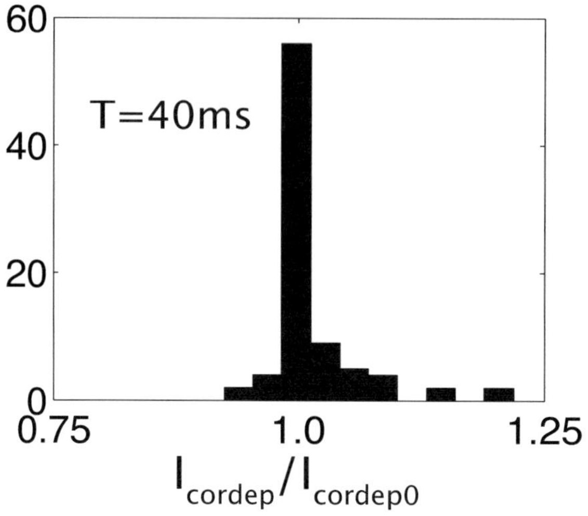

Figure 2: The effect of Δ (at $T = 40$ ms for 102 pairs of neurons). $I_{cor-dep} = I_{cor-dep0}$ corresponds to $\Delta = 0$.

We conclude from the histogram that the contribution from Δ will be reasonably small. In practice, we included Δ in our calculations (without bias correction).

Having expressed all of the quantities necessary to perform the information component analysis in terms of entropies of particular (in some cases conditional) distributions, these entropies are computed by inserting the respective distributions into the NSB entropy estimation routine. We found that, for our dataset, this procedure resulted in substantially better estimation performance than we were able to achieve with techniques that we have used previously [54]. In addition, the NSB approach provides a convenient error estimate for the entropies, which can be propagated into error estimates for the information quantities. An alternative method to estimate $I_{cor-dep}$ has been also recently developed by [65].

In the next section will will illustrate first the performance of the method NSB method with some simple examples. Then, in the following sections we will present a detailed discussion of the application of the Information breakdown method to investigate the role of correlations in neural spike train.

2.2 Example 1: Information calculation for a pair of V1 neurons

Mutual Information [50, 52] quantifies how well an ideal observer can discriminate between all members of the stimulus set based on the neuronal responses of a single trial. All our

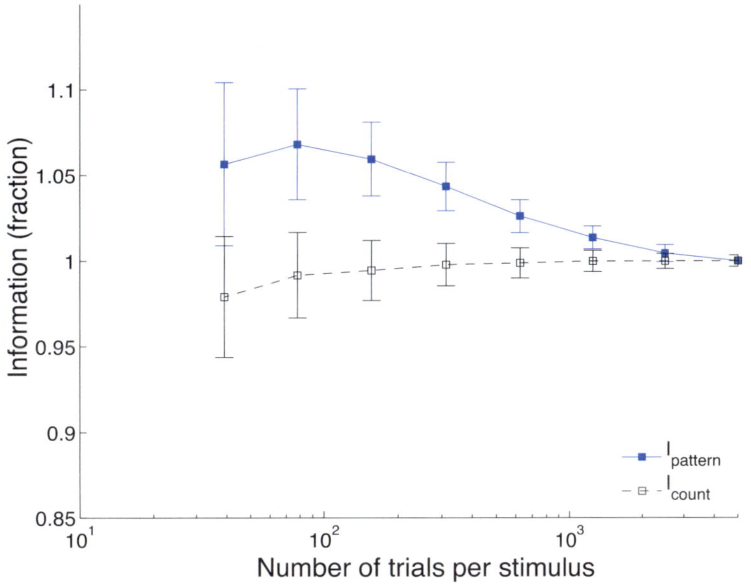

Figure 5: Sampling bias of NSB-corrected information estimates from a Poisson-generated surrogate dataset. The dataset was generated to match two-photon calcium imaging data from an ensemble of 7 cerebellar Purkinje neurons. Spike count and spike pattern codes are compared - the latter being more vulnerable to sampling bias due to higher dimensionality.

sampled data-set consist of creating a new "shuffled" response data-set, with shuffled responses R_{sh}. The shuffling is performed to remove correlations, while preserving the single bin statistic and the pseudo-response statistic will have a "shuffled" noise entropy.

In Figure 4 we show the pairwise Mutual Information of Poisson simulated responses. The NBS [61] entropy estimator is contrasted to QE [56]. Shuffled Information quantities are included for both entropy estimators. Notice, that the NSB method has certainly a great advantage in terms of a much smaller entropy bias deviation with respect to any another standard method to remove bias deviation. The shuffling method works well in combination with other entropy estimation techniques such as NSB and QE (Figure 4).

2.3 Example 2: Neural ensembles recorded optically from the cerebellum

In this study, a multiphoton microscope was used to image the activity of neural populations in the cerebellum of the anaesthetized rat [71]. Calcium indicator dye (Oregon Green BAPTA-1 AM) was injected into the cerebellar cortex, and fluorescence emission monitored using a laser-scanning microscope and photomultiplier tubes. As calcium influxes relate to

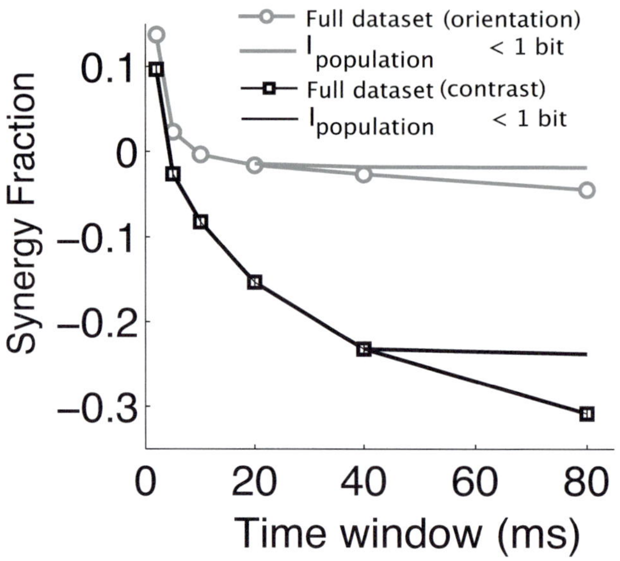

Figure 6: The synergy fraction (defined as $1 - I_{sum}/I_{ensemble}$ and indicating the fraction of information beyond that expected if the cells are independent) is plotted against the time window used for counting spikes. The lines show the average (+/− s.e.m.) over the entire dataset, and over only those pairs showing ensemble information lower than 1 bit (this latter curve excluding potential artefactual redundancy effects due to limited stimulus entropy). Orientation information shows weak synergy at short timescales, shifting to independence at longer timescales. The degree of redundancy in the coding of contrast quickly becomes substantial as the time window is increased. The results for contrast coding are indicated by squares and a grey line; for direction coding by the black line and circles.

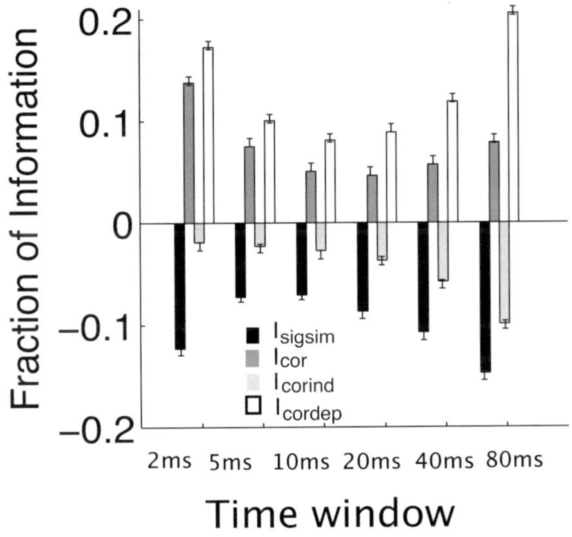

Figure 7: The breakdown of information about stimulus direction carried by pairs of V1 neurons. Bar charts report the fraction of the information accounted for by the different components of the information breakdown plotted as function of the length of the post-stimulus time window used to compute the spikes (average +/− s.e.m. over the entire data-sets).

action potentials, the fluorescence can be used to monitor neural spiking, from a population consisting of essentially every neuron in an area of tissue. This powerful new neural recording technique offers many new opportunities for studying the neural code, but also provides many technical problems, one of which is the sampling problem, which becomes dramatically worse the larger the population that is monitored.

In this study, information about the presence or absence of an airpuff stimulus was calculated for two codes, a population spike count (or 'pooled') code, and a spatial pattern code. Around 200 trials per stimulus were available. We used the NSB estimator to correct for sampling bias, and examined the residual sampling bias using surrogate data as described in Figure 5. For ensembles of 7 neurons, the amount of information available from the spatial pattern code was 44% greater than that from the population spike count code, a difference substantially greater than the 5% or so expected due to residual finite sampling bias.

2.4 Numerical Implementation

Several different implementations of the NSB method are available from the authors http://nsb-entropy.sourceforge.net/. The original Octave code was converted to a Matlab compatible version. The Matlab/Octave version has the advantage of cross-platform portability and simplicity of use, but suffers from some limitations, notably speed and wordlength restrictions. For this reason, we recommend the C++ version available at the same site instead. Note that NSB, together with code for some other estimators such as QE, is also now available as part of some other software packages, notably the Spike Train Analysis Toolkit of [72], and the Python toolbox of [73].

3 The role of correlations in the neural code

In this section, we review how to use an information theoretic approach to investigate the neuronal code. We discuss how to quantify the information carried by spike correlations in the neural population, and apply this formalism to simultaneous recordings of multiple spike trains.

3.1 The role of correlations in encoding stimulus information in monkey visual cortex

One simple way to study the effect of correlations on the neural population code is to compute the information available from the entire ensemble (considered together), and compare it with the sum of the information from each individual neuron making up the ensemble [66, 74, 75]. As we mentioned previously a code in which the whole is greater than the sum of the parts can be referred to as "synergistic", and one in which the whole is *less* than the sum of the parts can be considered to be "redundant" (note that this is a slightly different use of the term redundancy to that used in classical communication theory). Such an analysis for pairs of neurons, showing weakly synergistic orientation coding at short timescales, is shown in Fig. 6.

The simple comparison between ensemble and summed individual responses conflates several forms of correlation, however, making it impossible to judge the relative importance of synchronous spiking, basic tuning properties, and stimulus (in)dependent correlation. In this section we have applied an information theoretic approach to this question, using the responses of pairs of neurons to drifting sinusoidal gratings of different orientations and contrasts, recorded in the primary visual cortex of anesthetized macaque monkeys. Our approach allows us to break down the information provided by pairs of neurons into a number of components. The results we present in this chapter are part of a work we have carried out in collaboration with Kohn and Smith [69].

The picture over the entire population is one of a small amount of synergy: with a time window of 5 ms, the ensemble code does on average 2.4% better than the sum of single cell contributions (see Figure 6). To compute the degree of synergy in the population for a range of time windows T, we calculated a synergy index (which we call the synergy fraction) as ($I_{ensemble}$-I_{sum})/ $I_{ensemble}$: values greater than zero indicates synergistic coding, values less than zero indicate redundancy. The synergistic effect of ensemble coding is more substantial at shorter time windows, where the fine temporal precision at which the spikes may synchronize has a significant effect – the maximum effect we observed was 13.7% at 2 ms (for technical reasons related to convergence of the numerical integral involved in the entropy computation, the smallest time window we were able to use), tailing off sharply at around 10 ms, where the synergy fraction is close to zero. Thus, being very precise about spike timing leads to synergy rather than information independence [75]. The short time scale over which synergy is present agrees well with the time scale of synchrony in the population: the mean full width at half maximum of the CCG peaks was 9 ms [70]. At longer time windows ($T\geq$ 20 ms), the information became comparable in size to the stimulus entropy ($\log_2$ 5 bits), and thus ceiling effects may reduce the raw information values somewhat. We corrected for this by removing pairs from the population that presented a higher information amount than 1 bits at a given time window, an arbitrary but conservative threshold. It is apparent that in the absence of ceiling effects, the trend for long time windows is for the synergy fraction to asymptote to a slightly negative value, indicating that the mutual information adds across the neurons as if they were roughly independent.

Notice that Figure 6 shows that the different neurons convey redundant information about contrast, as might be expected give that all neurons have relatively similar contrast tuning curves. This provides a useful comparison for the weakly synergistic coding of stimulus orientation. Whether other variables related to stimulus structure are encoded synergistically has not been studied to date.

We then applied the information breakdown approach to investigate of how groups of neurons in primary visual cortex (V1) encode contrast and direction of motion of visual stimuli. To address this question, Kohn and Smith [70] recorded, from area V1 of anesthetized macaques, responses of pairs of neurons to drifting sinusoidal gratings of different directions and contrasts. Using cross-correlation measures, they showed that an orientation sensitive component of the response correlation dominated at short time scales, and an orientation insensitive component that was prevalent at longer time scales. An important difference between these experiments and some others [76] was the care taken to map receptive field and tuning properties of the neurons, and use stimuli which drove both cells well, thus ensuring the greatest chance to observe stimulus dependent correlations, which may be particular vulnerable to false negative issues (also see [77]). In a subsequent analysis of the same data, we used the information breakdown to gain a more quantitative insight into the role of noise correlations in the encoding of contrast and orientation [69].

Figure 7 summarizes the results for the whole population, of the breakdown of information about direction carried by pairs of neurons [69]. For this analysis, the neural response of each pair was quantified as the spike count carried over a post-stimulus window of fixed length. Different window lengths (in the range 2-80 ms) were used. The contribution of the information components is strongly dependent of the length of the response window: while the negative contribution of $\mathrm{I}_{cor-ind}$ grows with the time window, the effects of the other components follow a 'U' or inverted "U" shape curve. For time windows lower than 10 ms, the preservation of temporally precise noise correlations [70] resulted in correlational information taking up a substantial fraction of the total information. At time windows longer than 10 ms the contribution from $I_{cor-dep}$ was significantly curtailed, and was effectively canceled by the signal similarity term, $I_{sig-sim}$, and also by $I_{cor-ind}$, which provided a greater negative contribution as the time window increased.

We found that for direction coding the information available by being precise about spike timing for pairs of neurons in V1 adds super-linearly. But as the timescale becomes more coarse, information about direction adds slightly sub-linearly across pairs of cells in agreement with a previous information theoretic study of V1 cells [78]. We have revealed the origin of these effects by the information component analysis, which shows a redundant (negative) contribution to the information due to the overlap in tuning and to the average level of correlation (*i.e.*, stimulus independent correlation). However, these components are balanced (or even slightly over-compensated for) on average by a contribution from the stimulus-dependence of synchrony.

To overcome the statistical inefficiency of high activity ratio distributed representations, which are the typical product of redundancy reduction code, one needs representations with minimum overlap. This is, representations with the minimum number of elements active in both inputs in pair of cells that need to be distinguished. However, our finding suggest that stimulus dependent correlations can help the neural code to overcome this problem. The fact that the synchronization depends on the stimulus direction is critically important for coding: it serves to reduce the redundancy caused by cells having similar tuning.

It is important to point out that this comes about because of a balance between the strong synergy provided by stimulus-dependent correlation and redundancy arising from the similarity in tuning of the neurons, and not because there are no signal and noise correlations. Although noise correlations are informative, the additional information they provides is frequently offset by the redundancy arising from the similar tuning properties of the two cells. An interpretation of this result is that this cancelation this would allow cortical circuits to enjoy the stability provided by having similarly tuned neurons without suffering the penalty of redundancy as the associated information transmission deficit is compensated for by stimulus dependent synchrony [69].

The coding of contrast was far more redundant than that of direction at all time windows considered, because of the strong signal correlations in the coding of contrast [69]. The results of this analysis warn against making general statements about the role of correlations in neural coding – as demonstrated here, the effect of correlations can be quite sensitive to both the timescale and the nature of the stimulus parameter that is being studied.

4 Conclusions

A critical problem in neural coding is to address how important spike correlations are. The Information breakdown method has the unique advantage of capturing all nonlinear dependencies of any statistical order that may be present in the neuronal data. Several studies found evidence of weak synergy in the primary visual cortex (V1)[48, 69, 78, 79, 80] or retina [39]. The amount of $I_{cor-dep}$ typically resulted to be typically about the 10 % of the total mutual information. Thus, one could ignore stimulus dependent correlations and lose at most 10 % of the information. Even if the impact of correlations is "relatively small" at the level of neuron pairs they could have a much stronger effect at the level of larger populations.

In a recent study we analysed a model based on physics that evaluates whether the probability of synchronous discharge can be described by interactions up to any given order. We showed that high-order interactions could play a role in shaping the dynamics of neural networks, and that they should be taken into account when computing the representational capacity of neural populations [81]. Thus, amongst the many open challenges in this area of Theoretical Neuroscience are two which are particularly important: extending the techniques applied in this work in order to account for correlations of all orders, and extending the Information breakdown method to much larger ensembles of neurons, potentially numbering in the hundreds [82].

Conflict of interest: None.

Acknowledgments

Research supported by PIP 0255/11 (CONICET). FM would like to thanks Dr. O. A. Rosso, and Dr. A. Kowalski for very useful discussions.

References

[1] Adrian, E. D. and Zotterman, Y. Impulses from a single sensory cell end-organ. J. Physiol. (London), 61,465-483 (1926).

[2] Baylor, D. A. ,Photoreceptor signals and vision, Invest Ophthalmol Vis Sci, 28,34-49,Google Scholar (1987).

[3] Ringach, Dario L.,Spontaneous and driven cortical activity: implications for computation.,Curr Opin Neurobiol,19,(4),439-44 (2009).

[4] Toyama, K. and Kimura, M. and Tanaka, K. ,Cross correlation analysis of interneuronal connectivity in cat visual cortex,J. Neurophysiol.,46(2),191-201 (1981).

[5] Michalski, A., and Gerstein, G.L., and Czarkowska, J., and Tarnecki, R. , Interactions between cat striate cortex neurons,Exp. Brain. Res.,51,97-107 (1983).

[6] Toyama, K. and Kimura, M. and Tanaka, K. ,Organization of cat visual cortex as investigated by cross-correlation technique,J. Neurophysiol.,46(2),202-214 (1981).

[7] Ts'o, D. Y. and Gilbert, C. D. and Weisel, T. N.,Relationships between horizontal interactions and functional architecture in cat striate cortex as revelated by cross-correlation analysis,J. Neurosci.,6(4),1160-1170 (1986).

[8] Aiple, F. and Krueger, J. , Neuronal synchrony in monkey striate cortex: interocular signal flow and dependency on spikes rates,Exp.. Brain Res.,72, 141-149 (1998).

[9] Ts'o, D. Y. and Gilbert, C. D. ,The organization of chromatic and spatial interactions in the primate striate cortex,J. Neurosci.,8,1712-1727 (1988).

[10] Gochin, P.M, and Miller, E.K., and Gross, C.G., and Gerstein, G.L. , Functional interactions among neurons in inferior temporal cortex of the awake macaque,Exp. Brain res.,84, 505-516 (1991).

[11] Hata, Y., Tsumoto, T., Sato, H., Hagihara, K. and Tamura, H. (1998),Inihibition contributes to orientation selectivity in visual cortex of cat,Nature,335,815-817.

[12] Alonso, J.M., and Martinez, L.M. , Functional connectivity between simple cells and complex cells in cat striate cortex, Nat. Neurosci., 1, 395-403 (1998).

[13] Molotchnikoff, S., and Shumikhina, S.,Synchronization, gamma oscillations and spatial integration of line segments in the cat visual cortex,Vis. Res.,36,2037-2046 (1998).

[14] Perkel, D. H., and Gerstein, G. L., and Moore, G. P.,Neuronal spikes trains and stochastic point processes II. Simultaneous spikes trains ,Biophys. J. , 7,419-440 (1967).

[15] Gerstein, G.L., and Perkel, D.H. , Simultaneously recorded trains of action potentials: analysis and functional interpretation,Science,164(881), 828-30 (1969).

[16] Moore, G.P., and Segundo, J.P., and Perkel, D.H., and Levitan, H. ,Statistical signs of synaptic interaction in neurons,Biophys. J.,10,876-900 (1970).

[17] Kirkwood, P.A.,On the use and interpretation of cross-correlation measurement in the mammalian central nervous system,J. neurosci. Methods,1,7152-7161 (1979).

[18] Aertsen, A.M.H.J., and Gerstein, G.L., and Habib, M.K., and Palm, G., Dynamics of neuronal firing correlation: modulation of 'effective conectivity', J. Neurophysiol.,61, 900-917 (1989).

[19] Surmier, D. J. and Weinberg, R.J. ,The relationship between cross-correlation measures and underlying synaptic events,Brain Res.,331,180-184 (1985).

[20] Bryant, J., and Marcos, A.R. and Segundo, G.H. , Correlations of neuronal spike discharges produced by monosynaptic connections and by common inputs,J Neurophysiol,36,205 - 225 (1973).

[21] Peters, A. ,Synaptic specificity in the cerebral cortex. In: Synaptic function , Edelman, G., and Gall, W., and Cowan, W., , eds., New York: Wiley,373-397 (1987).

[22] Braitenberg, V. and A. Schüz ,Cortex: Statistics and Geometry of Neuronal Connectivity, Springer,249,ISBN: 3-540-63816-4,Berlin, Germany (1998).

[23] Desimone, R. , Face-selective cells in the temporal ortex of monkeys ,J. Cogn. Neurosci.,3,1-8 (1991).

[24] Andersen, R.A. ,Encoding of intention and spatial location in the posterior parietal cortex, Cereb. Cortex,5, 457-469 (1995).

[25] Churchland, P. S., and Ramachandran, V. S., and Sejnowski, T. J. (1994), A critique of pure vision, Large-Scale Neuronal Theories of the Brain, MIT Press, Cambridge, MA, Koch, C. and Davis, J.L. Editors.

[26] Cobb, S.R., and Buhl, E. H., and Halasy, K., and Paulsen, O., and Somogyi, P., Synchronization of neuronal activity in hippocampus by individual GABAergic interneurons,Nature,378,75-78 (1995).

[27] Azouz, R. and Gray, C. M. , Dynamic spike threshold reveals a mechanism for synaptic coincidence detection in cortical neurons *in vivo*, Proc Natl Acad Sci U S A, 97, 14, 8110-5, organization The Center for Neuroscience, University of California, Davis, CA 95616, USA (2000).

[28] Azouz, Rony and Gray, Charles M. , Adaptive coincidence detection and dynamic gain control in visual cortical neurons *in vivo*, Neuron,37, 3, 513-23 (2003).

[29] Bruno, Randy M. and Sakmann, Bert ,Cortex is driven by weak but synchronously active thalamocortical synapses.,Science,312(5780),1622-7 (2006).

[30] Uhlhaas, P. J. and Singer, W., Neural synchrony in brain disorders: relevance for cognitive dysfunctions and pathophysiology,Neuron,52(1), 155-68(2006).

[31] Eckhorn, R., and Bauer, R., and Jordan, W., and Brosch, M., and Kruse, W., and Munk, M.,and Reitboeck, H.J. ,Coherent oscillations: a mechanism of feature linking in the visual cortex? Multiple electrode and correlation analyses in the cat,Biol Cybern,60,121-130 (1998).

[32] Gray, C.M. and Singer, W. ,Stimulus specific neuronal oscillations in orientation columns in cat visual cortex,Proc. Natl. Acad. Sci. ,86,1698-1702 (1989).

[33] Gray, C.M., and Koening, P., and Engel, A.K., and Singer, W. , Oscillatory responses in cat visual cortex exhibit inter-columnar synchronization which reflects global stimulus properties, Nature,3888, 334-337 (1989).

[34] Meister, M., and Lagnado, L., and Baylor, D.A.,Concerted signaling by retinal ganglion cells,Science,270,1207-1210 (1995).

[35] Vaadia, E. and Abeles, M. and Bergman, H. and Prut, Y. and Slovin, H. and Aertsen, A. , Dynamics of neuronal interactions in monkey cortex in relation to behavioural events,Nature,373,515-518 (1995).

[36] De Charms, R.C., and Merzenich, M.M. ,Primary cortical representation of sounds by the coordination of action-potential timing,Nature,381,610-613 (1996).

[37] Dan, Y., Alonso, J.-M., Usrey, W.M. and Reid, R.C. ,Coding of visual information by precisely correlated spikes in the lateral geniculate nucleus,Nature Neurosci.,1(6),501-50 (1998).

[38] Steinmetz, P. N. Roy, A. and Fitzgerald, P. J. and Hsiao, S. S. and Johnson, K. O. and Niebur, E.,Attention modulates synchronized neuronal firing in primate somatosensory cortex,Nature 404,187-190 (2000).

[39] Nirenberg, S., and Carcieri, S. M., and Jacobs, A. L., and Latham, P. E. , Retinal ganglion cells act largely as independent encoders,Nature,411,698-701 (2001).

[40] Oram, M.W., Hatsopoulos, N.G., Richmond, B.J., Donoghue, J.P. (2001),Excess synchrony in motor cortical neurons provides redundant direction information with that from coarse temporal measures ,J Neurophysiol,86,1700-1716 (2001).

[41] Petersen, R.S., and Panzeri, S., and Diamond, M.E.,Population coding of stimulus location in rat somatosensory cortex,Neuron,32(3),503-14(2001).

[42] Levine, M.W., and Castaldo, K., and Kasapoglu, M.B.,Firing coincidences between neighboring retinal ganglion cells: inside information or epiphenomenon?,Biosystems,67,139-146 (2002).

[43] Panzeri, S., and Pola, G., and Petroni, F., and Young, M.P. and Peterson, R.S. , A critical assessment of different measures of the information carried by correlated meuronal firing,Biosystems,67,177-185 (2002).

[44] Panzeri, S., and Golledge, H.D.R., and Zheng, F., and Pola, G., and Blanche, T.J., and Tovee, M.J., and Young, M.P.,The role of correlated firing and synchrony in coding information about single and separate objects in cat V1, Neurocomputing,44-46,579-584,(2002).

[45] Averbeck, B.B., and Crowe, D.A., and Chafee, M.V., and Georgopoulos, A.P. , Neural activity in prefrontal cortex during copying geometrical shapes. II. Decoding shape segments from neural ensembles, Exp Brain Res.,150(2),142-53 (2003).

[46] Averbeck, B.B., and Lee, D. , Neural Noise and movement-related codes in the macaque supplement area,J Neurosci.,23,7630-7641 (2003).

[47] Averbeck, B.B., and Lee, D. , Coding and transmission of information by neural ensembles, Trends Neurosci.,27(4), 225-30 (2004).

[48] Golledge, H. D., and Panzeri, S., and Zheng, F., and Pola, G., and Scannell, J. W., and Giannikopoulos, D. V., and Mason, R. J., and Tovee, M. J., and Young M. P. ,Correlations, feature-binding and population coding in primary visual cortex,Neuroreport,14,7,1045-50 (2003).

[49] Gregoriou, Georgia G. and Gotts, Stephen J. and Zhou, Huihui and Desimone, Robert , High-frequency, long-range coupling between prefrontal and visual cortex during attention.,Science,324 (5931),1207-10 (2009).

[50] Shannon, C. E.,A mathematical theory of communication AT and T Bell Labs Tech. J., 27,379-423 (1948).

[51] Pola, G. and Schultz, S. R. and Petersen, R. S. and Panzeri, S., A practical guide to information analysis of spike trains,Neuroscience Databases. A Practical Guide, 139-153 (2003).

[52] Cover, T.M., and Thomas, J.A. ,Elements of Information Theory: Books: Wiley Series in Telecommunications and Signal Processing (1991).

[53] Paninski, L.,Estimation of entropy and mutual information,Neural Computation,15, 1191-1254 (2003).

[54] Panzeri, S. and Treves, A. ,Analytical estimates of limited sampling biases in different information measures,Network,7,87-107(1996).

[55] Ma, S. K. ,Calculation of entropy from data of motion,Journal of Statistical Physics,26 (1981).

[56] Strong, S. P. and Koberle, R. and de Ruyter van Steveninck, R. R. and Bialek, W.,Entropy and information in neural spike trains,Phys Rev Lett,80, 197-200 (1998).

[57] Schultz, S. R., and Panzeri, S.,Temporal correlations and neural spike train entropy, journal=Physical Review Letters,86,5823-5826 (2001).

[58] Nemenman, I., and Shafee, F., and Bialek, W.,Entropy and inference, revisited, Advances in Neural Information Processing 14, Dietterich, T.G., and Becker, S. , and Ghahramani, Z., eds., MIT Press,471-478 (2002).

[59] Wolpert, D. H. and Wolf, D. R.,Estimating functions of probability distributions from a finite set of samples,Physical Review E,52,6841-54 (1995).

[60] Samengo, I. Estimating probabilities from experimental frequencies. Phys. Rev. E 65:046124 (2002).

[61] Nemenman, I., and Bialek, W., and Steveninck, R.,Entropy and information in neuronal spike trains: Progress on the sampling problem,Phys. Rev. E,69,056111-6 (2004).

[62] Carlton, A. G.,On the bias of information estimates ,Psychological Bulletin,71, 108-109 (1969).

[63] Panzeri, S., and Schultz, S. R. ,A unified approach to the study of temporal, correlational, and rate coding,Neural Comput.,13,1311-1349 (2001).

[64] Rust, N. and Schultz, S. R. and Movshon, J. A.,A reciprocal relationship between reliability and responsiveness in developing cortical neurons,Journal of Neuroscience,22,10519-10523.

[65] Montemurro, M.A., and Senatore, R., and Panzeri, S.,Tight data-robust bounds to mutual information combining shuffling and model selection techniques,Neural Computation,19(11),2913-2957 (2007).

[66] Panzeri, S., and Schultz, S. R., and Treves, A., and Rolls, E. T., Correlations and the encoding of information in the nervous system,Proc R Soc Lond B Biol Sci,266,1001-1012 (1999).

[67] Pola, G., Thiele, A., Hoffmann, K-P., and Panzeri, S. (2003). An exact method to quantify the information transmitted by different mechanisms of correlational coding. Network Computation in Neural Systems 14, 35-60 (2003).

[68] Latham, P. E., and Nirenberg, S.,Synergy, redundancy, and independence in population codes, revisited,J Neurosci., 25(21),5195 - 206 (2005).

[69] Montani, F., and Kohn, A., and Smith, M. A., and Schultz, S. R., The role of correlations in direction and contrast coding in the primary visual cortex, J Neurosci, 27, 9,2338 - 48 (2007).

[70] Kohn, A., and Smith, M. A.,Stimulus Dependence of Neuronal Correlation in Primary Visual Cortex of the Macaque,J Neurosci,25,14 , 3661-3673 (2005).

[71] Schultz, Simon R. and Kitamura, Kazuo and Post-Uiterweer, Arthur and Krupic, Julija and Håusser, Michael,Spatial pattern coding of sensory information by climbing fiber-evoked calcium signals in networks of neighboring cerebellar Purkinje cells., J Neurosci, 29 (25),8005-15 (2009).

[72] Goldberg, David H. and Victor, Jonathan D. and Gardner, Esther P. and Gardner, Daniel ,Spike train analysis toolkit: enabling wider application of information-theoretic techniques to neurophysiology.,Neuroinformatics,7,3,165-78 (2009).

[73] Ince, Robin A. A. and Petersen, Rasmus S. and Swan, Daniel C. and Panzeri, Stefano , Python for information theoretic analysis of neural data.,Front Neuroinformatics,3,4 (2009).

[74] Gawne, T.J., and Richmond, B. J. ,How independent are the messages carried by adjacent inferior temporal cortical neurons?,J Neurosci,13, 2758-2771 (1993).

[75] , Schneidman, E., Bialek, W., Berry, M.J. (2003). Synergy, redundancy, and independence in population codes. J Neuroscience 23 , 11539-11553 (2003).

[76] Ecker, Alexander S. and Berens, Philipp and Keliris, Georgios A. and Bethge, Matthias and Logothetis, Nikos K. and Tolias, Andreas S. (2010),Decorrelated neuronal firing in cortical microcircuits.,Science,327 (5965),584-7.

[77] Samonds, J. M., and Bonds, A. B., Gamma oscillation maintains stimulus structure-dependent synchronization in cat visual cortex. J. Neurophysiol 93, 223-236 (2005).

[78] Reich, D.S., Mechler, F., Victor, J.D. (2001). Independent and Redundant Information in Nearby Cortical Neurons. Science 294, 2566-2568 (2001).

[79] Rolls, E. T. and Franco, L. and Aggelopoulos, N. C. and Reece, S., An information theoretic approach to the contributions of the firing rates and the correlations between the firing of neurons, J Neurophysiol,89,2810-2822 (2002).

[80] Kayser, C., and Konig, P. ,Population coding of orientation in the visual cortex of alert cats an information theoretic analysis,Neuroreport,15(18),2761-2764(2004).

[81] Montani, F., and Ince, R. A., and Senatore, R., and Arabzadeh, E., and Diamond, M. E., and Panzeri, S. ,The impact of high-order interactions on the rate of synchronous discharge and information transmission in somatosensory cortex,Philos Transact A Math Phys Eng Sci,367(1901),3297-310 (2009).

[82] M.T. Schaub, S.R. Schultz, The Ising decoder: reading out the activity of large neural ensembles, J Comput Neurosci, 32, 101-118 (2012).

Index

A

B

C

Andres M. Kowalski, Raul D. Rossignoli and Evaldo M. F. Curado (Eds)

D

E

F

G

H

I

J

K

L

Q

R

S

T

V

W

Z

Made in the USA
Las Vegas, NV
03 November 2026

57424414R00257